"Transforming our world into one that is sustainable and desirable and that achieves the UN Sustainable Development Goals is the overarching challenge for humanity today. This book is an important and innovative guide to sustainability science as a transformative, whole system learning process that is essential to achieving this goal."
— Professor Robert Costanza, *The Australian National University, Australia*

"This book will certainly promote the debate on the role of scientific knowledge in sustainability transitions. It convinces by its enlightened perspective on appropriate forms of knowledge production in the 21st century."
— Professor Uwe Schneidewind, *Wuppertal Institute for Climate, Environment and Energy, Germany*

"This book provides a powerful set of concepts and methods that have proved very effective for the participatory development of new perspectives and actionable knowledge on sustainability in practice, on campus and beyond. Whether for course-work or for practice – this book gives clear guidance. It is a gem!"
— Associate Professor Maki Ikegami, *Hokkaido University, Japan*

Sustainability Science

Sustainability Science: Key Issues is a comprehensive textbook for undergraduates, postgraduates, and participants in executive trainings from any disciplinary background studying the theory and practice of sustainability science. Each chapter takes a critical and reflective stance on a key issue or method of sustainability science. Contributing authors offer perspectives from diverse disciplines, including physics, philosophy of science, agronomy, geography, and the learning sciences.

This book equips readers with a better understanding of how one might actively design, engage in, and guide collaborative processes for transforming human–environment–technology interactions, whilst embracing complexity, contingency, uncertainties, and contradictions emerging from diverse values and world views. Each reader of this book will thus have guidance on how to create and/or engage in similar initiatives or courses in their own context.

Sustainability Science: Key Issues is the ideal book for students and researchers engaged in problem and project based learning in sustainability science.

Ariane König is a Senior Researcher in the Research Unit for Education, Cognition, Culture and Society at the University of Luxembourg.

Jerome Ravetz is an Associate Fellow at the Institute for Science, Innovation and Society, University of Oxford.

Key Issues in Environment and Sustainability

This series provides comprehensive, original and accessible texts on the core topics in environment and sustainability. The texts take an interdisciplinary and international approach to the key issues in this field.

Low Carbon Development
Key Issues
Edited by Frauke Urban and Johan Nordensvärd

Sustainable Business
Key Issues
Helen Kopnina and John Blewitt

Sustainability
Key Issues
Helen Kopnina and Eleanor Shoreman-Ouimet

Ecomedia
Key Issues
Edited by Stephen Rust, Salma Monani and Sean Cubitt

Ecosystem Services
Key Issues
Mark Everard

Sustainability Science
Key Issues
Edited by Ariane König in collaboration with Jerome Ravetz

Sustainability Science

Key Issues

Edited by
**Ariane König in collaboration
with Jerome Ravetz**

 Routledge
Taylor & Francis Group

LONDON AND NEW YORK

First published 2018
by Routledge
2 Park Square, Milton Park, Abingdon, Oxon OX14 4RN

and by Routledge
711 Third Avenue, New York, NY 10017

Routledge is an imprint of the Taylor & Francis Group, an informa business

British Library Cataloguing-in-Publication Data
A catalogue record for this book is available from the British Library

Library of Congress Cataloging-in-Publication Data
A catalog record for this book has been requested

ISBN: 978-1-138-65927-8 (hbk)
ISBN: 978-1-138-65928-5 (pbk)
ISBN: 978-1-315-62032-9 (ebk)

Typeset in Bembo
by Apex CoVantage, LLC

Printed and bound in Great Britain by
TJ International Ltd, Padstow, Cornwall

For Laurine

Contents

Figures

Tables

Boxes

Contributors

Julia Affolderbach is a lecturer in human geography at the University of Hull. Her research interests lie in the field of environmental economic geography involving concepts from political economy and governance studies. Her research primarily consists of actor centred inquiries of greening processes both in urban and rural contexts including innovations in green building as urban climate change mitigation strategy, environmental justice and green entrepreneurship and environmental conflicts in the forest sector. She holds a degree in Geography from the University of Cologne, Germany and a Ph.D. in Geography from Simon Fraser University, Canada.

Julien Bollati is a Financial Engineering Manager at the Innovation and Network Executive Agency (established by the European Commission) where he deals with financing and funding of trans European transport and energy infrastructures. Until 2016 he was a statistical officer in Eurostat working in the Macroeconomic Imbalances Procedure (MIP) Task Force, on the quality monitoring frameworks and reporting structures to ensure EU policy makers are provided with high-quality statistical indicators in support of the MIP macroeconomic surveillance mechanism. He participated to course 'Science and Citizens Meet the Challenges of Sustainability' first as a student and then as a lecturer. He has a Master of Science in International Economic and Finance from Univerista degli Studi di Milano.

Vladimir Broz worked in banking and finance for over thirty years, where he held a number of different positions, first as a treasurer in risk management and subsequently as a private client advisor. He holds a joint degree in Economics and Business Administration from Sophia University in Tokyo, as well as a postgraduate degree in international Studies from the Institut Européen des Hautes Etudes Internationales (IEHEI) in Nice. Throughout his career he nurtured an interest in ethical investments and renewable energy development. This motivated him to pursue the course work for the Certificate in Sustainable Development and Social Innovation.

Kim Chi Tran worked as a petroleum chemist in Australia, research scientist in France, program specialist in UNESCO (Thailand, Indonesia and France) and associate professor at Kwansei Gakuin University, Japan. During her appointment in Japan, she developed the research project on "Ecosystem Health Monitoring involving Local Communities of Yucatan, Mexico". This project focused on transformative capacity-building approaches to involving local community participation in sustainable development. In 2005, Dr. Tran moved to Norway where she worked as research scientist at the Norwegian Water Research Institute (NIVA), and as senior adviser for several public authorities. She joined Petrad as project director in 2014 where she works with capacity building for the public institutions and NGOs in developing countries in sustainable development of their energy sector. She holds a BSc (First Class Honours) and a PhD in Chemistry from Murdoch University, Perth, Australia.

Emilie Crouzat is currently a post-doc researcher at the French Lab of Alpine Ecology (LECA – CNRS). Her research explores social-ecological systems through the prism of ecosystem services, both in inter- and trans-disciplinary ways. She is currently leading the national ecosystems and ecosystem services assessment for mountain ecosystems in France. As the composite information gathered through biophysical and socio-cultural assessments can foster dialogue and learning opportunities between stakeholders, she aims at mobilizing the research process as a way to support a collaborative management of environmental resources. She holds a Ph.D. degree in Ecology and a Master degree in Agronomics.

Federico Davila is a Doctoral Scholar at the Fenner School of Environment and Society, at the Australian National University. His research and work is focused on using a diversity of critical social science methods to explore food systems across regions. Some of his current work includes smallholder participation in policy and research Philippine food systems, agricultural research impact assessment in Asia and the Pacific, and the use of systems thinking and human ecology for teaching and project design. His broader interests include human ecology, sustainability science, transdisciplinary research, and research impact on policy. He is a visiting research fellow at the Southeast Asian Regional Center for Graduate Study and Research in Agriculture. He holds a Bachelor of Interdisciplinary Studies (Sustainability), and a Master of Environment (Research), both from the Australian National University.

Nicolas Dendoncker is Professor of Geography at the University of Namur, Belgium. His main research interests relate to the sustainable management of landscapes. He is particularly concerned by the sustainability of agriculture and the needed transition to agroecology. He is a specialist in integrated valuation of ecosystem services and tries to use this approach to help foster sustainable landscape management. He is involved in several action research projects related to the transition to sustainable agriculture. He recently co-edited the book "Ecosystem Services: Global issues, Local Practice" (Elsevier, 2013).

Gerard Drenth is an Associate Fellow at Saïd Business School, University of Oxford and Senior Partner and Managing Director at NormannPartners, a strategy consultancy. At Oxford, he teaches on degree courses and executive education programmes. He has over 20 years' industry experience, having previously worked as Vice President in the European Strategy Group for Morgan Stanley, and for Shell as the Scenario Manager in Shell's scenario team in London. Gerard holds the CFA UK Certificate in Investment Management, an Executive MBA (London Business School) and a M.Sc. in Computer Science with Operational Research from the VU University Amsterdam.

Robert Dyball is a Senior Lecturer in the Fenner School of Environment and Society at the Australian National University. He is also Visiting Professor at the College of Human Ecology, University of the Philippines Los Baños. He is author of the text book "Understanding Human Ecology" (Routledge, 2015). Positions that Robert currently holds are President of the international Society for Human Ecology (SHE), Chair of the Human Ecology Section of the Ecological Society of America (ESA), and editor of the journal Human Ecology Review. His research involves the application of systems thinking to the justice and sustainability dimensions of human-environment interactions, with a primary focus on urban and rural interdependencies. He has been awarded an Australian Learning and Teaching Council citation for Outstanding Contribution to Student Learning. His PhD is from the Australian National University.

Boris Eisenbart is a Senior Lecturer in Product Design Engineering at Swinburne University of Technology in Melbourne and External Research Associate to the Product Innovation Management Department at the Delft University of Technology. Before his current role, Boris has been an Assistant Professor in Product Innovation Management at TU Delft and worked as Postdoctoral Researcher in Design Thinking and Strategy at the University of Sydney. Boris focuses on modelling and managing complex product development processes. A second thread of his research addresses behavioural strategies and design thinking in innovation management. He holds a Diploma in mechatronics engineering from Saarland University and a Ph.D. in engineering sciences from the University of Luxembourg.

Shirin Elahi is a scenario architect. She started her professional career as an architect designing and constructing buildings and public spaces around the world. Today, she applies that practical knowledge and understanding of the creative process to build scenarios for the future as a tool for strategy and innovative change. She has worked for a diverse number of public, private and civil society organisations. She has written and lectured widely on scenarios, risk and innovation. Shirin lives in London, UK.

Kilian Gericke is Research Scientist at the University of Luxembourg leading the Engineering Design and Methodology research group. Since 2010 he works at the University of Luxembourg. Before coming to Luxembourg he worked at the Technische Universität Berlin. His main research interests are in the area of product development with a focus on design methodology and design process management, including the methodical support of designers during the early stages of product development. Kilian Gericke is elected member of the Advisory Board of the Design Society, and is chair of the Design Society's Special Interest Group on Modelling and Management of Engineering Processes (MMEP). He studied Mechanical Engineering in Berlin.

Paulina Golinska-Dawson is an Assistant Professor in engineering and management at Poznan University of Technology (PUT) in Poland. Recently she is visiting scholar at Luxembourg University. Her current research includes: remanufacturing in the automotive industry, sustainability in the supply chain, circular economy in production and logistics with focus on products' recovery. She acted as an expert to the European Commission in the working group of STRIA (Strategic Research and Innovation Agenda) in the field of vehicle design and manufacturing. She is editor in chief of the books' series "EcoProdution – Environmental Issues in Logistics and Manufacturing" by Springer International. She received her PhD degree from the Faculty of Computing and Management at PUT, in the field of "construction and exploitation of machines", specializing in the organization of production systems in the automotive industry. She graduated from the joint programme "Executive Master of Business Administration" at the University of Warsaw and the University of Illinois at Urbana Champaigne (USA).

Paula Hild is working since 2016 on her Ph.D. at the University of Luxembourg. Her work looks at circular economy practices in companies in Luxembourg. She is particularly interested in organizational and institutional dimensions that matter for the circular economy. Before joining the University of Luxembourg, she worked at the Luxembourg Institute of Science and Technology (LIST). There, she was an R&D Engineer in the competence fields of environmental assessment and management, including material flow analyses; implementation of environmental management systems in companies; Footprint methodologies such as the Carbon Footprint, Ecological Footprint; and Life Cycle Assessment (LCA). She holds Master degrees in Civil Engineering from the University of Portsmouth (GB) and in Adult Education from the University of Kaiserslautern (D).

Kristina Hondrila As Ph.D. candidate at the University of Luxembourg, Kristina Hondrila currently works on the transdisciplinary project "Anticipating future challenges at the nexus of Luxembourg's water and food systems". Under the supervision of Ariane König, she develops a framework to evaluate actionable knowledge in the context of transformative research. Kristina Hondrila has been a "cooperation facilitator" throughout her career, supporting research projects and cross-border university collaborations at the universities of Luxembourg and Saarland as well as, prior to that, citizen and stakeholder involvement in EU politics. In 2016 she completed the Certificate in Sustainable Development and Social Innovation of the University of Luxembourg, including a peer group project, the main findings of which are presented in chapter 13 of this book. She holds a MA in European Studies from KU Leuven (2002), a MA in Modern Culture from the University of Copenhagen and a BA in Philosophy from Roskilde University in Denmark.

Dr Ariane König is Senior Researcher at the University of Luxembourg. Her research focuses on facilitating and evaluating collaborative systems thinking and transformative social learning for sustainability in practice. She has developed the Certificate in 'Sustainability and Social Innovation' that served as a platform for collaboration to develop this book. She edited the book "Regenerative Sustainable Development in Universities and Cities: The role of Living Laboratories" (Edward Elgar, 2013). Before coming to Luxembourg she worked at the Universities of Harvard and Oxford, focusing on governance of new technologies and risk. Prior to this, she worked as regulatory affairs manager in a leading multinational life science corporation, and later as independent scientific consultant for the OECD, the European Commission, and EU research consortia. Ariane König holds a Bachelor, Master and Ph.D. Degree in Biochemistry from the University of Cambridge, Emmanuel College.

Sebastian Manhart was appointed to the junior professorship for Educational Science at the University Trier in 2006, and then held the role of acting chair of Organizational Education at the University of Trier. His research interests include history of science and education; history of semantics with a focus on the development of a networks of central concepts in Modernity such as life, state, market, history, and education. Another key area is the meaning and use of numbers and the development of measurement regimes and their consequences for individual and organizational learning and the development of modern society. He is a member of the core team who have developed the Certificate in Sustainability and Social Innovation. He holds a Magister (History/Economics/Sociology) from the University of Bielefeld and obtained his PhD in History from the Universities of Bielefeld and Trier in 2006.

Jules Muller just retired from his position as Director of British Telecom Global Services Luxembourg. Jules held similar positions as technical and operations manager at Infonet Services Corporation, AT&T and Reuters Ltd. Jules also lectures in the lifelong learning programmes in the field of IT applications and telecommunication. Jules is member of the consultative commission 'energy' of the commune of Junglinster, and cofounder and president of EquiEnerCoop, the first cooperative in Luxembourg in the energy sector. "equisolar 2012" is the flagship project to demonstrate how citizenship may be implicated in electrical production plants and invest in those plants. Jules holds a maîtrise in electrical engineering.

Barry Newell is an Honorary Associate Professor in the Fenner School of Environment and Society at The Australian National University (ANU). He is also a Visiting Research Fellow at the International Institute for Global Health at the United Nations University. Barry is a physicist with a focus on the dynamics of social-ecological systems. He is involved in action research aimed at developing practical transdisciplinary approaches to the co-production of knowledge and adaptive policy making. This work has led to the systems thinking and modelling approach called Collaborative Conceptual Modelling (CCM) that he and Katrina Proust have developed in the context of many multi-stakeholder workshops. He is co-author, with Robert Dyball, of the textbook *Understanding Human Ecology: A systems approach to sustainability*. Barry holds Bachelors and Masters Degrees in Physics (University of Melbourne) and a PhD in Astronomy and Astrophysics (ANU).

Simon Norcross is retired from paid employment but nevertheless manages to keep himself very busy. He takes an interest in broad environmental issues, including renewable energy and the relationships between politics, economics and environmental degradation, both locally and globally. He is involved in environmental activism on a voluntary basis. His professional career was mainly spent in information technology, developing and maintaining bespoke applications for commercial companies and international financial institutions and managing teams doing so; he also had brief spells in both marketing and development finance. He successfully completed the Certificate in Sustainable Development and Social Innovation at the University of Luxembourg; the project described in Chapter 13 of this book was carried out as part of that course. He holds an MA in Pure Mathematics from Cambridge University.

Isabel Page consults on organisational strategy and performance, and coaches senior professionals wishing to develop sustainable integrated leadership and deep communication skills. In collaboration with the core team at the Presencing Institute, MIT, she conducts action research interests focus on the design and evaluation of o2o leadership development in the context of massive open online courses (MOOCs). She also was Associate Lecturer on MBA and MSc Human Resource Management for the Open University Business School. A career with The European Parliament included effecting the 2004 EU enlargement transitioning on a zero-based budget as a member of the Secretary-General's internal organisation team. In 2010, voluntary sustainability field work in Mexico fostered an introduction to Dr C. Otto Scharmer and team at MIT's Presencing Institute. From this emerged the Certificate session on social technology and Theory U, and offline tutorial support for the Luxembourg Hub of MIT's innovative U.Lab MOOC, hosted by the University as a Certificate antenna event. Isabel holds an MBA from the Open University Business School, UK.

Michael A. Peters is Professor of Education at the University of Waikato (NZ) and Emeritus Professor in Educational Policy, Organization, and Leadership at the University of Illinois at Urbana–Champaign (US). He is the executive editor of the journal, Educational Philosophy and Theory, and founding editor of five international journals, including The Video Journal of Education and Pedagogy (Springer), Open Review of Educational Research (T&F). His interests are in philosophy, education and sustainability, and he has written some eighty books, including: Wittgenstein and Education: Pedagogical Investigations, (2017) with Jeff Stickney, Paulo Freire: The Global Legacy (2015) with Tina Besley, Environmental Education: Identity, Politics and Citizenship (2008) with Edgar González-Gaudiano. He has acted as an advisor to governments and UNESCO on these and related matters in the USA, Scotland, NZ, South Africa and the EU. He was made an Honorary Fellow of the Royal Society of NZ in 2010 and awarded honorary doctorates by State University of New York (SUNY) in 2012 and University of Aalborg in 2015.

Bérénice Preller is currently in the final stage of her PhD at the University of Luxembourg. Her works looks at sustainable building policies and projects in two European cities: Luxembourg and Freiburg in Germany. She is particularly interested in argumentation lines and justifications for a change towards green building. Before joining the University of Luxembourg, she held several positions in the public and private sector, including in a consultancy in Luxembourg, a German local authority. Bérénice studied Political Sciences as well as Regional and Urban Policies at the Institut d'Etudes Politiques de Paris, France and the London School of Economics in the UK.

Katrina Proust is a Visiting Fellow in the Fenner School of Environment and Society at the Australian National University in Canberra, and Visiting Research Fellow at the United National University International Institute for Global Health in Malaysia. Her research interests address the historical factors and the feedback dynamics that shape relationships between humans and our environment. With Barry Newell she has developed Collaborative Conceptual Modelling, an approach which provides a conceptual framework to support studies of social-ecological systems. Katrina has contributed to the work of the

International Council for Science (ICSU) in the Regional Office for Asia and the Pacific (ROAP). She has contributed to the ROAP Science Plan on Health and Wellbeing in the Changing Urban Environment: a systems approach (2012). She has a PhD from the Australian National University in environmental and applied history, with a focus on complex human-environment relationships.

Jerome Ravetz is an Associate Fellow at the Institute for Science, Innovation and Society at Oxford University. His main research interest is in the social and ethical problems of contemporary natural science, together with sustainability studies and the management of uncertainty and quality in science. He is a member of the core team who have developed and implemented the course SCCS over the last five years. In 1971, he published his seminal work, Scientific Knowledge and its Social Problems. (Oxford U.P. 1971, Transaction 1996). With Silvio Funtowicz he published Uncertainty and Quality in Science for Policy (Kluwer 1990), where the 'NUSAP' notational system is developed. Also in collaboration with Silvio Funtowicz, he has developed the theory of 'post-normal science', which is the appropriate form of inquiry when 'facts are uncertain, values in dispute, stakes high and decisions urgent'. He has a PhD in Pure Mathematics from Cambridge University, and built his academic career in History and Philosophy of Science at Leeds University.

Aydeli Rios is a pedagogue who leads her own firm as global consultant and corporate trainer in management, leadership, and women empowerment. Since 1994, she has occupied positions in education and corporate. Professor of Humanities at the Universidad Autónoma de Nuevo León and in Entrepreneurship at the Instituto Tecnológico y de Estudios Superiores de Monterrey, in Mexico. In 2010, she worked for LEGO Operaciones de Mexico as e-Learning supervisor where she was involved with the TWI Job Instruction methodology (USA) in the moulding department's plants of America and Europe. In 2017, she founded a social organisation that supports gender equality worldwide. Aydeli has a B.A. in International Relations from ITESM/University of Wisconsin-STOUT; a master degree in Psycho pedagogy from the Escuela de Ciencias de la Educación (MX); certificates in Women's Empowerment from UN Women; and the Certificate of Sustainability and Social Innovation at the University of Luxembourg.

Christian Schulz is Professor of European Sustainable Spatial Development and Analysis (since 2006) and currently Head of the Institute of Geography and Spatial Planning at the University of Luxembourg. His research foci are in the fields of regional governance in Europe and environmental economic geography. With Boris Braun (Cologne), he co-authored a textbook on Economic Geography (UTB, Stuttgart, 2012). Current projects include green building transitions, circular economy policies and alternative economies, as well as the challenges and opportunities of post-growth dynamics for economic geography. He studied geography at the Universities of Saarbrücken/Germany, Québec/Canada and Metz/France. He holds a PhD (Dr. phil.) from the University of the Saarland (1998) and completed his habilitation at the University of Cologne/Germany in 2004 with a study of environmental producer services and their impact on the environmental performance of manufacturing firms.

Susanne Siebentritt is a Professor in physics and heads the laboratory for photovoltaics at the University of Luxembourg. Her research interest is twofold: the development of new thin film solar cells and the semiconductor physics of the materials used in these cells. In 2014 she received the FNR Outstanding Publication Award, together with three co-authors. In 2015 she was awarded the "Grand Prix en Sciences Physique – Prix Paul Wurth" of the Luxembourgish Institut Grand Ducal. She is vice chair of the scientific council of Science Europe and a board member for the Kopernikus projects, a 10 years research programme for the energy transition of the German Ministry of Education and Research. She studied physics at the University of Erlangen and received her doctoral degree from the University of Hannover.

Philipp Sonnleitner is Head of Test Development at the Luxembourg Centre for Educational Testing where he is applying his expertise to the Luxembourg school monitoring program. His personal mission in research and teaching is to develop valid, high-quality assessment methods and to give guidance concerning their correct and fair application in educational and psychological contexts. In the course of his career, Philipp Sonnleitner has been involved in developing tests for the school monitoring programs in Austria and Luxembourg, as well as the OECD's PIAAC-, and PISA assessments. He also gathered extensive experience developing instruments to capture competencies ranging from college aptitude to complex problem-solving behavior. The latter research project resulted in a freely available and internationally recognised test: www.assessment. lu/GeneticsLab. He received his Master's degree in Psychology from the University of Vienna in 2007 and is a licensed work- and occupational psychologist. His thesis received the German Society of Psychology's best master thesis award. In 2015, he was awarded his PhD in Psychology with honors from the Free University of Berlin.

Olivier Thunus is senior statistician at the National Statistical Institute of Luxembourg. He is responsible for Energy and Environmental accounts. In 2012, he was invited by the Ministry of Sustainable Development to participate in the experts group requested to select a new set of SD indicators for Luxembourg. Since 2014, he is the president of the "environment statistics" group of the Official Statistics Commitee. Since 2016, he is a member of the SDG indicator Task Force for Luxembourg. After studies of Bio-engineer, he obtained his PhD in Environmental Science at University of Liege. He also took a Master of Economy specialized on energy and sustainable development from the University of Grenoble.

Arjen Wals is Professor of Transformative Learning for Socio-Ecological Sustainability at Wageningen University in The Netherlands. Furthermore, he is the Carl Bennet Guest Professor in Education for Sustainable Development at IDPP, Gothenburg University and he holds the UNESCO Chair of Social Learning and Sustainable Development. His teaching and research focus on designing learning processes and learning spaces that enable people to contribute meaningfully sustainability, by creating conditions that support (new) forms of learning which take full advantage of the diversity, creativity and resourcefulness that is all around us, but is rarely tapped? He is editor and co-editor of a number of popular books including more recently the "Routledge International Handbook on Environmental Education Research" (2013) and "Envisioning Futures for Environmental and Sustainability Education" (Wageningen Academic, 2017). Recently he also contributed as a senior policy advisor to UNESCO's Global Education Monitor 2016 Report 'Education for People and Planet' on the role of education in helping realize the UN's Sustainable Development Goals. He maintains a popular blog at www.transformativelearning.nl

Gregor Waltersdorfer is a Ph.D. candidate in Engineering Design at the University of Luxembourg since 2013. His topic is about harnessing the role of meaning in products and services to foster eco-sufficient user behaviour. He received his Master degree in Mechanical Engineering and Business Economics from the Graz University of Technology in 2011. After that, he worked as an energy-efficiency business consultant at Stenum GmbH for 2.5 years in Graz.

About this book

Purpose: How can we transform society towards greater sustainability? This book presents arguments that any such endeavour will rely on sustainability science. Sustainability science is a collaborative and future-oriented inquiry in very diverse groups, whose goal is a better understanding of and action upon complex and dynamic social-ecological-technological systems. This sort of inquiry will only allow co-creation of actionable knowledge if contradictions and incongruences from diverse expertise, worldviews, and associated values are embraced as creative space by those engaged. This requires a critical mind-set about how we know and what we can know and the ability to put yourself into someone else's shoes to understand their perspective even if it differs from yours. The book presents guidance and resources to build capacity for engagement in sustainability science within the community of science and in the broader society. It is based on an original conception of sustainability science as a transformative process that builds on established concepts of social learning, post-normal science, complex systems, and uncertainty. Sustainability science has the potential for transforming human–environment interactions by enabling the emergence of an innovative social practice.

Use: Accordingly, the book is designed for use in diverse settings. For use in higher education, this book provides a comprehensive coverage to serve as a primary text for university courses on the theory and practice of sustainability science. These include courses in traditional master's-level or doctoral school programmes or part-time certificates that target mixed audiences of students and professionals. Elective courses that can be chosen by bachelor's programmes from different study courses in their later semesters can also be designed based on this book, provided that groups of enrolled students will represent a fair diversity of perspectives and experiences. For all such courses, it has to be ensured that the learning environment in which they are staged will allow for the active engagement of course participants. The textbook is designed also to give an appropriate basis for problem- and project-based learning in sustainability, as this is a key auxiliary activity that is closely associated with the course at the University of Luxembourg on which this book is based. The book will also prove useful in the design of executive training programmes or other professional trainings, as well as workshops organised by governments or firms or organized civil society groups.

The target audience includes persons who consider designing and leading such a course, as well as course participants. Courses will be most successful that can recruit diverse participants, for example, composed of advanced undergraduate and post-graduate students as well as professionals with diverse backgrounds and interests, together with engaged citizens. The intention is to equip engaged readers for connecting learning, research, and practice in order to foster transformative learning for the co-creation of knowledge, science, technologies, and social practices for more sustainable societies.

Content and organisation of the book

Content: The book provides recent insights on research related to sustainability science and the most salient issues associated with transformation for sustainability from diverse fields of knowledge, such as geography, human ecology, physics, and the educational sciences. We have taken care to draw on internationally salient issues and examples. It also provides the conceptual tools and methods required to collectively question this knowledge and reframe the issues to adapt them to different contexts. In sum, the book equips readers with a better understanding of how one might actively design, engage in, and guide processes by which society and the environment interact. This will require embracing complexity, interdependencies, uncertainties, and contradictions emerging from diverse perspectives, values, and worldviews. Each reader of this book thus will have guidance on how to create and/or engage in similar courses in their own context and to apply their lessons in practice.

The introductory chapters on sustainability science and transformative learning set out core concepts. The remainder of the book is organised in three parts:

Part I. Embracing complexity and alternative futures: conceptual tools and methods

Part II. What can transformations look like? Sectoral challenges and interdependence

Part III. Tracking, steering and judging transformation

Each chapter in these parts offers a reflective account of a particular challenge, an approach by which this challenge is being researched and/or addressed, a critical discussion of the merits and limitations of the approach, and a conclusion on the significance of the insights gained. For improved accessibility, chapters can include text boxes to highlight classic studies that have contributed to a sea change leading to the perspective presented in the chapter, particular cases in which the approach has been leveraged, or suggestions on how to bring the described approach to life in class in an engaged learning activity.

Pedagogical material: In this book and the underlying course we provide and critically discuss conceptual tools for the collaboration of scientists from diverse disciplines, professionals, and students. Complex systems are explored from diverse or even conflicting perspectives. The ability to hold diverse conflicting perspectives in one mind to offer a new creative space is trained; only with this transformative ability can we start to develop new shared expectations and social practices, with agreed goals and actions to realise these.

Transformative learning must rely on collaborative learning in diverse groups, organizations, or networks. We therefore provide two chapters that provide insights on theory from the learning sciences on transformative learning for sustainability, the design of learning environments, and tools and artefacts for problem-based experiential learning in groups and team teaching. The challenge of evaluating transformation from learning and in practice from diverse perspectives and developing metrics for success is addressed in the concluding chapters.

The book also provides useful guidance on how to establish group projects in a higher education setting: the complementarity of theory presented and discussed in the form of lectures, meetings, and experiential problem-based learning in peer groups is deemed central to successful transformative learning. This approach enables the provision of practical guidance on how to integrate education and research and thereby to effect change in practice.

To work productively in groups on points of dissonance, contradictions, and knowledge gaps in terms of both known and unknown unknowns depends on learning environments that are carefully designed and guided so that these can be seen as safe spaces for creativity and innovation of expectations and practice.

How was this book developed?

The book is based on a model course taught at the University of Luxembourg that has been running successfully for over four years and that is now embedded as a centrepiece in an innovative study programme in 'Sustainability and social innovation'. This programme is designed as a platform for universities to engage with citizens in fostering social and technological change for sustainability. The courses in this programme are unique in that they are designed for part-time study in conjunction with disciplinary degree programmes for advanced undergraduate and post-graduate students, as well as for professionals in full-time employment. The study programme also requires engagement in collaborative scientific inquiry in peer groups in order to address salient problems of Luxembourg's transition to a more sustainable society. The content and approach of this book thus have been developed with extremely diverse groups of course participants and contributors, drawing on a wide range of disciplinary perspectives and experience. Furthermore, course participants usually came from at least four continents.

Most of the authors have worked together over a period of four years to develop a matured shared language, with common conceptions of sustainability and of the roles of integrative science and citizens in social learning processes. Through this process, authors share the new vision of a science that draws on diverse logics of knowledge creation. These include, respectively, the empirical/theoretical science prevailing in natural science, the interpretative science prevailing in humanities, the practice-based sciences such as architecture and engineering, and finally the 'citizen science' of concerned collectives that face specific local sustainability challenges.

Peer group projects: Peer groups are encouraged to experiment with methods for systematically tapping into collective intelligence by engaging experts and practitioners, largely based on methods presented in the chapters of Part I of this book. The projects were co-created together with academics and stakeholders who are active in transition practice in the public sector, private sector, or organized civil society in Luxembourg and some of whom provide guidance to the projects. Peer group topics over the past three years have included 'Democratizing renewable energy', 'Developing local sustainability indicators', 'Contributing towards a sharing economy', and 'Creating actionable knowledge from community based water monitoring'. Chapters presenting sectoral perspectives in Part II of the book have also been developed to serve as primers to start off peer group projects.

Participants in each group are selected based on their motivation and diversity within the groups with respect to age and expertise/disciplinary backgrounds. Peer group projects are participant led in that the group remit initially has a broad scope and the group has to develop their own problem framing and specific approach. For example, peer groups on democratizing renewable energy first had to agree on an angle of interest to all group members, such as how the group may develop resources relevant to the establishment of citizen cooperatives in the energy sector. The groups are asked to draw together the more abstract academic literature resources on social innovation and the structure of the energy sector, together with legal and regulatory documents, as well as practical information on local emerging citizen cooperatives, specific local problems, and finally information from direct interaction and guidance on needs by active members of three energy cooperatives in Luxembourg. The main challenge encountered in organizing peer group projects is the resource intensity of guiding groups to establish a social learning process with stakeholders on complex problems.

Evaluation of sustainability science, including peer group projects, is not trivial (König, 2015). In the study programme we distinguish four overlapping objectives of evaluation: self-evaluation to reflect on what has been learned at the individual level, diagnosis of what has not been learned, evaluation for the certification of learning towards a degree, and and diagnosis of areas for improvement of the course and programme at large. Accordingly, the evaluation of peer group projects in the certificate relies on judgments at several levels: self-evaluation by participants; peer group self-assessment in a group discussion; and evaluative feedback by the 'peer project steering group', that is, the facilitators, together with key stakeholders in the project. Evaluation through multiple perspectives is deemed more valuable for learning than just drawing on one perspective. First, at the end of each semester a 360-degree feedback is organized, where each participant is asked to evaluate their level of engagement and that of all other members in the group. The peer group then reflects on their work, pooling individual reflections in a group judgment of strengths and weaknesses of their work. These reflections are included in the peer group presentation and in the final written report. These in turn are evaluated by group facilitators and the course coordinator. The external stakeholders' part of the evaluation through the project steering group is essential for a social learning platform.

Acknowledgements

First I would like to thank all contributing authors for the rich exchanges leading to this book. Each author introduces concepts and tools in order to critique them by pointing out their merits and limitations. Exploring boundaries to meaning-making in this way will often be somewhat uncomfortable, especially when one's own favourite conceptual tools to which one has dedicated years of one's research are put under scrutiny. I am grateful to everyone in our book team for engaging at this level. Without Tea Sikharulidze's dedication, reliability, and attention to detail, assembling this book would not have been possible. Tea has not just been fulfilling her role as student assistant, but she has become a core member of the book team.

Special thanks are also due to all who have at any point in time joined our teaching team in the certificate. All lecturers are always asked to start with a reflection on their personal motivations and interests and how these shaped their approach to understanding and acting on sustainability challenges. Writing the book was only possible after five years of learning from these sets of lectures for the certificate – every contribution counts.

I would also like to highlight the key role that the Conseil Supérieur pour un Développement Durable (CSDD) played, by opening the doors to the practice of transformative sustainability science in Luxembourg. This advisory group was open to my suggestion to stage Luxembourg's first scenario project for education, which was then developed together with Gerard Drenth, following the Oxford Scenario Planning Approach (OSPA) described in this book. Without the trust and readiness to experiment, in particular by Francis Schartz, its president, and Marguy Kohnen of the Luxembourg Ministère pour un Développement Durable et Infrastructures (MDDI), the practice of transformative sustainable science would not have gained its first foothold in Luxembourg. This project offered invaluable experience that also served as the foundation to writing this book. It is also thanks to this successful start in the CSDD, that now Minister of the Environment Carole Dieschbourg, Secretary of State of the MDDI Camille Gira, and Vice President for Research of the University of Luxembourg Ludwig Neyses have decided to contribute to fund the first larger five-year project for the practice of transformative science towards fostering sustainable water governance in Luxembourg (this is described in Chapter 19).

Jerry Ravetz has been the most steady travel companion on the seven-year-long journey, from the inception of the first course for the certificate, to its launch, and then writing of this book, one could wish for. He has been there, coming across once to twice a semester from Oxford, always ready to listen, critique, provide advice, and be immensely motivating as he saw the programme mature. I could not wish for a better schooling in critical thinking about science, democracy, and styles of thinking we may need to propagate in order to live up to new existential and political challenges of the twenty-first century. As mentors, I would also like to thank Sheila Jasanoff and John Holdren; my time spent with them at Harvard's Kennedy School of Government was deeply formative. My time at Oxford's Institute for Science and Civilisation working with Steve Rayner (just before my daughter was born) was short, but full of inspirational impulses stimulating creative thinking.

To my daughter, Laurine König, I am grateful for her understanding that on some days the book's completion had to take precedence; she has taught me many things that have gone into making this book. Like Oscar Wilde, I think one's best education for life can be received from one's children. Last but not least, I owe more than just a mere thank you to my parents, Brigitte and Dieter König, who always were ready to read, provide critical feedback, and engage in discussion, always there when needed and just always immensely encouraging.

Ariane König, 29 July 2017

Introduction

1   Sustainability science as a transformative social learning process

Ariane König

Existential challenges of the twenty-first century

Throughout history mankind has caused unintended changes in the natural environment, including phenomena such as extinction of vulnerable species and soil erosion from deforestation (Crosby, 2015). These changes, however, were confined to local or regional scales. Unprecedented in the twenty-first century is that science and technology associated with industrialization, whilst they have further transformed the conditions of life for humanity, now create environmental impacts at the global scale. Our socio-industrial-agricultural metabolism and land-cover changes are significantly altering the earth system's natural bio-geo-chemical cycles and threaten the integrity of the biosphere, which both when unperturbed contributed to maintain the relatively stable conditions under which humanity has thrived over the last 10,000 years.

The existential problems of the human civilization in the twenty-first century are complex as they relate to interactions between humans and their environment in a world that is experiencing accelerating changes in the technological, cultural, political, economic, and environmental spheres. Moreover, changes in all these spheres are globally interconnected and interdependent. Traditional disciplinary fields of 'normal' science and static approaches to management and governance relying on prediction, regulation, and control can play only a limited role in resolving such complex problems. Experts in different fields of knowledge in the natural and social sciences often fail to understand each other, and we are drowning in specialised expert reports that often do not sufficiently account for interdependencies and feedbacks between changes in these different spheres. New approaches to combining knowledge co-creation and distributed governance should be designed to enable the effects of human-environment interactions to be continuously and iteratively monitored, evaluated, judged, and acted upon, based on social norms that respond to locally determined sustainability issues.

This book's main premise is that one of the most fundamental challenges we face as we are affecting the functioning of the entire planetary system, is to develop new approaches to knowledge co-creation and governance that will enable us to relate to our environment and to each other in view of more complex interdependencies between local and global circumstances, in how we conceive of the natural and social worlds, now and in the future. The silo-based approach to science and expertise, government and practice, with strict separation of research in the natural sciences and social sciences and the humanities that has co-evolved with industrialization, is no longer adequate for our civilization to cope with twenty-first-century challenges. New approaches to combining research, governance, and learning in communities of public authorities, stakeholders, and scientists are required to complement our increasingly fragmented knowledge fields and to learn to change the unsustainable ways in which we relate to our environment and to each other. These new approaches should draw on new opportunities for new forms of knowledge co-creation and governance in a networked society; the hope is that such new forms of knowledge co-creation with broader public engagement might also contribute to address the current great risks of democratic disengagement and post-truth politics (as highlighted by Wals and Peters, Chapter 2 in this book). Not only may we have reached an end to a long period of stability in the biosphere of our planet, providing us with living conditions in which we humans thrived, but we may have also reached an end of a period of stability of the type of specialised expert knowledge that governments have drawn upon in modernity (or 'epistemic stability') (Maggs & Robinson, 2016). In the words of Michael Hulme, 'where science is practiced, by whom and in what era, affects the knowledge that science can produce' (Hulme, 2009, as cited in Maggs & Robinson, 2016).

In this book, we therefore introduce a new conception of science that emerges from dialogues on the need for a more future-oriented and systemic social learning process. The overarching goal of this 'transformative sustainability science' is for humanity to become better at embracing complex relationships and interactions (complexity) that depend on local settings and people (contingency); contradictions and conflicts between diverse worldviews; and associated sets of values, uncertainty, and ignorance when interacting with each other and with our planet's life support system which we are a part of. We need to become better at understanding the social and cultural dimensions of these challenges to knowledge creation (Maggs & Robinson, 2016; Castree, 2015; O'Brien, 2015). A starting point is to develop a critical and reflective mind-set towards what we can and cannot know, how we know, and why we want to know in an age of accelerating and interconnected change. Furthermore, we need to practice to understand different facets of complex situations through very different perspectives rather than just defending a viewpoint through a particular (organizational) lens associated with a particular set of interests (see also Chapter 2). Similar questions can be asked about learning – with a focus on how we might better link learning in individuals, organizations, and societies. In these processes, scientific analysis serves as a basis *not* for predication and control, but for social learning (e.g. Ison et al., 2007). Accordingly, such approaches should enable reflection, review, and transformation of the prevailing social structures and practices, technologies, research, and learning approaches in the light of new learning Digital technologies for networked societies have a role to play.

This implies fundamental changes in - or transformation of - currently prevailing mechanisms for social coordination, including organisation of the economy and governance in our society. At present this is largely accomplished through hierarchical government and through market rules that assume competing utility-maximising rational actors (e.g. Ostrom, 2010). Figure 1.1 provides an overview on transformations in prevailing patterns for social organisation, economy and communications over the course of human evolution. The emergence of the networked knowledge society in the twenty-first century presents huge opportunities and novel types of virtual spaces that can be leveraged in such a transition and is already significantly altering social coordination, including ways in which knowledge is co-created.

The remainder of this chapter introduces the practice of sustainability science as one promising new approach to organizing knowledge co-creation across silos of expertise and practice. First, we introduce sustainability science and different

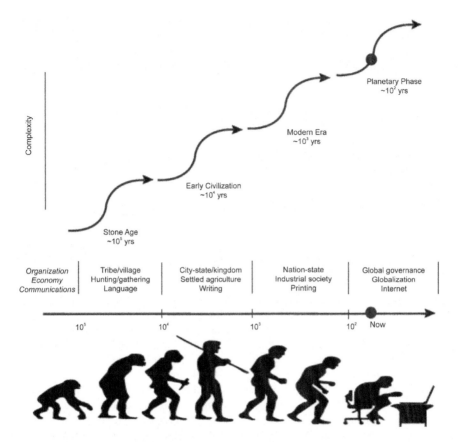

Figure 1.1 Phases from human history
Source: Adapted from Paul Raskin, 2016

meanings attributed to it in the academic literature. Second, we discuss why this approach to science may be considered problematic. Third, we explore a diversity of purposes and approaches that can be associated with sustainability science. Fourth, we focus on transformative sustainability science as one approach that is akin to a transformative social learning process and ask what exactly we envisage shall be transformed. The second part of this chapter presents the structure of the entire book and summarises the individual chapters.

What is sustainability science and what makes it transformative?

Sustainability science is an emerging approach to science that actively seeks improved ways for governing human–environment interactions. One main goal is to leverage social and natural science and technology for a transition towards sustainability (Jerneck et al., 2011). Some engaged scientists consider sustainability a field of knowledge that relies on integrating across diverse established fields of knowledge (Kates et al., 2001; Clark & Dickson, 2003; Clark, 2007). One of the first publications referring to 'sustainability science' describes it as an interdisciplinary field of research that takes a systems approach to understand dynamic interactions between nature and society (Kates et al., 2001; Clark & Dickson, 2003; Matson et al., 2016). Core research questions include: 'How can today's relatively independent activities of research, planning, observation, assessment and decision-support be better integrated into systems for adaptive management and social learning?' (Kates et al., 2001).

Some leading scholars distinguish between descriptive and transformative sustainability science (Wiek et al., 2011; Miller et al., 2013; Wiek & Lang, 2015). On the one hand, descriptive sustainability science that is largely practiced by scholars in the environmental and earth sciences states as a main objective the building of a body of knowledge on the functioning of the earth system to inform the traditional top-down policy-making processes. For example, a group of very successful researchers set out to determine planetary boundaries for creating a safe operating space for humanity that is to inform evidence-based policy making (Rockström et al., 2009; Steffen et al., 2015). On the other hand, in transformative (or sometimes also called 'transformational') sustainability science, in line with pragmatic views on knowledge and science, the stated goal is to fundamentally change human–environment interactions and all associated social practices and expectations in a process that will necessarily question prevailing values, worldviews, and ways of knowing and doing. Such research at its best achieves the combination of approaches that are critical and challenge driven. (This book refers to 'challenge driven' rather than 'solution oriented', as we consider sustainability challenges in highly dynamic worlds not as 'solvable', but as amenable to being addressed in an iterative learning process over time.) Sustainability in this case can be considered an emergent phenomenon, consisting of sets of new expectations and social practices from a societal conversation that is scientifically informed (Robinson et al., 2013).

Together with a range of other scholars in this book we thus take a procedural view and consider transformative sustainability science as a social deliberative process, a different approach to conducting science that engages experts and stakeholders with diverse perspectives on complex challenges (Miller et al., 2013; Wiek & Lang, 2015; Robinson et al., 2011; Schneidewind & Singer-Brodowski, 2014; Schneidewind et al., 2016; Grunwald, 2016; König, 2015). Similar to related calls for a new social contract for global change research within the field of geography (Castree, 2015), an approach to research is sought that takes the human dimension seriously. One requisite to this research is to clearly acknowledge the political dimension in the conduct of science – for example, by problematizing human agency in pluralist societies, in which diverse groups defend different sets of values. A monolithic view of the role of science as preaching simple truths to power is seen as inadequate by many scholars concerned with developing science for sustainability (O'Brien, 2015). We also require innovation in the ways that science can foster changes in social practice in as many diverse groups as possible, not merely by serving as expert-shaped evidence for policy makers.

Transformative sustainability science starts from the assumption that in order to change social practice affecting how we understand and relate to our environment and draw on environmental and social resources, research needs to be co-created in a collaborative process that connects diverse disciplinary perspectives with practice. Figure 1.2 describes a process for co-creation of four knowledge types that relies on rigorous participatory scientific inquiry and quality control. In the boxes shaded in grey relying on dialogue between diverse participants, science is on tap, not on top. Different methods can be used in different stages to ensure that a future oriented

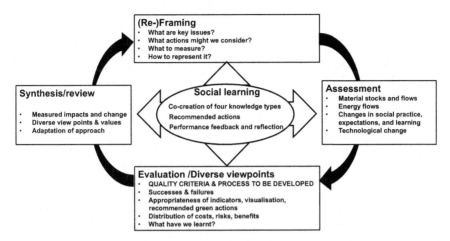

Figure 1.2 Transformative sustainability science: an iterative process with four stages **(Re-) Framing:** A dialogue for participatory consideration of the following questions: What are key issues? What actions might we consider? What should be measured? How shall we represent our findings? **Assessment:** Serves to characterize changes in the core variables over time and space in terms of material and energy stocks and flows; technology, its uses and users; and social practice, expectations, and learning and how these may be interrelated. **Evaluation:** A dialogue for participatory development of criteria and a process to determine diverse perspectives on successes and failures; distribution of costs, risks, and benefits; appropriateness of indicators; visualisation/representations; and recommended actions addressing the question 'What have we learnt'? This takes a humble attitude by all engaged about what and how we know and requires empathy (based on Jasanoff, 2003). **Synthesis/review:** Measured impacts and change are reviewed, as are diverse viewpoints and values, as well as joint judgments, leading to reframing the adaptation of our approach.

systems view that is sensitive to diverse perspectives is adopted. These can include collaborative conceptual systems mapping (CCM) and scenario practice for framing and visioning to structure evaluation. Such a participatory process can be conceived as an iterative cycle in four stages that include participatory framing (or re-framing) to jointly define a system of interest within a problem space and to understand which variables and relationships may matter most and how boundaries might be drawn; assessment; participatory evaluation; and synthesis, leading to a new round of reframing (see Figure 1.2). These stages, however, are not as cleanly separated in time in practice as suggested by this figure, but rather are overlapping and non-linear as they occur in time. Requisites to such knowledge co-creation processes are that they embrace (i) complexity by directing attention to the relationships between different parts of a system and engaging diverse disciplines, including critical and challenge-driven inquiry across the natural, social, and engineering sciences (see Chapters 3, 4, and 5); (ii) contingency by connecting theory and practice and drawing on and producing place-based knowledge – local and very diverse forms of knowledge count, as well as knowledge about different viewpoints of the present and possible future worlds (Chapter 6); (iii) contradictions and trade-offs by taking an actor-oriented perspective; and (iv) uncertainty and ignorance (see also Chapter 16 and the postscript by Jerome Ravetz). Concepts further differentiating uncertainty, such as 'technical' and 'methodological' uncertainty, as in the NUSAP system (Chapter 16, Funtowicz & Ravetz, 1990; van der Sluis, 2005), are deemed important so as not to entirely lose sight of what we don't know in terms of known and unknown unknowns. 'Ignorance', for example, can refer both to external realities and to our self-awareness. The distinction between 'surprise' (discovering unknown unknowns) and 'denial' (after Freud, refusing to be aware of the known) are also very useful. All such concepts are highly effective for analysing the states of knowing and knowledge, and of non-knowing and non-knowledge, that occur in meeting sustainability challenges and will be deployed where appropriate through the book.

The ambition of embracing complexity will often also facilitate directing attention to and deliberating what we do not know. The idea of 'complexity' has many roots in practice and many approaches. In this book, we present complex systems as a structured view of parts of a bounded whole that is developed for the purpose of investigating relationships between the parts. The structure of a system is designed as a function of the purpose of the analysis. Where to draw the system boundary for study is the first task of any inquiry; it depends partly on the complex system that the inquiry is part of in itself. Uncertainty is present to all degrees, and there is no privileged perspective among participants (see also Ravetz, 2006, and Chapters 3, 4, and 5, and the postscript). The focus on relationships with possible time delays will call into question many simplified models developed for the purpose of close investigation of specific cause–effect relations, in a particular sub-system at a particular scale such as the economy, an ecosystem, or a particular cell. Adopting a systems perspective whilst acknowledging uncertainty enables gaining an integrated understanding and repertoire of action on situated problems. Quality assurance is a key concern in these new approaches to knowledge production (Funtowicz & Ravetz, 2015; Haklay, 2015).

What knowledge for sustainability?

The procedural view of sustainability science developed in this book stems from a pragmatic philosophical perspective on science, technology, and knowledge. It presents questions on what it might mean to 'integrate' learning, research, and practice for the purpose of gaining an integrated understanding by drawing on insights from different theory and research approaches that have emerged from disparate disciplines.

The philosophical perspective of pragmatism further invites us to reconsider the relation of the human mind with the universe, not as representational but as causal. The languages and symbols we use to evoke reactions and convey meanings are not pictures of reality, but are rather part of the causal network that bind humans to their environment – tools of an intelligent animal. According to Richard Rorty and others, knowledge resides in interaction with others and the environment and in the contemplation thereof. It is fluid and dynamic; it is the basis for our beliefs, which we then make explicit in their justifications. Knowledge is simply what we are justified in believing, and justification is a social/relational phenomenon. Pragmatism seeks to avoid confusion between 'the human justification of knowledge claims' and 'causality in reality'. Pragmatists replace the 'appearance'/'reality' distinction with descriptions of ourselves and the environment that are 'more or less useful'. Similarly, science is conceived as not serving to represent, but rather to manage realities. And according to Rorty, 'the purpose of inquiry is to achieve agreement of what to do, with what end, by what means to make life better – any other inquiry is just word play'. Accordingly, the quality of science is revealed in its power on how to make life better (Rorty, 1999).

In this book, it is assumed that the more we are aware of and embrace the culture–technology–environment connection in our knowledge co-creation processes, the more successfully we will be able to transform prevailing ways of thinking and doing to make them sustainable. Technology, which significantly contributes in shaping human–environment interactions, can be seen as an expression of not only scientific knowledge, but also the prevailing cultural beliefs and worldviews that shape our relation to nature (see, for example, Parodi, 2008, p. 15; Dyball & Newell, 2015). In fact, technology is shaped by but also contributes to changing prevailing values and worldviews.

Building on Sir Geoffrey Vickers's work (1984) investigating the relation between (scientific) theory and practice in an 'appreciative setting', we distinguish three dimensions of judgments in societal deliberative processes that are to lead to more fundamental changes in prevailing social practice in the form of more stable norms and customs. He describes 'appreciation' as a deliberative process by which judgments are formalized in diverse groups of scientists and stakeholders. Accordingly, building on Vickers and Burt and van der Heijden (2008), we distinguish three types of knowledge that appreciative judgments are based on that will influence changes in social practice, technological innovation, and shared expectations and notions of progress that guide these changes:

(1) **Knowledge on perceived realities** requiring an understanding of human–environment relations as complex dynamic systems;
(2) **Knowledge on values** defended by diverse individuals and groups, from which a direction or purpose can be deduced and which can also present grounds for conflicts, contestation, and polarization to the extent of blocking learning and decision making in value pluralist societies; knowledge on values also facilitates cognitive switching and building empathy by looking at challenges from diverse perspectives;
(3) **Knowledge for action**, or 'actionable knowledge', which combines the two previous sets of knowledge to arrive at judgments on socially robust and acceptable and desirable courses of action, with insights on the functioning of organizations and institutions as sedimented forms of social practice that may pose barriers to change or lock-ins of undesirable practices and some strategic insights on how these might be overcome.

In line with Armin Grunwald's research on understanding sustainability (Grunwald, 2016), we add a fourth type:

(4) **Knowledge about emergent futures**, which can be gained from participation in participatory scenario approaches and visioning practices, which conveys a sense of shared imaginaries about the future (see also Jasanoff, 2015), together with their social robustness in terms of their aggregate desirability, acceptability, and plausibility. Joint deliberation on alternative futures also helps to make explicit uncertainties, areas of ignorance, and potentially disruptive events.

Such knowledge also helps cognitive switching between past, present, and alternative future worlds. The question of 'How can the future be scanned in a creative, rigorous, and policy-relevant manner that reflects the normative character of sustainability and incorporates different perspectives' is critical in seeking reorientation in discussions on the meaning of progress and development trajectories (Grunwald, 2011; Swart et al., 2004; Robinson et al., 2011). The conception of transformative sustainability science of this book offers a process that allows the emergence and explicit deliberations on these four forms of knowledge (see Figure 1.3), and to draw on all four to gain an integrated understanding of the complexity of the issues and socially robust courses of action from them (König, Chapter 19).

Core research questions that prevail in transformative sustainability science are concerned with better understanding from diverse stakeholder perspectives what is 'actionable knowledge' and further improving methods for stakeholder interaction and learning to change their social practices with methods such as systems thinking and alternative futures. They explore the design of more interactive processes that enable the exploration of highly complex and uncertain value-laden issues and

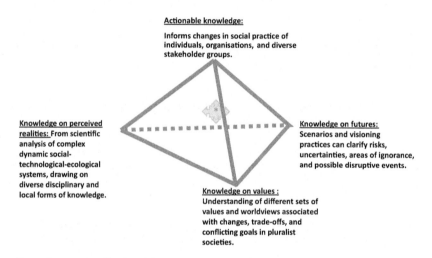

Figure 1.3 Four knowledge types in transformative sustainability science

Knowledge on perceived realities: From scientific analysis of complex dynamic social-technological-ecological systems, drawing on diverse disciplinary and local forms of knowledge.

Knowledge on values: Understanding different sets of values and worldviews associated with changes, trade-offs, and conflicting goals in pluralist societies.

Knowledge on futures: Imagined futures serve to motivate change in the present. Scenario practices can clarify risks, uncertainties, areas of ignorance and possible disruptive events, and alternative possible futures and their trade-offs. A vision provides a desirable future that can help orientation and concerted action in diverse groups; it is normative.

Actionable knowledge: Informs changes in social practice of individuals, organizations, and diverse stakeholder groups.

bring these processes closer to social learning and action in real life (Robinson et al., 2011; König, 2013, 2015; Wiek & Lang, 2015; but also Dyball & Newell, 2015; and Newell and Siri, 2016). One further essential challenge remains: to understand and try to track social change that allows for the emergence of new forms of production, consumption, and distribution with new forms of combinations of technologies, organization, institutions, and lifestyles (Jerneck et al., 2011). This will require directing more attention to how new knowledge, science, and technologies are actually co-produced with social practice that is then stabilized in the form of social values, norms, rules, and institutions (Jasanoff, 2003; Jasanoff, 2015).

Transformative sustainability science can thus also be understood as a transformative social learning process, which relies on scientific inquiry in diverse groups of stakeholders and experts. The relation of learning across different scales of social organization matters (Wals et al., 2014). Conceptions of social learning in environmental management and human ecology are closely aligned and suitable to build upon in further developing conceptions of transformative sustainability science (Keen et al., 2005; Dyball & Newell, 2015). This conception of learning is rooted in John Dewey's work (1938), who argues that knowledge is constructed for action and that learning can be mediated by iterative cycles of making an hypothesis, systematic inquiry testing hypotheses in practice, observation, and reflection. Transformative learning then emerges from dialogue between groups with diverse sets of values and worldviews, in which each group is brought to reflect on and creatively re-consider their own ways of thinking and doing (Lotz-Sisitka et al., 2015) (see the fourth section of this chapter on the diversity of purpose in sustainability science). There is an interesting convergence in the literatures on procedural requisites for transformative sustainability science and social learning, or transformative social learning for sustainability (Peters & Wals, 2013; Wals et al., 2014). Understanding diverse facets of challenges from distinct perspectives of different experts and stakeholders who engage in transformative social learning from each other is deemed necessary to better understand complexity (Wals et al., 2014; Newell & Doll, 2015; Dyball & Newell, 2015).

Why sustainability science may be seen as problematic

Associating the concept of sustainability with science raises a number of problems. First, the concept of sustainability is normative, as it suggests a direction in terms of 'good' or 'bad' ways in which society and environment interact. Furthermore, a focus on the human–nature interface requires us to embrace complexity, in the diversity of perspectives and commitments; situated knowledge, rather than abstract scientific knowledge that is expressed in formulae or models that claim universal validity; contradictions associated with contested expertise and interests; uncertainty; and ignorance. This is at odds with the social norms widely associated with science, including 'disinterestedness' and 'universalism' (Merton, 1973). Because these norms are still very much the basis for determining 'excellence' through peer review, they are then in turn reflected in terms of career rewards, even survival, in research organisations.

Building on Thomas Kuhn's observations of science as a social institution, most of 'normal science' that is conducted within disciplined groups of scientists can be seen to focus attention on the exclusive study of a tightly specified system, at specific spatial and temporal scales. Further it is customary in most disciplined fields to build on assumptions that reduce complexity, uncertainties, and value pluralism.

Quality criteria in most disciplines drive towards abstract and generalizable knowledge that claims quasi-universal validity, rather than situated knowledge that takes account of local contexts. Kuhn also points out that science education often implicitly conveys and enforces specific sets of beliefs and worldviews. Both education and research practice rely on review and reward systems that can systematically suppress disagreement, contradictions, and dissenting views. Kuhn refers to a situation when "the profession can no longer evade anomalies" (Kuhn, 1962, p. 6). The resulting fragmentation of fields of knowledge impedes the sense-making of complex systems, which are the matter of sustainability science. This phenomenon of fragmentation of disciplinary knowledge fields from the drive to further specialization also has the effect of undermining quality control through peer review, even in the most 'normal' of sciences. This problem is now publicly acknowledged, as we now see with the 'reproducibility crisis' in the natural and social sciences and 'endarkenment' in the humanities (Millgram, 2015).

Kuhn has demonstrated the inadequacy of his 'normal science' for engaging with the problems of the complex world of sustainability. He says (Kuhn, 1962, p. 37)

> It is no criterion of goodness in a puzzle that its outcome be intrinsically interesting or important. On the contrary, the really pressing problems – such as, a cure for cancer or the design of a lasting peace – are often not puzzles at all, largely because they may not have a solution ... A paradigm can, for that matter, even insulate the community from those socially important problems that are not reducible to the puzzle form, because they cannot be stated in terms of the conceptual and instrumental tools the paradigm supplies.

Building on these insights, 'Post-Normal Science' (PNS) was the first conceptualization that explicitly introduced uncertainty and value loading into the description of a type of scientific practice related to policy (Funtowicz & Ravetz, 1993). It introduced an element of democratization of science (see also Chapter 2 by Wals and Michaels) in the mention of the 'extended peer community'. As the concept evolved, complexity was incorporated into the scheme. Because most scientific practice, and all science education, is still 'normal' in Kuhn's sense of puzzle solving, PNS still has a very useful function as the first step in understanding the position and role of science in the contemporary world. It can also lead towards the comprehensive vision of sustainability science.

However, the classification of sustainability science in relation to 'normal' disciplinary fields of knowledge remains virtually impossible, for a defining attribute of sustainability science is that it draws from and builds on theory, methods, and practice of a wide range of sciences. Professional identities of scientists and the roles they play in society are conceived quite differently, depending on the fundamental worldviews in the various disciplines. This is reflected in distinct understandings of what science is, from whence it derives its authority and legitimacy, and how it is done. Furthermore, the *quality* of an 'integrated process or understanding' that is derived from drawing on diverse fields of knowledge cannot be judged based on criteria or review procedures from any single one of the relevant fields. Issues of quality criteria and processes are hence of central concern in sustainability science (Funtowicz & Ravetz, 2015).

Normal and post-normal science now need supplementing with another that we describe as 'transformative sustainability science'. Transformative sustainability science is a social process that draws on methods and practices to structure and systematize knowledge creation from diverse disciplines, as well as from other types of knowledge, including diverse forms of local knowledge. Participatory environmental sensing and sense-making technologies and virtual spaces on the Internet that are now fashioned for co-creative scientific inquiry in citizen science projects, such virtual spaces open up significant opportunities for scaling certain dimensions of transformative sustainability science. It also adds new spaces for shared representations for joint 'reflection' in such knowledge co-creation processes.

For the previously mentioned reasons, including the current lack of established methods for quality control, it is very challenging to establish sustainability science initiatives in more traditionally oriented research organisations. Systemic change in the university system and in prevailing ways to practice and legitimise science is therefore at the core of any strategy to foster and catalyse sustainability science-informed transformations in society at large (Sterling, 2013; Barth, 2015; Schneidewind et al., 2016).

Sustainability science: diversity of purpose and approaches

The possibility of juxtaposing different ways of describing problems and different ways of knowing helps to identify knowledge gaps and dissonances. This can contribute to a better understanding of how different facets of complex problems might be related to each other and acted upon, and thereby may open up creative new perspectives. Meanings of 'sustainability' that shape what 'purpose' is seen for sustainability science usually depend on the local context and people involved in addressing a specific challenge, such as water scarcity that can also affect local food production and electricity generation. This often also depends on diverse conceptions of actors and agency, which in turn affect more normative ideas on 'action fields and who the main actors are' that can emerge from such research (O'Brien, 2012; O'Brien, 2015).

Four broad discourses can be distinguished in academia (although this is by no means an exhaustive list) that have all different interpretations about meanings of sustainability and purpose of sustainability science. (Chapter 3 by König, comparing diverse systems approaches, provides a similar and more detailed comparison of how different disciplines direct attention to different problem framings and solution approaches). Research in environmental and earth system sciences that has gained great attention from policy makers worldwide aims to better understand the range of natural variation in, and anthropogenic impacts on, the earth system in order to deduce global and regional *'boundaries'* or limits to human activities that should be respected (Rockström et al., 2009; Steffen et al., 2015). Some scholars from the field of economics are usually interested in problems of, *'internalizing* externalities', for example by accounting for environmental damages caused in the production of goods and provision of services in pricing them, such that prices more appropriately reflect these 'externalities' in terms of collateral damages caused (Costanza et al., 2014). Other economists focus on conditions of 'market failure', as in the case of 'tragedy of the commons'. This tragedy describes conditions and types of goods and services to which basic market rules governing supply and demand of more typical goods do not apply. In these cases, groups of users may self-organize for sustainable management of vulnerable resources stocks, as is the case for *management of the commons*, including public forests, fish stocks, or ground water basins (Ostrom, 2009). By contrast engineering and urban planning often focus more on the technological dimension, engaging in discourses of *ecological modernization*, such as how public investment and market forces may be improved to reduce pollution at its source (Hajer, 1995). In philosophy and the learning sciences, the advocates of *deep ecology* argue that we require profound changes in human consciousness about our relations to nature (including non-human animals) and to each other. In consequence, the goal is not to be 'doing things better', but instead to be 'doing better things' (Peters & Wals, 2013). Last but not least, some researchers fear change may come too slowly in the absence of urgent crisis. This perspective of impending *'Doom'* is often the basis for research on resilience and adaptive management in crisis.

Research approaches associated with such diverse worldviews differ profoundly. They include disciplinary, interdisciplinary, or trans-disciplinary projects conducted by scientists alone or in collaborative processes with stakeholders, or are embedded in practice. Systems of interest include nature, society, or technology, at any scale. Research may be concerned with natural systems and planetary boundaries of the earth system or more concerned with social systems, problematizing actors and agency, and how sets of human values can serve as ordering principles in society (Castree, 2015). Research may be designed to better understand or to transform interactions between these systems. Ways of knowing can be conceived as inductive or deductive language-based reasoning, or knowledge can be seen as emergent from interactions between people and their environment, through practice or experience. Many research projects are aimed at the 'integration' across disciplines. Given very different sets of assumptions that shape concepts and methods used in different fields, this can, however, be a somewhat nonsensical endeavour. Armin Grunwald carefully discusses the challenges to and meaning of 'integration' in such 'integrative research projects' (Grunwald, 2016).

Sustainability science as a transformative social learning process

Transformative sustainability science can also be understood as a transformative social learning process that relies on scientific inquiry in diverse groups of experts and stakeholders (Peters & Wals, 2013; Wals et al., 2014; König, 2015). The process needs guidance and a safe space to ensure that everyone engaged is prepared to juxtapose diverse perspectives and to question their own assumptions. The relation of learning across different scales of social organization matters (Wals et al., 2014), building on prevailing conceptions of social learning in environmental management and human ecology (Keen et al., 2005; Dyball & Newell, 2015). The conception of learning advanced in this book is rooted in John Dewey's work (1938), who assumes that knowledge is constructed for action and that learning can be mediated by iterative cycles of making hypotheses, systematic inquiry testing hypotheses in practice, observation, and reflection.

In the face of scarce or damaged environmental resources, social learning is a form of social coordination that is more effective in translating research findings into real-world coping mechanisms than either hierarchical government regulation or market competition (see Table 1.1). Physical, technological, and social infrastructures for supporting the institutionalization of such community self-governance processes are key for their effectiveness. Platforms for analytic deliberation to deal with value conflicts are required (Dietz et al., 2003; Ison et al., 2007). Social learning focuses on the relation of learning across different scales of social organization: individual, group, organizational, or societal (Medema et al., 2014; Wals et al., 2014). Building on this research, learning is conceived of in this book as a process of developing enriched understandings and repertoires of action on complex problems as

Table 1.1 Forms of social coordination

Social coordination	Actors	Form of knowing	Taking account of complexity, uncertainty, management of commons
Regulation Education (hierarchical)	Government	Fixed form applied to a defined problem	X
Market-based Competition (individualistic)	Firms and consumers	Fixed form applied to a defined problem	X
Social learning (Non-coercive, egalitarian)	All stakeholders in a resource		– Iterative learning process towards shared understanding of a dynamic complex unknowable situation – Shared learning serves as basis for concerted action – Solutions are emerged properties of the social process

a result of open and iterative cycles of experimentation, observation, analysis, and judgment of results. The fact of knowing more and mastering knowledge in a different manner through experimentation, reflection and continual learning changes our relationship between the world and ourselves.

Reasons for the prevalence of particular ways to organize knowledge and learning include the ways we perceive our brain to function; how we learn to make conceptual representations in our minds, including of our identity and our surroundings; and how we manage our emotions and creativity. Sense-making can be seen as a process of matching perceptions to accumulated embodied experiences, which are organized in our minds with conceptual representations of our surroundings. These concepts can pre-structure and thus filter what we perceive and experience. As we learn, these conceptual representations can also be iteratively adapted or changed based on personal experiences from interactions with our physical and social surroundings that do not match previous organization in our mind. Hence, situations in which different groups of stakeholders consider disparate truths as self-evident can arise when each group relates the problems that are discussed to entirely different conceptual environments, considering problems at different scales and time frames. For example, even within the field of biology, molecular biologists and ecologists will often disagree whether genetically modified organisms present acceptable risks in agricultural or other industrial applications.

Stated goals of some collaborative projects to gain an *integrated* understanding from diverse perspectives on one system may therefore be hampered at the conceptual level. This is usually evidenced in many time-consuming and eventually failed attempts to develop common, precise, language-based definitions for key terminology to describe and structure a problem. Consequently, some collaborations just avoid the feat of developing shared sharp definitions; this, however, can prevent effective shared understanding and communication of new shared insights, in particular, if the main vehicle for sharing and exchanging perspectives remains just technical language (Newell & Proust, 2012). Language does not automatically convey meaning, as words stripped from their context do not convey meaning (Reddy, 1993). The use of metaphors, indicators, and pictures to link new conceptual representations to particular embodied experiences help to open new doors of perception, transformative learning, and new realities to be discovered and stabilized in our minds (Newell & Proust, 2012; Newell, 2012). This insight builds on research in cognitive linguistics, which posits metaphor as a logic-preserving mapping from concrete experiences to abstract concepts (Lakoff & Johnson, 1980). Once the limitations of language are recognised, efforts can be made to develop more powerful ways of communicating.

Accordingly, we infer that collaborative inquiry about complex systems performed in diverse groups can foster the emergence of new related phenomena, as expectations in participants change, at times converge, and become stabilized in changes in social practices and structures (König, 2015). Progress then can build on the evaluation of and passing judgment on a direction of development. Related competences to be developed in education for sustainability include systems thinking to embrace complexity and normative capacities to better understand diverse sets of values and worldviews in order to gain an enhanced judgment of courses of action deemed acceptable and feasible by many (Wiek et al., 2011).

This learning requires different conceptions of teaching and learning environments (see Table 1.2) from traditional approaches to teaching that still prevail – at least in most of higher education.

Research on learning processes identified citizen science–based community monitoring projects as an effective means to learn to improve natural resource governance in a range of diverse stakeholder groups and settings (Wals et al., 2014). Citizen science is 'scientific work undertaken by members of the public, often in collaboration with or under the direction of professional scientists' (Oxford English Dictionary, 2014). Public participation in scientific research can take diverse forms, ranging from

Table 1.2 Comparative table of two cultures of learning

	Transmissive learning	*Transformative learning*
Purpose and Scope	Understand defined cause and effect relationships	Personal transformation in contribution to systemic change
Process	Transfer of information from experts	Action-oriented development process
Teaching	Teacher defines meaning	Teaching facilitates negotiation and construction of meaning in diverse groups
Learning Environments	Classroom or laboratory	Emergence of new knowledge from interaction with complex real-world learning environments in diverse groups
Outcomes and Impacts	Efficient reproduction	Shared actionable knowledge, transformed perspectives, and environments
Assessment and Evaluation	Standardized testing	Self-evaluation and critical support

Source: Adapted from König, 2015b

scientists soliciting passive sensing activities or brain power from volunteers, to fully collaborative projects, in which citizens engage in problem framing, research design, observation, analysis, and interpretation (Haklay, 2015). Haklay defines the form of 'extreme' citizen science practiced by his research group as 'a situated, bottom-up practice that takes into account local needs, practices and culture and works with broad networks of people to design and build new devices and knowledge creation processes that can transform the world'.[1] Accelerating innovation in cheap sensor technologies, mobile computing tools, networking applications, and data aggregation and processing tools allows for crafting rich user interfaces and the storing and sending of time- and location-tagged data. Flexible open-source software tools that are easily adapted for diverse monitoring purposes to combine sensor-derived data, photographs, and the input of subjective data on environmental quality allow usage by citizens and researchers with little computing knowledge (Stevens et al., 2014).

Empirical evidence suggests that participatory monitoring projects can transform the relationship of ecosystems, local communities, and economies; reconnect people to the landscape and to each other; and achieve appreciation of complexity and renegotiation of what and how values are attributed in the community (Fernandez-Gimenez et al., 2008). Chapter 14 describes the beginnings of citizen science in the developing world for improved water governance. In the EU, several citizen science projects and citizen observatories, including one on flooding and a coordination platform, were financed under FP7 and the Horizon 2020 framework programme and the EU Digital Unit of the European Commission. The first eco-schools are adopting citizen science inquiry–based approaches to combine learning with the staging of community environmental projects (Hargreaves, 2008). And during the Obama administration in September 2015 a communication from John Holdren, the Director of the White House Office for Science and Technology Policy and Chief Science Advisor to the president, even invited all federal agencies to prepare for soliciting data for evidence-based policy making, including environmental monitoring data from new approaches to crowdsourcing and citizen science. Chapter 2 introduces reasons for the urgency to foster the combination of transformative learning from and with civic and citizen science in an era of post-truth politics stabilised by the Trump government in the United States. Citizen science projects can be designed as environments in which transformative learning for changing social practices across scales of social organisation takes place. These designs can include conceptual tools and associated social processes presented in Part I of this book, including systems, futures, social technologies, and human-centred design thinking. A recent project that exemplifies a research design based on these insights is starting in Luxembourg in 2017 and is described in more detail in this book's conclusion.

What systemic transformation?

Systemic transformation first relies on the emergence of new social practices. These in turn can be consolidated into new social structures such as institutions and prevailing sets of rules, social norms, and values. Systemic transformation for sustainability requires new forms and logics of collaboration. In systemic change, there is a complementarity of the changes at the individual level in terms of attributing new meaning and the various social levels. It is therefore helpful to consider the characteristic patterns of change at both levels (see also Chapter 16 by Manhart).

Individual meaning-making is linked to perceptions, conceptual representations, and experiences and expectations. Embodied experiences are matched to conceptual representations in one's mind, and these are continuously and iteratively changed to accommodate these experiences. And the iteratively refined conceptual representations in turn serve as filters to pre-structure subsequent perceptions. Transformative learning will result in a reframing and reconfiguration of those perceptions, conceptions, experiences, and expectations. This can more fundamentally change how we relate these to each other within ourselves and in dialogue with others. For these functions, the language is not merely descriptive and analytical, but it will also rely on metaphors (Newell, 2012, citing Lakoff, 1980). Beyond that, other vehicles for meaning will be used, including graphic representations of quantities, along with drama, music, and art. These can include diagrammatic techniques used in the approaches to low-order conceptual systems mapping described in this book (Newell and Proust, Chapter 5), which describe how we can step by step establish a shared visual language on which aspects of complex challenges may matter most from diverse perspectives as a basis from which to use focused dialogue to develop well-defined conceptual/theoretical frameworks. This is one of many evolving approaches and methods for making connections and creating new ways to arrive at shared perceptions, experiences, and concepts to allow the reframing and reconfiguration of the experienced reality.

The complementary level of transformation, the social, will require creativity resulting in entirely new expectations and their associated social practices and technologies (see also Chapter 16 by Manhart). Moreover, these new social practices must emerge from, and be fit for, the dynamic networked society of the future rather than the hierarchical industrial society of the past. Radically new concepts of knowledge and learning, and associated social processes and spaces, are required, and digital technologies have a role to play (Wegerif, 2007). The emerging generations just born in this new millennium, who seem to spend more time and thus gain more experiences in the networked virtual space, may engage in meaning-making not so much by relating concepts to embodied experiences from interacting in and with the physical world, but by seeking and comparing different perspectives from the virtual space. This can be much more pluralist and rely on cognitive switching between diverse perspectives. This is the perspective of the 'global village' first articulated by Marshall McLuhan, where diversity of customs, assumptions, and values is a familiar experience to citizens. When the practice of meaning-making is more sensitive to highly divergent assumptions of what constitutes a reality and how it may be experienced, new possibilities for systems approaches and new creative solutions may emerge. Knowledge gaps and contradictions may become more easily apparent and less troubling to embrace, and these can present spaces for creativity (Wegerif, 2007). A serious challenge here is the requisite for new types of quality assurance for legitimating and protecting this emerging new shared knowledge.

Contemplation of such a far-reaching transformation of society and consciousness might seem to be ambitious to the point of being utopian. But such transformations occur regularly in the history of civilisations. In Europe, we have three well-marked epochs in the cultural dimension that we call 'medieval', 'renaissance', and 'modern'; and in the technological and economic dimensions we have 'industrialisation' and 'capitalism'. We have even witnessed an attempted transformation that, at least in the short to medium term, has largely failed, namely 'socialism'. In each case there were, or were intended to be, profound transformations in some crucial aspect of social or cultural life at social and individual levels. Much of the deep theoretical reflection on society has been stimulated by the challenges and contradictions that were witnessed in the replacement of one epoch by another.

The pace and pattern of transformation to a new social-cultural order cannot be predicted, but there are many indicators of a rapid change in the style of social behaviour along lines towards co-creation. A mere century ago discrimination based on race was generally considered to be only right and natural in many societies. Arguably, Rachel Carson's book, *The Silent Spring*, and photos of planet Earth from outer space contributed to transforming our sense of 'the environment'. Such transformations will be complex and partial, and many practices that violate now-established normative concepts such as equity and environmental protection persist unchecked. But a direction can be discerned.

Decisions on alternative paths of action invited by just-discernible changes in social norms, however, are not often obvious, and what is considered obviously 'good' from one perspective can be considered 'an evil threat' from another. For some individuals, they promise liberation and self-realisation; for others, they threaten a loss of integrity and livelihood. Institutions that are endangered will respond with all possible means. 'Good' causes and campaigns can be caught up in power struggles in which they are mere pawns and risk corruption or extinction. Campaigns for practices for environmental protection at a very local level have the potential to 'harm' some people and their environment.

A genuine education for transformation will prepare participants for such complexities and contradictions. Practicing putting yourself in other people's shoes is crucial to recognise why some people or institutions opposing such change may just be selfish or reactionary. People have the right to fundamentally disagree about what is good and in relation to the facts, the right to be wrong. For such leaders as Martin Luther King and Nelson Mandela, dialogue with malevolent opponents did not come easily; but they knew that that is the only true path to systemic transformation (see Chapter 6 on scenarios).

How can transformation through the creation of new shared meanings and expectations be achieved? As stated earlier (see page 7, Figure 1.2) sustainability science conceived as a transformative social learning process will have four main procedural attributes it helps to embrace: (i) complexity, by adopting a systems perspective and developing the practice of cognitive switching between diverse perspectives for meaning-making; (ii) contradictions and learning to dialogue for collaboration across different worldviews; (iii) contingency and the need to produce situated knowledge in a systematic manner; and (iv) recognising gaps, unknowns, and ignorance for making creative spaces that enable new ways of seeing and new social practices. Last but not least in this era of accelerating change, we need to learn a new way to relate to the past, present, and future. These are fundamental requisites for transformational change.

Judging transformative learning

Transformative learning will be assessed based on observable changes in communication and behaviour by individuals and organizations over time. One approach is to define and access relevant competences, usually identified as systems thinking, a future orientation, and normative competences (Wiek et al., 2011). The content of learning may also be assessed based on Vickers's (1984) conception of appreciation of complex systems in three dimensions: reality judgement that is reflected in the enriched systems understanding; value judgement relating to perception of the diversity of values and worldviews; and instrumental judgement on which actions may be acceptable and feasible to diverse stakeholders. Another approach is

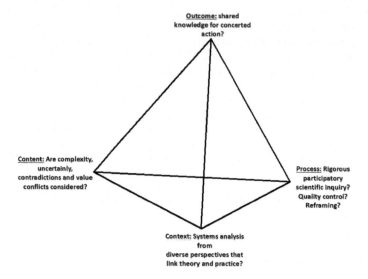

Figure 1.4 Evaluation questions for transformative sustainability science

to seek to track empowerment through education (as described by Wals and Peters in Chapter 2). The assessment of creative thinking in diverse groups can be assessed simply by observing the types of dialogue they engage in (Wegerif et al., 2015). Three types of talks are distinguished, disputational, cumulative, and exploratory talk, in order to observe how participants enter into dialogue in group interviews and workshops.

Apart from observations of transformative learning at the individual and organizational level, the evaluation of impacts and outcomes of transformative sustainability science is a complex field. This would require the evaluation of the quality of the social process for knowledge co-creation, as well as the evaluation of visible changes that may have been affected through the process, such as changes in behaviour and communication; institutions; material and energy stocks and flows of natural and anthropognic nature; and the mergence of new technologies, or new uses or users. There is an important and growing literature on quality criteria and the evaluation of transformative sustainability science, which is beyond the scope of this chapter, but in part is addressed in Chapters 14 to 18.

Structure of the book

In order to provide guidance and resources for capacity building in communities of scientists and citizens to engage in transformative sustainability science, we have divided this book into three parts. Each chapter in these parts offers a reflective account of a particular challenge, an approach by which these challenges can be explored, a critical discussion of the merits and limitations of the approach, and a conclusion on the significance of the insights gained. All contributors have been encouraged to state their assumptions in theory and methods they refer to as well as personal convictions and motivations to engage in their research or profession. These are requisites to allow for critical inter-disciplinarity, which directs attention to limitations and contradictions between specific disciplinary approaches to generating new knowledge. Because this book records experiments in a new methodology, a fully matured, critical inter-disciplinarity is not always achieved. In sum, the book equips readers with a better understanding of how one might actively design, engage in, and guide processes by which society and the environment interact.

The second chapter of the introductory part, Chapter 2 by Arjen E.J. Wals, a learning scientist, and Michael Peters, a political philosopher, builds on John Dewey's sense that 'the cure for ailments in democracy, is more democracy'. Risks of coercive forms of sustainability implemented by elite 'expert-driven' technocratic governments, based on top-down goals and indicators, and counter-movements supporting post-truth politics, are to be countered by combining education and transformative social learning with collaborative inquiry in civic and citizen science projects. Recent insights on necessary changes in how we conceive of research excellence and education in academia and policy making are critically discussed. The organisation of knowledge co-creation should take into account concerns of social fractures and risks of reinforcing inequity. The importance of developing spaces and education for transgressive thought that questions existing orders and boundaries is highlighted, without which no democracy can learn to live up to the promises of the democratic principles across changes in time.

Part I: Embracing complexity and alternative futures: conceptual tools and methods

Part I provides an overview on diverse conceptual tools and methods that are useful in structuring transformative social learning processes that draw on diverse types of expertise and practical experience. These methods are required to collectively question prevailing knowledge and problem framings and reframe the issues to adapt them to different contexts.

The first three chapters focus and provide ways to overcome our own limitations in the face of complexity in dynamic social-technological-ecological systems that are rooted in the ways our human brains function, but also in our education system that generally reinforces breaking down complexity into too simple direct cause–effect relations. Chapter 3 by Ariane König discusses diverse approaches to conceptualise human–environment interactions as embedded in complex and dynamic social-ecological systems, in which interactions between different elements in the system play a greater role in determining the behaviour of the overall system than do interactions within the different sub-parts of the system. Chapter 4 by Philipp Sonnleitner, a cognitive psychologist, starts by describing cognitive challenges in the face of complexity. Self-reflection on personal emotions in this process is an important dimension of the learning experience. In-class exchanges on different cognitive approaches and emotional responses in the face of complexity are also important learning dimensions in order to build the empathy required to work in diverse groups on complex problems. Chapter 5 by Barry Newell and Katrina Proust provides guidance on how to implement a step-by-step process to characterise what matters most in complex, dynamic human–environment systems in diverse groups. This approach to Collaborative Conceptual Modelling (CCM) of dynamic human–environment systems directs attention to interactions and feedbacks that shape overall patterns of behaviour within such systems. This enables groups to collectively decide which leverage points to address to develop feedbacks from policies and associated learning by design that counteract unsustainable trends. CCM approaches are usefully embedded in scenario approaches that allow the same groups to get a better understanding of drivers of change in the past, present, and future and what is certain and uncertain about these.

The next two chapters describe methods to explore and sense alternative futures in diverse groups. Chapter 6 by Gerard Drenth, Shirin Elahi, and Ariane König introduces the readers to participatory scenario approaches in order to better understand uncertainty in complex conditions and potential disruptive changes and human choice and constraints. The role of diverse imaginations of the future in deliberations and motivations to change current practices is uncontested. There are diverse methods to structure explorations of alternative futures. Exploratory scenarios are plausible stories describing future worlds that illustrate alternative outcomes of developments. The approach blends qualitative and quantitative analysis in order to explore alternative outcomes of global change and associated implications locally in the transactional environment, where some changes might be brought about if a critical mass of stakeholders engages. A set of scenarios usually serves to highlight things we can or can't know about the future; uncertainties that matter but are rarely talked about; and inter-dependencies in alternative future development paths, human choices and constraints, and differential power distribution in society. Sets of scenarios may also be designed to sketch the interdependence of culture and values prevailing in society and how these are interdependent with technological choices; this may also be related to experienced quality of life and environment and how distributional issues might play out in different futures. Visions, by contrast, offer a desirable future to serve as an orientation for concerted action in diverse groups – they are normative. Chapter 7 by Isabel Page introduces Theory U, a step-by-step approach to exploring the emerging future in diverse groups. It is necessary for transformational change for members of a group to relate to each other in different ways in order to suspend their own ideas, fears, and judgments. Then the group can enter a special space in which all identify with the problem and try to switch between diverse perspectives on this problem rather than defending their own viewpoints and interests. In that way they can conjointly sense the merging future possibilities that might be realized in collective action.

Chapter 8 by Kilian Gericke and Gregor Waltersdorfer introduces human-centred design thinking as one approach to structure processes to develop a particular form of 'actionable knowledge' as specific solutions to design problems that have been identified in scenario and conceptual systems mapping exercises. The iteration between ideation and prototyping following the motto of implementing, failing, learning, and improving fast is the core of this method.

Part II: What might transformations look like?
Sectoral challenges and interdependence

The seven chapters in Part II ask what is transformed and how, both in theory and in practice, providing more disciplinary and sectoral perspectives on particular challenges of sustainability. We have taken care to draw on internationally salient issues and examples. Examples are taken from agriculture, energy use, and water management. In each of these sectoral chapters we discuss changing prerequisites to scientific research and practice, as well as challenges and pitfalls in the discourses on change that are prevailing in Luxembourg, in the EU, and in other world areas, including in the developing world. In each theme we consider challenges of human–technology–environment interactions and useful bridging concepts to consider these, such as 'ecosystem services'. Systematic critical discussions in each chapter of the role of science and the merits and limitations of scientific concepts and methods help to convey the uncertainties in our knowledge and set the foundations for critical interdisciplinarity and future-oriented dialectics.

The two first chapters in this part explore the transformation of agricultural systems to greater sustainability from two different perspectives. Chapter 9 by Nicolas Dendoncker and Emily Crouzat, geographers, characterises three past agricultural revolutions with attention to changes in land use and its consequences; subsequently the chapter maps out requisites for a fourth revolution towards a more sustainable agricultural system that relies on agroecology. The concept of ecosystem services plays a central role as a bridging concept to structure collaborative inquiry on stocks and flows in nature and to better understand diverse sets of values associated with these in diverse communities. Chapter 10 by Federico Davila and Robert Dyball introduces a conceptual tool from the field of human ecology, the cultural adaptation template (CAT), to explore in more detail the relations between the state of the environment, human well-being, human culture and human institutions, and unintentional as well as policy-based 'designed' feedbacks between them. This tool is then applied in support of a discourse analysis of two prevailing discourses in food politics: one focussing on technological innovation to solve food security issues, and one directing attention to self-reinforcing equity issues in the global food production system. These juxtaposed insights on the merits and limitations of problem framing in each of these discourses open a creative space to think about alternative futures and possible policy design afresh.

The three subsequent chapters engage with challenges in transformations of our energy system. Building on Samadi et al. (2016), three types of complementary measures are distinguished: those aiming at consistency by replacing fossil fuel–based energy sources with renewables, efficiency gains allowing us to do more with less, and sufficiency measures with the ultimate aim of reducing demand for energy-intensive products and services, such that efficiency-related gains are not merely reinvested into more energy demand–generating activities. In Chapter 11 Susanne Siebentritt, a physicist, introduces the readers to basic principles of physics, placing emphasis on the fact that energy is only ever transformed from one form into another, not actually consumed or used up. Some forms are more accessible and storable and versatile for human use than others. On this basis some key design requisites for future sustainable energy systems are characterised. Chapter 12 by Julia Affolderbach, Bérénice Preller, and Christian Schulz provides insights from research in the field of geography on transitions in the green building sector and drivers of change and reasons for lock-ins in unsustainable practices in four case studies distributed across three continents. Chapter 13 by Hondrila et al., which is actually written by a peer group based on a practice-embedded project during one academic year of the certificate in sustainability and social innovation, explores challenges to democratizing renewable energy in Luxembourg, with reference to recent related developments in Germany.

The last chapter in this section, Chapter 14 by Kim Chi Tran and Ariane König, then explores the role community-based monitoring and citizen and civic science might play in reconfiguring the problematic science–policy–practice interface for sustainable water governance.

Part III: Tracking, steering and judging transformation

Part III discusses the role of measurement regimes and sets of indicators in defining and tracking progress. We are surrounded by quantitative indicators. They are one of the main channels whereby the citizen interacts on a day-to-day basis with science. How can the citizen make decisions in this jungle of indicators? Indicators are not simple facts about the world, but are the products of design, constructed from data, assumptions, and conventions. For the citizen, the question is not so much about the truth of some indicator, but about its quality. Moreover, measurement regimes act as devices for steering development across various levels of social organization and governance from diverse perspectives, including sociology and practice. The misuse and abuse of numbers in society and the neglect of representation of associated uncertainties is thematised, drawing on prior writings of Ravetz. Chapter 15 by Jerome Ravetz, Paula Hild, Julien Bollati, and Olivier Thunus identifies and critically discusses quality criteria for indicators in such measurement regimes. On this basis we deliberate on strengths and limitations of some statistical indicators, as well as other measures developed to motivate societal transition to sustainability – notably the Ecological Footprint. The assessment of the quality of an indicator, particularly one that is compounded from data relating to diverse fields of activity, is a demanding task from which value commitments cannot be kept separate. The NUSAP notational system, representing the range of attributes of scientific information from the most quantitative to the most qualitative, can be very useful in aiding the assessment of indicators. The Ecological Footprint, one of the most popular and influential sustainability indicators, performs a diversity of functions; the discussion of its merits and limitations is correspondingly complex.

Chapter 16 by Sebastian Manhart argues that in order to reflect and engage in complex transformative processes for sustainability, it is necessary to better understand how humankind approaches complexity on an individual and societal level, using numbers as well as other signs such as letters, language, and pictures. This chapter presents a conceptual framework to analyse how different categories of signs are used to represent complexity, using language, numbers, and pictures independently and in combinations. Each category of signs is associated with limiting rules on their assembly and interpretation for sense-making and communication. From this perspective, transformative learning requires a synthesis of diverse subjective and social constructions. The argument in this chapter is that most effective for this purpose are representations that draw and combine diverse logics of diverse sign categories, including numbers, language, and pictures. Different ways of representing complexity will result in drawing different productive boundaries around systems of interest. Reflection about merits and limitations of each sign category will make us more effective in using them in combination to better understand, communicate, and act upon complex circumstances.

Given the prevailing logics for governance in our liberal democracies that are rooted in neo-classical economics, the penultimate Chapter 17 by Jerome Ravetz asks what a reformed science of economics might contribute to sustainability. The use of numbers and language in the form of concepts and paradigms in the disciplined science of economics is critically discussed.

Chapter 18 by Jerome Ravetz is a post script that presents the attributes of sustainability science conceived as transformative social learning process as a basic set of heuristics in more detail and relates why just this set of heuristics is significant in view of his analysis of the current crisis of legitimacy in science and expertise, in particular when drawn upon in technocratic forms of government.

Chapter 19 by Ariane König offers an outlook on the theory and practice of future transformative sustainability science. It briefly draws together insights across all chapters and concludes with more general guidance for designing and implementing research and learning initiatives for the practice of transformative sustainability science.

Questions for comprehension and reflection

1 Why are traditional approaches to knowledge production that have evolved over the past 200 years inadequate to address existential challenges of the twenty-first century?
2 What distinguishes descriptive and transformative sustainability science?
3 What can be learnt from the practice of transformative sustainability science?

Note

1 www.ucl.ac.uk/excites/home-columns/full-what-is-extreme-citizen-science/

References

Barth, M. (2015). *Implementing sustainability in higher education: Learning in an age of transformation*. London and New York: Routledge.

Burt, G. & van der Heijden, K. (2008). Towards a framework to understand purpose in futures studies: The role of Vickers' appreciative system. *Technological Forecasting and Social Change* 75 (8): 1109–1127. doi:10.1016/j.techfore.2008.03.003

Castree, N. (2015). Geography and global change science: Relationships necessary, absent, and possible. *Geographical Research* 53 (1): 1–15. doi:10.1111/1745-5871.12100

Clark, W. C. (2007). Sustainability science: A room of its own. *PNAS* 104 (6): 1737–1738.

Clark, W. C. & Dickson, N. M. (2003). Sustainability science: The emerging research program. *PNAS USA* 100 (4): 8059–8061. doi:10.1073/pnas.1231333100

Costanza, R., Kubiszewski, I., Giovannini, E., Lovins, H., McGlade, J., Pickett, K. E., . . . Wilkinson, R. (2014). Time to leave GDP behind. *Nature* 505: 283–285.

Crosby, A. W. (2015). *Ecological imperialism: The biological expansion of Europe, 900–1900*. Cambridge: Cambridge University Press.

Dewey, J. (1938). *Education and experience*. New York: Simon and Schuster.

Dietz, T., Ostrom, E. & Stern, P. C. (2003). The struggle to govern the commons. *Science* 302 (5652): 1907–1912.

Dyball, R. & Newell, B. (2015). *Understanding human ecology: A systems approach to sustainability*. Oxon: Routledge.

Fernandez-Gimenez, M., Ballard, H. L. & Sturtevant, V. E. (2008). Adaptive management and social learning in collaborative and community-based monitoring: A study of five community-based forestry organizations in the Western USA. *Ecology and Society* 13: 4–19.

Funtowicz, S. & Ravetz J. R. (1990). *Uncertainty and quality in science for policy*. Kluwer: Dordrecht.

Funtowicz, S. O. & Ravetz, J. R. (2015). 'Peer review and quality control', in Wright, J. D. (Ed.) *International encyclopedia of the social & behavioral sciences, 2nd edition*. Oxford: Elsevier. pp. 690–694.

Funtowicz, S. O. & Ravetz, J. R. (1993). Science for the post-normal age. *Futures* 25 (7): 739–755.

Grunwald, A. (2011). Energy futures: Diversity and the need for assessment. *Futures* 43 (8): 820–830.

Grunwald, A. (2016). *Nachhaltigkeit verstehen: Arbeiten an der bedeutung nachaltiger entwicklung*. München: Oekom Verlag.

Hajer, M. A. (1995). *The politics of environmental discourse: Ecological modernization and the policy discourse*. Oxford: Oxford University Press.

Haklay, M. (2015). *Citizen science and policy: A European perspective*. Washington, DC: Woodrow Wilson Center for International Scholars.

Hargreaves, L. G. (2008). The whole-school approach to education for sustainable development: From pilot projects to systemic change. *Policy and Practice: A Development Education Review* 6: 69–74.

Hulme, M. (2009). *Why we disagree about climate change: Understanding controversy, inaction and opportunity*. Cambridge: Cambridge University Press.

Ison, R., Roling, N. & Watson, D. (2007). Challenges to science and society in the sustainable management and use of water: Investigating the role of social learning. *Environmental Science and Policy* 10 (6): 499–511. doi:10.1016/j.envsci.2007.02.008

Jasanoff, S. (2015). 'Future imperfect: science, technology and the imaginations of modernity', in Jasanoff, S & Kim S.-H. (Eds.), *Dreamscapes of modernity: Sociotechnical imaginaries and the fabrication of power*. Chicago: University of Chicago Press. pp. 1–33.

Jasanoff, S. (2003). Technologies of humility: Citizen participation in governing science. *Minerva* 41 (3): 223–244. doi:10.1023/A:1025557512320

Jerneck, A., Olsson, L., Ness, B., Anderberg, S., Baier, M., Clark, E., . . . Persson, J. (2011). Structuring sustainability science. *Sustainability Science* 6 (1): 69–82. doi:10.1007/s11625-010-0117-x

Kates, R. W., Clark, W. C., Corell, R., Hall, J. M., Jaeger, C. C., Lowe, I., . . . Svedin, U. (2001). Sustainability science. *Science* 292 (5517): 641–642. doi:10.1126/science.1059386

Keen, M., Brown, V. A. & Dyball, R. (Eds.) (2005). *Social learning in environmental management: Towards a sustainable future*. London and New York: Earthscan.

König, A. (Ed.) (2013). *Regenerative sustainable development of universities and cities: The role of living laboratories*. Cheltenham, UK: Edward Elgar.

König, A. (2015a). Towards systemic change: On the co-creation and evaluation of a study programme in transformative sustainability science with stakeholders in Luxembourg. *Current Opinion in Environmental Sustainability* 16: 89–98.

König, A. (2015b). Changing requisites to universities in the 21st century: Organizing for transformative sustainability science for systemic change. *Current Opinion in Environmental Sustainability* 16: 105–111.

Kuhn, T. S. (1962). *The structure of scientific revolutions*. Chicago: University of Chicago Press.

Lakoff, G. & Johnson, M. (1980). *Metaphors we live by*. Chicago: University of Chicago Press.

Lotz-Sisitka, H., Wals, A.E.J., Kronlid, D. & McGarry, D. (2015). Transformative, transgressive social learning: Rethinking higher education pedagogy in times of systemic global dysfunction. *Current Opinion in Environmental Sustainability* 16: 73–80. doi:10.1016/j.cosust.2015.07.018

Maggs, D. & Robinson, J. (2016). Recalibrating the Anthropocene: Sustainability in an imaginary world. *Environmental Philosophy* 13 (2): 175–194. doi:10.5840/envirophil201611740

Matson, P., Clark, W.C. & Andersson, K. (2016). *Pursuing sustainability: A guide to the science and practice.* Princeton, NJ: Princeton University Press.

Medema, W., Wals, A. & Adamowski, J. (2014). Multi-loop social learning for sustainable land and water governance: Towards a research agenda on the potential of virtual learning platforms. *NJAS – Wageningen Journal of Life Sciences* 69: 23–38.

Merton, R. K (1973). *The sociology of science: Theoretical and empirical investigations.* Edited and with and introduction by Storer, N.W. Chicago: University of Chicago Press.

Miller, T. R., Wiek, A., Sarewitz, D., Robinson, J., Olsson, L., Kriebel, D. & Loorbach, D. (2013). The future of sustainability science: A solutions-oriented research agenda. *Sustainability Science* 9 (2): 239–246. doi:10.1007/s11625-013-0224-6

Millgram, E. (2015). 'The great endarkenment', in Millgram, E. *The great endarkenment: Philosophy for an age of hyperspecialization.* Oxford: Oxford University Press. Chapter 2. pp. 21–53.

Newell, B. (2012). Simple models, powerful ideas: Towards effective integrative practice. *Global Environmental Change* 22 (3): 776–783. doi:10.1016/j.gloenvcha.2012.03.006

Newell, B. & Doll, C. (2015). *Systems thinking and the cobra effect.* Tokyo: United Nations University.

Newell, B. & Proust, K. (2012). *Introduction to collaborative conceptual modelling.* Working paper, ANU Open Access Research. Retrieved from https://digitalcollections.anu.edu.au/handle/1885/9386

Newell, B. & Proust, K. (2017). 'Escaping the complexity dilemma', in König, A. (Ed.) *Sustainability science: Key issues in connecting learning, research, and practice.* Abingdon: Routledge. Chapter 5.

Newell, B. & Siri, J. (2016). A role for low-order system dynamics models in urban health policy making. *Environmental International* 95: 93–97.

O'Brien, K. (2012). Global environmental change II: From adaptation to deliberate transformation. *Progress in Human Geography* 36 (5): 667–676. doi:10.1177/0309132511425767

O'Brien, K. (2015). Political agency: The key to tackling climate change. *Science* 350 (6265): 1170–1171. doi:10.1126/science.aad0267

Ostrom, E. (2009). A general framework for analyzing sustainability of social-ecological systems. *Science* 325: 419–422. doi:10.1126/science.1172133

Ostrom, E. (2010). Beyond markets and states: Polycentric governance of complex economic systems. *Transnational Corporations Review* 100 (3): 641–672.

Oxford English Dictionary. (2014). *OED Online: The definitive record of the English language.* Oxford: Oxford University Press. www.oed.com

Parodi, O. (2008). *Technik am fluss.* München: Oekom Verlag.

Peters, S. & Wals, A.E.J. (2013). 'Learning and knowing in pursuit of sustainability: Concepts and tools for transdisciplinary environmental research', in Krasny, M. E. & Dillon, J. (Eds.) *Trading zones in environmental education: Creating transdisciplinary dialogue.* New York: Peter Lang. pp. 79–104.

Ravetz, J.R. (2006). Post-Normal Science and the complexity of transitions towards sustainability. *Ecological Complexity* 3: 275–284.

Raskin, P. (2016). *Journey to Earthland: The great transition to planetary civilization.* Boston: Tellus Institute.

Reddy, M. J. (1993). 'The conduit metaphor: A case of frame conflict in our language about language', in Ortony, A. (Ed.) *Metaphor and thought.* Cambridge: Cambridge University Press. pp. 164–201.

Robinson, J., Berkhout, T., Cayuela, A. & Campbell, A. (2013). 'Next generation sustainability at the University of British Columbia: The University as societal test-bed', in König, A.

(Ed.) *Regenerative sustainable development of universities and cities: The role of living laboratories.* Cheltenham, UK: Edward Elgar. pp. 27–48.

Robinson, J., Burch, S., Talwar, S., O'Shea, M. & Walsh, M. (2011). Envisioning sustainability: Recent progress in the use of participatory backcasting approaches for sustainability research. *Technological Forecasting and Social Change* 78 (5): 756–768.

Rockstrom, J., Steffen, W., Noone, K., Persson, A., Chapin, F. S., Lambin, E. F., . . . Foley, J. A. (2009). A safe operating space for humanity. *Nature* 461 (7263): 472–475. doi:10.1038/461472a

Rorty, R. (1999). *Philosophy and social hope.* London: Penguin Books.

Samadi, S., Gröne, M.-C., Schneidewind, Uwe, Luhmann H.-J., Venjakob, J. & Best, B. (2016). Sufficiency in energy scenario studies: Taking the potential benefits of lifestyle changes into account. *Technological Forecasting & Social Change* – available online 3 October 2016. doi:10.1016/j.techfore.2016.09.013

Schneidewind, U. & Singer-Brodowski, M. (2014). *Transformative wissenschaft.* Marburg: Metropolis Verlag.

Schneidewind, U., Singer-Brodowski, M., Augenstein, K. & Stelzer, F. (2016). *Pledge for a transformative science: A conceptual framework.* Wuppertal Papers, No. 191. Retrieved from http://nbn-resolving.de/urn:nbn:de:bsz:wup4-opus-64142

Steffen, W., Richardon, K., Rockström, J., Cornell, S. E., Fetzer, I., Bennett, E. M., . . . Sörlin, S. (2015). Planetary boundaries: Guiding human development on a changing planet. *Science* 347 (6223): 736–746. doi:10.1126/science.1259855

Sterling, S. (2013). 'The sustainable university: Challenge and response', in Sterling, S., Maxey, L. & Luna, H. (Eds.) *The sustainable university: Progress and prospects.* Abingdon: Routledge. pp. 17–50.

Stevens, M., Vitos, M., Altenbuchner, J., Conquest, G., Lewis, J. & Haklay, M. (2014). Taking participatory citizen science to extremes. *IEEE Pervasive Computing* 13 (2): 20–29. doi:10.1109/MPRV.2014.37. Retrieved from http://ieeexplore.ieee.org/stamp/stamp.jsp?tp=&arnumber=6818498&isnumber=6818495

Swart, R. J., Raskin, P. & Robinson, J. (2004). The problem of the future: Sustainability science and scenario analysis. *Global Environmental Change* 14: 137–146. doi:10.1016/j.gloenvcha.2003.10.002

van der Sluis, J. (2005). Uncertainty as a monster in the science–policy interface: Four coping strategies. *Water Science & Technology* 52 (6): 87–92.

Vickers, G. (1984). *Judgement: Chapter 16 in the open systems group: The Vickers Papers.* London: Harper & Row. pp. 230–245.

Wals, A. E. J., Brody, M., Dillon, J. & Stevenson, R. B. (2014). Convergence between science and environmental education. *Science* 344 (6184): 583–584. doi:10.1126/science.1250515

Wegerif, R. (2007). *Dialogic education and technology: Expanding the space of learning.* New York: Springer.

Wegerif, R., Li, L. & Kaufman, J. C. (Eds.) (2015). *The Routledge international handbook of research on teaching thinking.* New York and London: Routledge.

Wiek, A. & Lang, D. J. (2015). 'Transformational sustainability research methodology', in Heinrichs, H., Martens, P. & Michelsen, G. (Eds.) *Sustainability science – An introduction.* New York: Springer. pp. 1–12.

Wiek, A., Withycombe, L. & Redman, C. L. (2011). Key competencies in sustainability: A reference framework for academic program development. *Sustainability Science* 6: 203–218. doi:10.1007/s11625-011-0132-6

2 Flowers of resistance

Citizen science, ecological democracy and the transgressive education paradigm

Arjen E. J. Wals and Michael A. Peters

> . . . who are you, then? I am part of that power which eternally wills evil and eternally works good.
>
> —— Goethe, *Faust*

> We have every reason to think that whatever changes may take place in existing democratic machinery, they will be of a sort to make the interest of the public a more supreme guide and criterion of governmental activity, and to enable the public to form and manifest its purposes still more authoritatively. In this sense the cure for the ailments of democracy is more democracy.
>
> —— John Dewey, *The Public and Its Problems* (1927)

Prelude

When the editors of this book asked one of us (Arjen) to contribute to this collection in June 2016, it was a different time. Even though Donald Trump had become the presidential candidate for the Republican Party, not many people believed he would actually become president of a major world power: the United States of America. This chapter was to be mainly about the transition movement that seemed to be gaining strength in science and society. In science, sustainability science, post-normal science and science-in-transition was gaining strength, representing a recognition of multiple ways of being (ontological pluralism) and knowing (epistemological pluralism) and the inevitability of persistent complexity, ambiguity and uncertainty. In society the call for living more lightly and equitably on the Earth was leading to transitions in energy (away from fossil fuel and centralized energy systems), food (away from agri-business and industrial farming towards more localized and sustainable food systems), economics (away from profit and growth oriented capitalist systems towards economies of sharing and meaning) and health care (away from centralized and privatized care systems towards localized cooperatives). Using a recently ISSC-funded project on T-learning[1] as a backdrop, the chapter would present these transitions as 'learning-based transitions' that would also require a transition in education. It would focus on what such education might look like in terms of its processes, design and outcomes.

But then the seemingly improbable happened: Donald J. Trump was elected president of the United States on November 8, 2016. His election was a major disruption in the sense that it unveiled serious shortcomings in 'modern' democracy, the role of social media, the press and, indeed, the use and misuse of science in general and facts in particular. The chapter needed to be re-considered in light of this historic world event that could become what we might call a negative tipping point that could trigger a spiralling towards the abyss where *Homo sapiens* will join many other species in becoming endangered. The political events in the United States make clear that this chapter would also have to address the idea of democracy and to discuss the political nature of change. By inviting critical education philosopher Michael Peters to join in the writing of the chapter, the chapter brings these twin concerns to the fore. Still the chapter focuses on transformative and even transgressive education and research in the context of sustainability transitions and possibilities that did not die with Trump's rise to power, and may in fact crystallise into new forms of green solidarity once the world has recovered from the inadvertent and improbable election of the oil and gas, anti-environment, climate change–denying U.S. president.

Post-Paris challenges

After the diplomatic achievement of the Paris agreement in 2016 when, against the odds, a number of accords were put in place acknowledging the desperate need for immediate global action on carbon transmissions, the international green agenda has suffered a major setback with the election of Donald Trump. The recent U.S. withdrawal from the Paris agreement signals how much painstaking effort is required to engineer a small concerted global step forward and the asymmetries of power involved when one man exercising executive power can overturn twenty-five years of effort in a single day. Trump's anti-environmentism has unfolded quickly with a 30 per cent cut of the Environmental Protection Agency (EPA)'s budget and the appointment of Scott Pruitt, a climate-change sceptic, to oversee its downgrading. Already the EPA has been directed by an Executive Order to review and then rescind the 2015 Clean Water Rule. Staff at the US Department of Agriculture (USDA) have been told to avoid using the term climate change in their work, with the officials instructed to reference "weather extremes" instead. "Climate change adaptation" is to be replaced with "resilience to weather extremes (The Guardian, 2017). Trump's Executive Order on climate change directed the Environmental Protection Agency to start the process of withdrawing and rewriting the Obama-era Clean Power Plan, which would have closed hundreds of coal-fired power plants, frozen construction of new plants and replaced them with vast new wind and solar farms (The New York Times, 2017). There is little doubt that Trump's ascendancy represents the triumph of the world oil and gas industrialism that makes sworn buddies of old Cold War enemies to threaten pristine environments like the Arctic region in a regime of unrelenting oil exploration, drilling and oil production.

Economically *and* politically, the shift represents a reassertion of oil and gas global capitalism where Trump's presidency provides the perfect vehicle for the robber barons to take back what they consider they have lost under the reign of global liberal internationalism. It is a calculated assault on the institutions of liberal environmentalism, although the courts still remain a vehicle of maintaining the green public interest. This represents a fundamental axial shift that makes oil buddies of Trump and Putin, signified by the appointment of ex-CEO of ExxonMobil Rex Tillerson as Secretary of State.[2]

Despite Obama's late ban of oil drilling in the federally owned areas of the Arctic and Atlantic to protect the unique Arctic ecosystem, a Republican-controlled Congress may yet be able to rescind the law, viewed by some as a last-ditch attempt by Obama to protect his climate-change legacy against Trump's climate-change–denying cabinet and his promise to renege on the U.S. commitment to the Paris agreement.

Trump's plans also have huge consequences for the coming 'environmental wars' at home and abroad when he lifted the moratorium of coal and began to dissemble the EPA. Hydraulic and gas fracking will be encouraged; in addition Trump eliminated by decree all controls on exploitation of oil, gas and coal. This already led to the cancelation of U.S. commitment to the Paris Climate Agreement and ended the curtailment of U.S. carbon emissions, which are likely to jump significantly, encouraging China and India also to ignore the new Paris 2 percent reduction protocols, although there are signs that China won't take the bait. His policies set back the environmental cause a generation and empower the oil and gas multinationals to more brazenly drill for oil in environmentally pristine areas. The best outcome for those concerned about the well-being of people and the planet is that, like the era of the Vietnam War, we may experience a flowering of global resistance against Trump's anti-environmentalism that works to galvanise and consolidate a variety of groups in a rainbow coalition to strike new values, 'soft skills' and forge a global action agenda.

A major problem is the lack of formal accountability structure in global civil society linking agencies to the publics they directly affect, especially in a system where accountability is derived from the consent of states. The international system of nation-states seems somewhat outmoded in dealing with global problems that spill over national boundaries and does not effectively recognise either sub-state actors or differentiated and emerging global publics such as the world's indigenous peoples. The new protectionism that is the heart of the rise of far-right inspired national populism is a wild contagion against all outsiders, refugees and migrants that militates against the liberal international order that prevailed in the last 70 years.

The structural imbalance in global governance between democracy and the market, especially capital markets, is part of the frustration felt by those scientists who believe that the sustainability paradigm has failed because although the science of climate change has firmed up against organized and well-funded climate deniers, the governance of climate change is painfully slow, cumbersome and open to 'buy-off'. Thus, in terms of grassroots democracy, local participation within the nation-state has been compromised by neoliberal reforms that substitute the market for the state, forcing green politics increasingly outside the realm of electoral democracy. This badly compromises the capacity of public schools, which are now under the threat of privatization and 'charterism', to critically engage students on issues of sustainability. One of the effects of globalization has been to advantage global markets, granting movement of capital and goods certain privileges and a head start over incipient global forms of democracy and the development of new international environmental agencies.

This fragile situation – some would say, 'inherent contradiction' – has led the likes of Ingolfur Blühdorn and Ian Welsh (2007, p. 185) to talk of the era of *post-ecologism* and *the politics of unsustainability*. They made this assessment before the appointment of a Republican Cabinet that is collectively anti-environmental and pro-oil and gas. Post-ecologism is reminiscent of diagnoses of the 'end of nature' (e.g. Carson, 1962; Merchant, 1980; McKibben, 1990) and earlier announcements of 'post-environmentalism', the 'fading of the Greens' and the 'death of the environmental movement' (Young, 1990; Bramwell, 1994; Shellenberger and Nordhaus, 2005). The question in this new political environment is how to name this historical moment and in face of a concerted attack on the environmental movement how to imagine a rallying resistance that harnesses all global forces, including education for sustainability, future green decades, green litigation and rainbow green coalitions.

Blühdorn and Welsh (2007) go on to argue 'the compatibility and interdependence of democratic consumer capitalism and ecological sustainability has become hegemonic' and 'faith in technological innovation, market instruments and managerial perfection is asserted as the most appropriate means for achieving sustainability' (p. 186). Is there a form of sustainability education that can chart a course that helps unhinge the easy accommodation between consumerism and sustainability, to encourage a more critical mode of education of market solutions to environmental problems?

Where many prominent environmental scientists see the failure of the sustain-ability paradigm as a failure of democratic governance at the global level – that is, a failure to act on the basis of strong and increasingly incontrovertible evidence, Blühdorn and Welsh (2007) emphasize the incompatibility of sustainability with the dominant neoliberal economic system and the culture of mass consumption it generates. What is more, they take the argument into the realm of subjectivity when they hypothesize that 'western practices of individualized, consumption-oriented identity formation' and 'the axiom of individual self-responsibility' cascading through 'the institutions of market-oriented governance', condition citizens to accept an environmental precariousness on the basis of a popular hegemonic set of relations between the fruits of consumer capitalism and 'feel-good' sustainability.

Hence, it is no surprise that living unsustainably has become the default ('nor-malized') on the planet: unsustainability is made easy; sustainability is made hard. There are clear trends and manifestations representing global systemic dysfunction, including rising inequality; loss of biodiversity and top soils; changing climates and weather patterns; and the continued toxification of water, air, soil and bod-ies. *Education as usual, much like business as usual, is no longer an option.* Education and science have become an extension of the globalizing economic system. We are preparing young people to become hard-working, lifelong learning, flexible workers who want to consume all the time. Education and science are both at risk of being hijacked for instrumental purposes that do not serve the well-being of people and the planet. Education and science are not for profit, to paraphrase Martha Nussbaum (2010). When education and science become tools to prescribe how people should live their lives, both become fundamentally undemocratic and, indeed, unsustainable. And yet in a post-truth world there are important issues that yoke science as empirical truth with democracy that we might christen *ecological democracy* that provides the warrant and justification for concerted civil action and demonstrates the new power of citizen science groups that can act autonomously in the interest of their local communities.

Counter trends in education and research?

DESD, GAP and GEM – *some responses from the UN*

UNESCO, the lead agency for the Global Action Programme (GAP) on Education for Sustainable Development (ESD) – the official follow-up to the UN Decade of ESD (DESD) which ran between 2005 and 2014 – is seeking to address the challenges facing the future of ESD in relation to current changes taking place in the world. Although the work done under this umbrella, certainly during the early years of the DESD, can be critiqued for its lack of historical awareness of social movements and other 'planetary' educations, for being 'top down', neo-colonial and undemocratic, for a-critically embracing 'development' as something to con-tinuously focus on, ESD also triggered and stimulated some important alternative approaches to education.[3] The major question concerns action-based pedagogies to develop sustainability as a way of life that rests on empowerment, broad-based citizen engagement, the significance of community models and education programs that help students become change agents. Arguably, with time, some manifestations of ESD also brought to the fore the importance of addressing systemic dysfunction and generating more systemic responses that address structure, agency, leadership, governance and content of education with full consideration of the context in which it takes place.

The 2016 Global Education Monitor Report (GEM) 'Education for people and planet: creating sustainable futures for all' (UNESCO, 2016), for instance, highlights the so-called 'whole school approach to sustainability' which simultaneously seeks to re-orient curriculum, school-community, relationships, professional development, management, pedagogy and a school's own sustainability performance (in terms of democratic processes, energy use, food served, waste reduction and management and use of school grounds). The 2016 GEM report focuses on 2 of the 17 UN-adopted Sustainable Development Goals (SDGs) – SDG 4 (quality education) and SDG 17 (partnerships) as mechanisms or processes that can help realize the 15 others which focus on topics related to the three commonly distinguished pillars of SD (people–planet–prosperity).

Does the 2016 GEM report signify a change from the dominant neoliberal agenda that sees education as an extension and a driver of the globalizing economy and its push for infinite growth, innovation and expansion?[4] If you are looking for confirmation of replication and affirmation of this agenda, you will find it; however, if you look for a shift in the common discourse, you will also find it. Reading the text through a 'transition lens', a counter-narrative and a potential shift away from business as usual can be detected. Here we list our own selection of quotes, normally not found in UN reports, that seem to support this (UNESCO, 2016):

- Current models of economic growth cause environmental destruction.
- For education to be transformative in support of the new sustainable development agenda, 'education as usual' will not suffice.
- Education cannot fight inequality on its own. Labour markets and governments must not excessively penalize lower-income individuals. Cross-sectoral cooperation can reduce barriers to gender equality.
- A whole-school approach is needed to build green skills and awareness. Campaigns, companies and community and religious leaders must advocate for sustainability practices.
- Expand education on global citizenship, peace, inclusion and resilience to conflict.
- Emphasize participatory teaching and learning, especially in civic education.
- Invest in qualified teachers for refugees and displaced people, and teach children in their mother language.
- Incorporate education into the peace-building agenda.
- Distribute public resources equitably in urban areas, involving the community in education planning.
- Mobilize domestic resources, stop corporate tax evasion and eliminate fossil fuel subsidies to generate government revenue for fundamental needs such as education and health.
- Include education in all discussions on urban development. Improve and fund urban planning programmes and curricula to include cross-sector engagement and develop locally relevant solutions.

- Promote the value of indigenous livelihoods, traditional knowledge and community-managed or community-owned land through actions such as land conservation and locally relevant research.
- Engage community elders in curricular development and school governance, produce appropriate learning materials and prepare teachers to teach in mother languages.
- Incentivize universities to produce graduates and researchers who address large-scale systemic challenges through creative thinking and problem solving.
- Promote cooperation across all sectors to reduce policy-related obstacles to full economic participation by women or minority groups, as well as discrimination and prejudice that also act as barriers.
- Support multi-stakeholder governance for the sustainable management of natural resources and of public and semi-public rural, urban and peri-urban spaces.

Additionally the 2016 GEM report has a somewhat different take on Sustainable Development compared to previous UN reports, recognizing that there are different perspectives, including ones that critique the notion of continuous development (Box 2.1).

Box 2.1   Excerpt from UNESCO's 2016 Global Education Monitor Report: change of narrative?

The different perspectives of sustainable development include viewing it as a model to improve current systems (endorsed by those focusing on viable economic growth), a call for major reforms (supported by those who advocate for a green economy and technological innovation) and an imperative for a larger transformation in power structures and embedded values of society (supported by transition movements). Some ecologists, such as deep ecologists, believe present-day human development focuses too much on people and ignores the plant, animal and spiritual parts of this world (Leonard and Barry, 2009). They believe humans must learn to be less self-interested and place the needs of other species alongside their own. Transformation advocates say societies should go back to ways of living that are locally sustainable – consuming and wasting less, limiting needs to locally available resources, treating nature with respect, and abandoning polluting technology that has become an integral part of modern society. Culture advocates believe sustainable living can happen only if communities truly embrace it as part of daily culture (Hawkes, 2001) so that it affects decisions about what to eat, how to commute to work and how to spend leisure time.

The South American *buen vivir* movement rejects development as materialistic and selfish, implying that living sustainably means finding alternatives to development (Gudynas, 2011). The *buen vivir* belief system comes directly from traditional values of indigenous people, and posits that collective needs are more important than those of the individual. In Ecuador, this concept is called sumak kawsay, the Quechua term for fullness of life in a community.

It involves learning to live within boundaries, finding ways to reduce use or to do more with less, and exploring non-material values. Ecuador and the Plurinational State of Bolivia have incorporated buen vivir into their constitutions.

Most definitions of sustainable development challenge the status quo, believing human development lacks meaning without a healthy planet. This view requires people, communities and nations to reconsider basic values of daily living and change the way they think. Understanding one's own values, the values of one's community and society, and those of others around the world is a central part of educating for a sustainable future. This means education systems need to continuously evolve and change in order to identify what practices work best within a given context and how they need to change over time. Indeed, for many of its advocates in education, sustainable development is best understood as a journey, rather than a destination.

Source: UNESCO, 2016, p. 4

The GEM2016 report seems to advocate the idea of sustainable development as a journey, a collaborative search for and engagement in sustainability, one that is not limited to small elites in society, but rather one that is accessible to all stakeholders, including those who are marginalized. The processes of searching and engaging are as important as their outcomes, as they enable a reflexive and 'learning' society that is capable of responding to setbacks, crises, challenges and systemic dysfunctions that may benefit some in the short term but harm all in the long term. Viewed as such, education and science for sustainable development require space for social learning. Such space includes space for alternative paths of development, but also alternatives to 'development' itself; space for new ways of thinking, valuing and doing; space for participation minimally distorted by power relations; space for pluralism, diversity and minority perspectives; space for deep consensus, but also for respectful *dissensus* (Koppen et al., 2002); space for autonomous, deviant, disruptive and counter hegemonic thinking; and space for self-determination.

UNESCO'S (2017) report *Education for Sustainable Development Goals* focuses on the capacity of education to pursue the 17 SDGs conceived as universal, transformational and inclusive provided within a unified global framework. The SDGs are pictured as global challenges for the survival of humanity as a whole that 'set environmental limits and set critical thresholdsfor the use of natural resources' while recognizing the interconnectedness of poverty and economic development. David Griggs and his colleagues (2013, p. 305) recognised that 'planetary stability must be integrated with United Nations targets to fight poverty and secure human well-being' and provided a unified framework that linked SDGs with MDGs. Such a unified framework could facilitate connecting what Murray Bookchin (2005, orig. 1982) called 'social ecology' with issues of 'environmental ecology' and 'political ecology', thereby raising the issue of democratic governance that is at the heart of 'ecological democracy'.

Although still ignoring the non-human and more-than-human, this connection could also represent a step toward an 'ecological' appreciation of the interconnectedness of systems linking questions of subjectivity and epistemology with environmental ecology and political governance (Bateson, 1972). The mind and human subjectivity has been part of the evolution of the planet, albeit rather recent in the 4.6-billion-year history of the Earth. One of the problems, as Guattari (2000) argues, is that 'Post-industrial capitalism – which Guattari calls Integrated World Capitalism – is delocalized and deterritorialized to such an extent that it is impossible to locate the source of its power' (Translators' Introduction, p. 6). Well before the era of post-truth he understood how *mental ecology* can be the manipulation of mass media–produced subjectivity – a critical issue for education.

Science in transition

The call for boundary crossing, partnerships, joint learning and alternatives *for* and even *to* development is not only limited to education in the context of sustainability, but also can be found in science in the context of sustainability. With the bankruptcy of science for impact factors nearing and the loss of public trust in science rising, the call for greater connectivity between science and society is growing louder (e.g. Gibbons et al., 1994; Martens et al., 2010). Echoing the visionary work of Funtowicz and Ravetz (1993), these voices call for the act of science to become a more open and democratic activity that invites public engagement and enables society to reflexively respond to urgent sustainability challenges. Inevitably, higher education (HE) needs to be closely attuned to these changing times.

The European Commission–funded EnRRICH (Enhancing Responsible Research and Innovation through Curricula in Higher Education) network concluded in its final report that higher education is at the crossroads, having to choose between the business-as-usual path of commodification of knowledge and learning focused on the well-being of the economy and the innovative path of socio-ecological transition requiring new responsible forms of collaborative research and learning and alternative capabilities and values that contribute to the well-being of the planet and its people (Tassone & Eppink, 2016). These collaborative forms of research and learning suggest a shift away from 'research as mining' grounded in empirical analytical, positivist and mechanistic traditions towards 'research as co-learning' rooted in hermeneutic interpretivist traditions and 'research as activism' rooted in socially critical transformative and even transgressive traditions (Dillon & Wals, 2006). The latter perspective is one that is quite controversial in science as it invites academic researchers to become advocates of a particular cause where they essentially are explicitly biased as opposed to neutral or objective (which from this vantage point is an illusion to begin with).

In light of global systemic dysfunction and structural favouritism towards exploitation and inequity, the idea of transgression or disruption of routines and systems that are inherently unsustainable but often highly resilient becomes an option worth pursuing. This pursuit will lead to questions about what methodologies, methods and competencies are needed for operating in a good way within such a transgressive activist framework. As far as methodologies and methods are concerned, a whole set of reflexive, participatory and transformative approaches is available (see e.g. Regeer et al., 2009; van Mierlo et al., 2010).

The EnRRich project referred to earlier created a so-called responsible research and innovation competence framework that acknowledges the complex and wicked nature of sustainability challenges, which is not meant to be prescriptive but rather as a guide or heuristic for higher education (Figure 2.1).

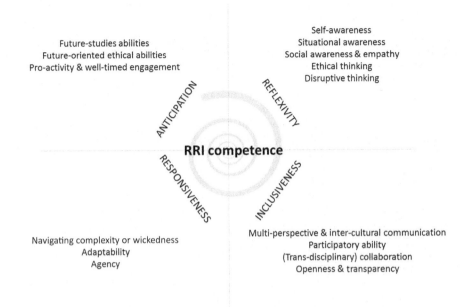

Future-studies abilities
Future-oriented ethical abilities
Pro-activity & well-timed engagement

Self-awareness
Situational awareness
Social awareness & empathy
Ethical thinking
Disruptive thinking

ANTICIPATION

REFLEXIVITY

RRI competence

RESPONSIVENESS

INCLUSIVENESS

Navigating complexity or wickedness
Adaptability
Agency

Multi-perspective & inter-cultural communication
Participatory ability
(Trans-disciplinary) collaboration
Openness & transparency

Figure 2.1 A competence framework for responsible research and innovation

Source: Tassone & Eppink, 2016

Although we think relational thinking and/or systems thinking is critical as well, we do believe all the competence areas in Figure 2.1 are crucial – as many of the other chapters in this volume seem to confirm – but of particular interest to us is the recognition of participation and collaboration and of 'disruptive thinking'. The first two link up with our concern for democracy; the latter with our concern for resilient unsustainable structures as it suggests that responsible research and innovation requires people who are capable to break with the ordinary, question the taken for granted and transgress stubborn routines. In the next two sections we discuss two such possibilities that relate to these capacities: citizen or, rather, civic science as a means of promoting ecological democracy; and transgressive education as a means to disrupt, transgress and transform unsustainable practices.

Ecological democracy

From its development in the 1980s and 1990s, Green Political Theory (GPT), or ecopolitics, founded on the work of John Dryzek (1987), Robyn Eckersley (1992), Val Plumwood (1993) and Andrew Dobson (1980), participatory democracy has been viewed as a central pillar and key value often associated with descriptions of decentralization, grassroots political decision making, citizen participation and 'strong democracy' (Barber, 1984; 2003, p. 18) and increasingly with conceptions of deliberative democracy. The value of participatory or grassroots democracy also seemed to gel with a new ecological awareness, non-violence and the concern for social justice. Green politics favoured participatory and more recently deliberative democracy because it provided a model for open debate and direct citizen involvement and emphasized grassroots action over electoral politics.

John Dewey (1916), as perhaps the arch defender of participatory democracy, proposed an 'ecological' system over 100 years ago that was based on a form of Darwinian naturalism that understood that knowledge arises from the experience of the human organism in the process of adapting to its environment. For Dewey, democracy is not just a means of protecting our interests or expressing our individuality, but also a forum for *determining our interests*. It was above all an account of democracy *as social inquiry* that emphasized the importance of discussion and debate as a mechanism of decision making with the institution of education at its heart. Democracy is a form of 'organized intelligence', and 'education is a regulation of the process of coming to share in the social consciousness', as he says in *My Pedagogical Creed* (Dewey, 1897, p. 15), as the only sure means of social reconstruction and reform. Dewey is the foremost philosopher of education in the twentieth century and perhaps also the most concerned for developing an account of education and democracy – of education as an essential democratic institution in building civil society and citizenship. As such it might be argued that Dewey proposed the ideal 'ecological' model of grassroots participation that cultivates green citizenship.

What is worthy of consideration is the ready acceptance by Dewey of the argument of social ecology when it comes to a democracy of 'organized intelligence' (what we might call 'collective intelligence' today). There are also good grounds for interpreting Dewey's naturalistic theory of experience as a fundamental ecological perspective, and that takes us in the direction of the definition of ecological democracy as *sustainability in action* – not merely a set of biological processes but simultaneously an orientation of grassroots social and political forces shaping the ecosphere.

The paradox is that although the global spread of democracy in the post-war era has been remarkable, there has been considerable 'backsliding' and the creation of a 'democratic deficit' in the long-established democracies as neoliberalism has come to prevail, leading to public scepticism of electoral politics and loss of faith in parliamentary authority. Economic liberalism has crowded out political liberalism and reduces democracy to market principles: policies as products, voters as passive consumers, politician as producers and elections as markets.

As Justin Turner (2015, p. 8) argues, 'being young in the age of globalization' is increasingly precarious in a world where 'over one billion children [are] living in poverty, 400 million lacking clean drinking water, and 165 million under the age of five experiencing stunted growth because of malnutrition'. Neoliberalism as an economic project relies increasingly on privatization of not just land, water, air, minerals and seeds, but also of education (e.g. through 'school choice', vouchers and charter schools)[5] as a political project tends to involve the deliberate shrinking of the government's role in the development and protection of civil society:

> The state experiences a redefinition of its responsibilities, shifting away from the interventionist and welfarist models toward a system more conducive to the accumulation of capital on a global scale ... and more reliant on privatization. As a pedagogical model, this historical project reshapes the social, not simply by influencing economic policies or modes of governance but by manufacturing a specific neoliberal subject.
>
> (p. 6)

Turner's conclusion, in our arguably 'elitist' words, is that 'neoliberalism submits youth to the logic of hyper-individualism and disengages them from community and society in general' that in turn makes them less prepared and less able to cope collectively with the consequences. So youth are the most critically affected 'by this exclusionary form of governance', and ultimately their privatized and individual schooling experience disbars them from natural participation or engagement in civil society or even of developing faith in collective decision making or direct forms of political action.

Foucault's now-classic concept of governmentality, applied to neoliberalism, subjectivity conditions the new generation as 'rational utility maximisers' in all of their behaviour, leading to the decline of social democracy and the erosion of the participatory ethos underlying 'thick' conceptions of green citizenship (Peters & Marshall, 1996; Peters, 2011).

Against the climate-change deniers, the scientific consensus is now almost complete and the evidence seems unassailable, yet political action is slow and open to policy reversals such as the ones initiated by Donald Trump. Achieving reductions in target emissions is notoriously difficult to police. Democracy is painfully slow and open to manipulation: the question must be asked whether it is up to the task in the new global environment where action is through agreement of interest-based states. Does it provide the appropriate decision-making mechanism and vehicle for political action at either the level of the nation-state or at the extra-state level?

Nico Stehr (2016) asks under exceptional circumstances 'Does Climate Change Trump Democracy?' He refers to a number of climate change scientists and political commentators who have lost faith in the ability of the democratic process, such as James Hansen who sounded the alarm on global warming in 1988 before Congress, and James Lovelock (2009), who in his book *The Vanishing Face* warns that we need to go on an emergency war footing with temporary suspension of democracy in order to cope with the challenges of climate change. It also turns to Dale Jamieson, a professor of environmental law at New York University and author of *Reason in a Dark Time: Why the Struggle Against Climate Change Failed – and What It Means for Our Future* (2014), who warns that climate change presents us 'with the largest collective action problem that humanity has ever faced, [but] evolution did not design us to deal with such problems, and we have not designed political institutions that are conducive to solving them', adding 'Sadly, it is not entirely clear that democracy is up to the challenge of climate change' (cited in Stehr, 2016, pp. 38–39).

The 'anti-democracy' claims of some scientists who suggest that democracy is too slow, too open to corruption and too irrational seems to be supported by a general public scepticism and disenchantment with democracy, not to mention historic referenda that have rendered deep national divisions and been open to ideological manipulation based on fear tactics. The 2015 report of the Fourth World Conference of Speakers of Parliament[6] speaks of the emergence of 'a toxic combination of adversarial politics, broken promises and a perceived inability to bring about positive change [undermining] public confidence in political processes' (IPU 2015, p. 23). The report also flags the demise of democracy on two fronts: an institutional failure to perform properly and a diminishing sphere of influence under global financial capitalism.

At the heart of public scepticism, perhaps, is a judgment about parliament's capacity to perform its functions effectively and to **embody key democratic values**.

The environment in which parliaments operate is changing, and in some ways very fast. **Much decision-making power no longer resides at the national level**, where parliaments can exert the most influence. Global financial markets increasingly shape our national policies, and international agreements can constrain a State's ability to regulate the economy independently. More decisions are taken within intergovernmental forums where parliaments typically have little influence – for example regarding the rules of international trade – and national politics are seen as powerless to influence developments.

(www.ipu.org/splz-e/speakers15/rpt3.pdf, p. 2. Bold in original)

The report concludes by suggesting a series of recommended actions to strengthen democratic culture and revitalize representative democracy, including 'investing more in civic and political education for children in schools' and 'making concerted efforts to encourage people, especially young people, to vote' (Stehr, 2016, p. 5). Perhaps of all the measures recommended, the last is the most important: 'democratizing the system of international relations, enhancing the role of parliaments vis-à-vis the issues that are high on the global agenda, and further developing the parliamentary dimension of the work of the United Nations' (Stehr, 2016, p. 5).

Stehr does not disagree with climate or political scientists about the challenges we face as a global society from climate change, nor, indeed, with the empirical evidence that indicates stable or increasing rates of deterioration, but he does draw the line when it comes to the inference that exceptional circumstances dictate that we turn away from democratic processes as a 'convenient' way to govern.

I will insist that there is no contradiction between democratic governance and scientific knowledge. Rather than lamenting the inconvenience of democratic governance, the need is to enhance democracy, not despite, but especially in light of, the massive challenges of a changing climate. We need to recognise our changing climate as an issue of political governance and not as an environmental or economic issue.

(Stehr, 2016, p. 39)

Rarely do the results of scientific inquiry provide such an unequivocal and overwhelming set of findings as climate science does in this case. One might argue that this provides an ideal Piercean community that has gathered confidence and also to all intents and purposes, achieved consensus on the interpretation of hard evidence as time has progressed. Stehr argues there is 'no contradiction between democratic governance and scientific knowledge', but the *political* arguments of climate scientists point to the need for stream-lined emergency decision making in our collective global interests, and they, also rightly in my mind, point to the encumbered nature of democratic governance at the global level. They are worried that time is running out and that the politics of climate change in the past 25 years reveals an inability to act decisively on the evidence indicating the structural incapacity of global governance working for consents of some 200 world states to achieve consensus or majority status in order to act both collectively and decisively. It is unimaginable that sovereign states would ever give up their power of consent, but it is certainly within the bounds of our technological and democratic capacities to develop continuous online voting, consensus building and rapid decision making.[7]

Tim Forsyth (2014), in an article 'Deliberative Democracy and Climate Change', reviews five significant contributions from leading scholars and emphasizes how the traditional international relations approach has been recently supplemented by approaches from other disciplines that focus

> more upon how different sub-state social actors such as citizens and businesses contest climate change politics, and how their actions are governed by underlying discourses, rather than on the analysis of national interests alone. A key theme of this analysis is deliberative democracy – or the achievement of political actions through open debate, and the consideration of differences between actors.
>
> (p. 1115)

These approaches that favour deliberative democracy explore 'how consensus might be achieved among different actors' who operate on the basis of different understandings and how developing countries can become involved in discussions. This is a process that depends upon clear communication of the findings of climate science to different public arenas and allows multiple levels of engagement (Dryzek et al., 2013). Other approaches tend to emphasize the normative and political dimensions of climate change and include it within the realm of deliberative rather than something given once and for all. Forsyth (2014) writes:

> Dryzek and Stevenson (2014) define deliberation as 'communication that is non-coercive, capable of connecting expression of particular interests or positions to more general principles, induces rejection on the part of those both speaking and hearing, in which participants strive to make sense to those who do not share their own conceptual framework'.
>
> (p. 12)

The authors usefully identify different spheres of deliberative systems, including the private sphere, public spaces, empowered space, transmission of influence, accountability, meta-deliberation and decisiveness. Deliberative democracy involves the articulation of reasoned argument in public spaces that is non-coercive, open and transparent as well as progressively transformative, that is, moving toward some accepted and effective resolution. What is important about this approach is that it enables us to focus on 'deeper origins of disagreements other than interests' and 'presents a framework for resolving differences through processes of discussion and engagement' (p. 1123).

The model of deliberative democracy philosophically springs from its pragmatist roots that suggest, over and above the mechanics of voting and representative democracy, that democracy needs to take a deliberative or participatory form as a social way of life that supports, builds and protects civil society (Florida, 2013). In *Deliberative Democracy and Beyond* John Dryzek offers this perspective:

> Deliberation as a social process is distinguished from other kinds of communication in that deliberators are amenable to changing their judgements, preferences, and views during the course of their interactions, which involve persuasion rather than coercion, manipulation or deception. The essence of democracy itself is now widely taken to be deliberation, as opposed to voting, interest aggregation, constitutional rights, or even self-government.
>
> (p. 1)

Some scholars embrace deliberative democracy for its educative power and its peda-gogical force in teaching students to reason about ecological issues and to accept responsibility for their daily practices and actions. The deliberative nature of ecological democracy has strong roots in grassroots participation in civil society. In philosophi-cal terms it is indebted to John Dewey's (1916) *Education and Democracy* and more recently to the German philosopher Jürgen Habermas's (1984) theory of communi-cative rationality based on the ideal of 'a self-organizing community of free and equal citizens', coordinating their collective affairs through their common reason. Free and open debate in society and the classroom is a necessary condition for the legitimacy of democratic political decisions based on the exercise of public reason rather than sim-ply the aggregation of citizen preferences as with representative or direct democracy.

Mauve Cooke (2000) puts forward five arguments for deliberative democracy, emphasizing (1) the educative power of the process of public deliberation; (2) the community-generating power of public deliberation; (3) the fairness of the pro-cedure of public education; (4) the epistemic quality of the outcomes of public education; and (5) the congruence of the ideal of politics articulated by deliberative democracy with our democratic subjectivities and values. Although criticisms can be made in terms of conceptions of radical democracy and insensitivities to minori-ties, the concept of an ecological deliberative democratic model offers substantial pedagogical benefits, especially when combined with school-based citizen science.

Entering citizen science, civic science

Muki Haklay (2015) provides a comprehensive report on how citizen science can significantly contribute to policy formation especially in environmental monitor-ing and decision making. He makes the case this way:

> The past decade has witnessed a sustained growth in the scope and scale of participation of people from outside established research organizations, in all aspects of scientific research. This includes forming research questions, record-ing observations, analyzing data, and using the resulting knowledge. This phenomenon has come to be known as citizen science. While the origins of popular involvement in the scientific enterprise can be traced to the early days of modern science, the scale and scope of the current wave of engage-ment shifts citizen science from the outer margins of scientific activities to the centre – and thus calls for attention from policymakers.
>
> (Haklay, 2015, p. 4)

An emerging challenge of citizen science is its deployment in education at all levels to promote participatory scientific practices integrating school, STEM education and environmental science and green studies at university to promote do-it-your-self (DIY) science for the benefit of both science and local communities. Often citizen science encourages committed and objective, disinterested research based on rigorous and systematic data collection, on the one hand, and, on the other, environmental responsibility for an action agenda – an indissoluble link carrying an ethical and political obligation to act on results. In light of the arguments made in this chapter so far, we might better characterize the action imperative as a result of the shift from the industrial science model to an ecological systems view that recognises the interconnectivity of all things and problematizes the disinterested scientist and spectator theories of knowledge.

The European Commission's *Green Paper on Citizen Science* entitled 'Citizen Science for Europe: Towards a better society of empowered citizens and enhanced research' (2014) puts the argument powerfully in terms that readily carry educational and pedagogical possibilities:

> ICT facilitates a shift of paradigm, with a more open research process sharing good and bad experiences through digital media and collaboration efforts. These new participative and networked relationships promote the transformation of the scientific system, allowing collective intelligence and new collaborative knowledge creation, democratizing research and leading into emergence of new disciplines and connections to study emerging research questions and topics. While doing this, participatory approaches contribute to long-term inclusive education, digital competences, technology skills and wider sense of initiative and ownership.
>
> (Haklay, 2015, p. 14)

We are at the beginning of a new era characterized by the cooperation of amateur and professional scientists where enhanced computing and computation power along with big and linked data signal an exciting mix of local and global, humans and machines, humans and nature in the transgressive pedagogical paradigm that moves beyond the industrial scientific model of applied science.

In the introduction to a special issue on citizen science in *Conservation Biology*, Dillon et al. (2016) introduce a particular strand of citizen science that fits well with the idea of ecological democracy. They speak of 'transition-oriented civic science' to emphasize that not the questions and concerns of scientist are the point of departure of collaborative inquiry, but rather those of concerned citizens. In other words, it is not so much about citizens supporting science but rather about science supporting citizens. The 'transition-oriented' suggests a normative stance towards a shift away from unsustainable routines and systems that tend to lead to the earlier global systemic dysfunction (e.g. planned obsolescence, built-in inequality, fossil fuel dependency, loss of identity and sense of place, etc.). This relatively new approach can be traced back to a post-normal science perspective (see for instance Ravetz, 2004), which assumes that citizens have or need to have agency, there are multiples ways of knowing and different types of knowledge that all are relevant (e.g. indigenous knowledge and local knowledge) and that improving a 'wicked' sustainability challenge requires social learning between the multiple stakeholders/actors affected by an issue (scientists being one of many). In their conclusion Dillon et al. write:

> [O]ur civic-science version of citizen science calls for expanding public participation beyond the volunteers who normally populate citizen science projects, shifting the role of scientists to one of the stakeholders (but with recognised important technical expertise), and engaging all stakeholders as co-creators and co-learners in a deliberate and systematic process of knowledge building. An important part of this process is treating emerging goals and knowledge as tentative and subject to revision based on ongoing critical and collaborative dialogue, inquiry, and action.
>
> (Dillon et al., 2016, p. 454)

The transformative power of conflict: transgressive education and research

As it is difficult not to, let us return to Trump again, albeit briefly. The people who voted for him are not a homogenous bunch, but turn out to be quite diverse. They include people who actually detest everything he represents but at the same time are deeply troubled by hegemonic structures of power and exploitation that are highly resilient (e.g. capitalism). These people, at least some of them, feel that the only way this resilience is broken is by creating a disruptive event that will create chaos and a new dynamic that might germinate a new society based on alternative principles, values and structures that they feel are more desirable. Any moderate consensus-seeking alternative will only strengthen the status quo. 'If not Bernie, then Trump to bring about real change', so to speak. The election of Trump can be viewed as a disruptive event in that it does upset conventional rules and norms and creates energy, disorder and confusion out of which something new might arise. Of course, the normative direction of the transition this disruptive event might take adds to the uncertainty: it could lead to transition towards fascism, for instance, or towards sustainability when counter-movements and grassroots sustainability niches might coalesce. We don't know at this point. What is of interest here is the potential power of disruption in bringing about fundamental change.

In sustainability discourse lots of emphasis is placed on the power of adaptation in responding to the manifestations of systemic global dysfunction such as climate change. The message is, simply put, that the world is changing, and we must learn to adapt and to become resilient to survive as a species. Ironically a focus on adaptation distracts from a critical exploration of the root causes of this systemic global dysfunction. Such exploration would reveal some serious flaws in the assumptions underlying modernity that in the end make unsustainability easy and almost our second nature and sustainability hard and almost counter-intuitive. Such exploration would also reveal that these assumptions and the structures, systems and lifestyles that result from them are very resilient themselves: changing them is hard, if not near impossible. From a sustainability point of view, disruptive capacity building and transgressive acts of transformation seem critical but are given little attention in education and research. This leads to questions like: What might disruptive capacity building entail? How can it be developed? What does transgressive learning look like? What is the role of schools and universities in supporting it?

Here we wish to connect to the ISSC-funded T-learning transformative knowledge network mentioned in the opening paragraph of this chapter. In this network learning has been identified as an important driver of change towards sustainability. Yet little is known about the type of transformative, transgressive learning (T-learning) that could potentially enable such change. The T-learning transformative knowledge network focuses specifically on transgressive social learning for social-ecological sustainability in times of climate change. The network seeks to uncover and enable T-learning processes at the climate–energy–food–water security and social justice nexus and aspires to generate, surface and describe qualities of such learning processes and their role and contribution to sustainability transformations. Specifically its objectives are to:

1   Investigate and expand the emergence and qualities of T-learning processes in selected food-water-energy-climate-social justice nexus contexts in diverse niche level settings,

2   Investigate and identify potential 'germ cell' sustainability activities and engage these in potential expansions within a multi-levelled perspective, and trace how this is done, and

3   Develop generative T-learning methodologies for informing social-ecological science research and praxis, and extend current theoretical work on T-learning within the social-ecological sciences.

(www.transgressivelearning.org)

Central in the T-learning network is the transformative potential of disruption and transgression in reframing dominant narratives in education and learning spaces (Lotz et al., 2015). The network seeks to strengthen commitment to the commons and the common good, to decolonisation, the good life, ecological economics and real sustainability and will seek to bring environmental and social justice into being.

Let us look at disruption and transgression a bit more closely. In a sense, a disruption represents a discontinuity of the ordinary of the usual of the expected. It forces people to leave one's comfort zone: a state of disequilibrium. Such discomfort can be generative when it invites people to explore other options, to build new alliances or to re-think what they always thought to be normal or true, for instance, but can also be regressive when it numbs them, makes them more susceptible to blindly follow others or leads to withdrawal. In other words the dissonance that arises out of acts of boundary crossing (leaving one's ordinary world) and being exposed to alternative ways of seeing and being in the world can both drive and block learning, co-creation and innovation. Dissonance and disruption entail conflict that might be destructive or constructive depending on when, where, how and with what intent. The crux is *how* conflict is dealt with. Too much conflict may lead to a break in interaction when people are too far out of their 'comfort zone' (and some people have bigger comfort zones than others with respect to being challenged), whereas too little of it is just as likely to prevent any significant learning.

In transgressive learning people are learning on the edge of their comfort zones with regard to dissonance. When facilitators of interactive processes manage to strike a balance between comfort and tension, creating 'optimal dissonance' by skilfully stretching comfort zones as needed (Schwarzin & Wals, 2012, p. 27), transformative disruptions can occur that push participants away from the 'comforting bubbles' of their own (potentially privileged) position and perspective and challenge them to view the world from the vantage point of (perhaps marginalized) others (van Gorder, 2008). When dissonance is addressed as 'oppositional discourse', in which participants embrace tensions between different positions and seek to uncover and probe paradoxes and contradictions, it can play a critical role in realizing transitions towards sustainability. The is not new: constructive approaches to conflict have long been shown to play a key role in individual learning (Berlyne, 1965; Festinger, 1957; Piaget, 1964) but from the perspective of transitions and social movements, they need to be studied and understood at the collective level as well. When dissonance is introduced carefully and dealt with in a proactive and reflexive manner, it can help participants reconsider their views and invite them to co-create new ways of looking at a particular issue and generate new thinking that can thaw frozen mind-sets and break deeply entrenched systems and routines.

Akkerman and Bakker's (2011) theory on learning across boundaries adds a new dimension to existing learning theories. Whereas conventional learning theories, such as social constructivism, focus on a person's development of knowledge or capabilities within a specific domain and in a specific context, a boundary perspective adds the dimension of two-sided actions and interactions between learners anchored in different contexts. Whereas in conventional theories of learning, diversity is often perceived as problematic, in trans-boundary learning this diversity is appreciated. Boundaries can be defined as socio-cultural differences leading to discontinuity in action or interaction (Cremers et al., 2016). 'Boundaries simultaneously suggest a sameness and continuity in the sense that within discontinuity, two or more sites are relevant to one another in a particular way' (Akkerman & Bakker, 2011, p. 133). In the context of sustainability the notion of socio-cultural differences probably needs to be expanded to include socio-ecological differences. Such differences refer not just to differences relating to physical and virtual locations or practices, but also to more abstract distinctions, such as different perspectives or perceptions of unfamiliar domains (Engeström, 1999). Cremers et al. (2016), referring to Kerosuo (2004), point out that these differences can be explicitly perceived by diverse actors, or they can be more implicit, but still empirically detectable by verbal markers. These markers, they suggest, can be references to synonyms of the word boundary (e.g. border, limit), metaphors (e.g. fences, walls), references to social relationships such as 'we versus them', or spatial references to different locations.

So where does this lead us when re-thinking education and research in light of transitions towards a more equitable and enjoyable world where people can live well together while respecting ecological boundaries and the non-human and more-than-human world? Earlier we referred to research as co-learning and research as activism as generative perspectives for transition-oriented forms of inquiry and change. Here we will expand this by introducing three types of work that seem necessary (Peters & Wals, 2013, p. 93). *The work of determining what is.* This includes naming, framing and setting problems; identifying, observing and documenting physical, social, cultural and political realities, phenomena and behaviours; identifying and documenting views, opinions and needs; and identifying and articulating ideals, values and interests. *The work of determining what should be and what can and should be done to close the gap between what is and what should be.* This includes public deliberation and debate; the production of public judgment; the running of experiments; and the development and testing of action plans, strategies and tools. *The work of determining, assessing and interpreting what happened and why and what to do next.* This is done both during and after taking action and running experiments, and it can include both quantitative forms of measurement and qualitative and narrative forms of evaluation and interpretive meaning-making.

Considering the importance of transgression and disruption, we propose to add some activities to the work around determining what should be and what can and should be done to close the gap between what is and what should be: determining what works with the changes that are desired change (enabling forces and conditions) and determining what works against these changes (forces and conditions that work to keep things the way they are). Adding this also implies that the work of determining, assessing and interpreting what happened and why would need to include work of determining, assessing and interpreting what did *not* happen and why. These additions to what Peters and Wals (2013) call *phronesis* invite critique of hegemony and expose systemic dysfunction and might lead to transgressive acts that open the door for deep transformation. Box 2.2 contains an excerpt from their 2013 chapter that seems quite relevant in the so-called post-truth era.

Box 2.2   Practical theory building as activism

How should educators working in New York State, above the Marcellus Shale (note: a geological formation that contains natural gas that can be extracted for commercial purposes), handle a controversial issue such as this in public classrooms and/or community settings? What position do they take? How do they bring in and treat scientific evidence coming from different sides? Do they take a stance? Do they actively engage in such a complex and existentially relevant issue? Are they able and willing to draw distinctions between rhetoric and reality, and to reduce, using Sandra Harding's language, the systemic bodies of ignorance created around highly politicized issues such as these? Or should they stay away from issues such as these altogether, and stick to school board – approved textbooks in order to keep controversy out of the classroom? We believe taking one side in the continuum between passive detachment and avoidance on the one hand ("playing it safe") and active involvement and engagement on the other ("taking a risk") will not be generative in taking education, learning, and research to a level where we can deal with issues such as these in a more satisfying, less polarizing way.

This is where practical theory building offers a third way forward, as it assumes that learners – including teachers, researchers, policy makers, etc. – do not accept facts as an external given; rather, it requires self-confrontation and joint fact finding as a starting point for learning.

Source: Peters & Wals, 2013, p. 97

Conclusion

In light of the earlier proposed need for boundary crossing, learning on the edge of one's comfort zones and the importance of discontinuities and plurality of ideas, it is also critical that education and research explore the phenomenon of blinding insights and lock-ins. It is well known that the way we frame experiences is closely connected to our cultural narratives and associated encultured and embodied ontological pre-dispositions. This framing gives us comfort, on the one hand, but can blind us to alternative ways of seeing and being that, from a sustainability point of view, might be more generative. Transformative processes are more likely to occur when those involved are or become aware of the frames or filters through which they perceive their reality and are able to deconstruct and reconstruct them in their joint pursuit of sustainability. This is no easy task, as people can become so stuck in their own often taken-for-granted and normalized ways of thinking and acting that they fail to see how this colours their judgment and narrows possibilities. The success of transformative learning lies in people's ability to transgress their way of seeing and being in the world. Such transgression is facilitated by the exposure to alternative ways of seeing and being in the world and participation in what Chaves et al. (2017) call ontological encounters. They write:

> When deployed in the 'transgressive' context of our own sustainability struggles, ontological encounters provide a treasure for learning that unsustainable realities are not destiny.
>
> (Chaves et al., 2017, p. 21)

Through these encounters people can become aware of their own frames, filters, values and assumptions, but also of their ideological underpinnings and the resulting tunnel vision that may blind them in pursuit of a more sustainable way of being. When this process takes place in a collaborative setting, where dissonance is properly managed, cultivated and utilized, allowing participants to be exposed to the deconstructed frames of others, old ideas can be challenged and new ones can be co-created (Wals, 2007). We can add to this the importance of creating an atmosphere that invites trust and empathy where people can be comfortably vulnerable, open up their minds and listen to others. Presently the culture in most schools and universities does not allow for such an atmosphere to emerge. Alternative leadership, management and forms of resistance to previously mentioned patterns of individualism, competition and accountability will be necessary. Education based on action pedagogies can play a significant role in joining up a deliberative ecological democracy with new forms of activist science and the rapidly growing forms of citizen science that encourage the use of empirical evidence and logic in a 'post-truth' world driving community-based science projects and encouraging linked-up international scientific agendas that promote collection of data and careful evaluation based on systematic observation and experiment.

Our final proposition, for now, on transgressive education and research, returning to 'flowers of resistance', is based on a philosophical understanding of 'dissident thought', bringing us back to the 'flowers of resistance, and hopefully poses that against the structures of power and repression

> engender sparks of dissidence that leads to a person, movement, literature, or a form of scholarship that actively challenges an established doctrine, policy, or institution to call out against unlawful violations of "human rights".
>
> Dissident thought has a kinship relationship with the ecology of concepts that proceed from the concepts of dissent and the very possibility of disagreement as an inherent aspect of discourse. It has taken many different forms in relation to discourse, thought and action, and encompassed and cultivated political norms associated with freedom of speech that allows the expression of opposition, protest, revolt and the expression of anti-establishment thought that takes the form of civil disobedience, non-violent protest and sometimes revolutionary activity.
>
> (Peters, 2016, p. 20)

'The flowers of resistance' in the age of post-truth – let a thousand flowers bloom!

Questions for comprehension and reflection

1   Is there a form of sustainability education that can chart a course that helps unhinge the easy accommodation between consumerism and sustainability, to

encourage a more critical mode of education that can also lead to question market solutions to environmental problems?

2   What is meant in this chapter by a 'transgressive research-as-activism framework'? What methodologies, methods and competencies are needed for engaging meaningfully in such a transgressive research-as-activism framework?

3   What action-based pedagogies are available or can be envisioned that allow us to develop sustainability as a way of life that rests on empowerment? How can we educate for broad-based citizen engagement? What is the significance of community models and education programs that help students become change agents?

4   Democracy is painfully slow and open to manipulation. The question must be asked: Is it up to the task in the new global environment where action is through agreement of interest-based states? Does it provide the appropriate decision-making mechanism and vehicle for political action at either the nation-state or extra-state level?

Notes

1   In 2015 the International Social Sciences Council (ISSC) established a research programme on Transformations to Sustainability. This programme seeks to strengthen social responses to climate change and environmental concerns. It has established three Transformative Knowledge Networks linking scientists, educators, civil society, policy makers, business and other stakeholders together. This chapter connects with one of those network: the T-learning transformative knowledge network (see: www.transformativelearning.org).

2   Tillerson developed close ties with Russia, overseeing the Exxon drilling project in 1996 and signing a production sharing agreement that became commercial in 2001. He brokered an agreement to drill for oil in the Russian Artic Ocean (even though Obama made sure Alaska remains off limits for the time being) with Rosneft, the massive Russian State oil company headed by Igor Sechin.

3   For critiques of ESD see: Huckle and Wals, 2015; Jickling and Wals, 2012; Helberg and Knuttson, 2016; Berryman and Sauve, 2016; Kopnina, 2016; Jickling and Sterling, 2017.

4   We should declare that one of us, Arjen, acted as an external consultant to the 2016 GEM report (UNESCO, 2016).

5   These are the exact policies that characterize U.S. Secretary of Education Betsy DeVos's program of reform under the Trump regime.

6   See www.ipu.org/splz-e/speakers15.htm and www.ipu.org/pdf/publications/speakers15-e.pdf. Reference in this paper is to the report 'Challenges Facing Parliaments Today'.

7   See, for instance, the UK Labour manifesto on digital democracy under Jeremy Corbyn that looks to a universal service network, an open knowledge library, a community media platform, platform cooperatives, a digital citizen passport, a people's charter of digital liberty rights and, most significantly, massive multi-person online deliberation – www.jeremyforlabour.com/digital_democracy_manifesto.

References

Akkerman, S. F. & Bakker, A. (2011). Boundary crossing and boundary objects. *Review of Educational Research* 81 (2): 132–169.

Barber, B. (1984, 2003). *Strong democracy: Participatory Politics for a New Age*. Berkeley: University of California Press.

Bateson, G. (1972). *Steps to an ecology of mind*. Chicago: Chicago University Press.

Berlyne, D. E. (1965). 'Curiosity and education', in Krumbolts, J. D. (Ed.) *Learning and the educational process*. Chicago: Rand McNally & Co. p. 67–89.

Berryman, T. & Sauvé, L. (2016). Ruling relationships in sustainable development and education for sustainable development. *The Journal of Environmental Education* 47 (2): 104–117. doi:10.1080/00958964.2015.1092934

Blühdorn, I. & Welsh, I. (2007). Eco-politics beyond the paradigm of sustainability: A conceptual framework and research agenda. *Environmental Politics* 16 (2): 185–205. doi:10.1080/09644010701211650

Bookchin, M. (2005, orig. 1982). *The ecology of freedom: The emergence and dissolution of hierarchy*. Oakland: AK Press.

Bramwell, A. (1994). *The fading of the green: The decline of environmental politics in the West*. New Haven, CT: Yale University Press.

Carson, R. (1962). *Silent spring*. London: Penguin.

Chaves, M., Macintyre, T., Verschoor, G. & Wals, A.E. J. (2017). Towards transgressive learning through ontological politics: Answering the 'call of the mountain' in a Colombian network of sustainability. *Sustainability* 9 (1): 21. doi:10.3390/su9010021

Cooke, M. (2000). Five Arguments for Deliberative Democracy. *Political Studies*, 48 (5): 947–969.

Cremers, P.H.M., Wals, A.E.J., Wesselink, R. & Mulder, M. (2016). Design principles for hybrid learning configurations at the interface between school and workplace. *Learning Environments Research* 19 (3): 309–334. doi:10.1007/s10984-016-9209-6

Dewey, J. (1897). My Pedagogical Creed. *School Journal* vol. 54 (January 1897), pp. 77–80 available at http://dewey.pragmatism.org/creed.htm

Dewey, J. (1916). *Democracy and education*. New York: Macmillan Company. Reprinted in Boydston, J.A. (Ed.) (1980). *The middle works of John Dewey*. Carbondale: Southern Illinois University Press. Volume 9, pp. 1899–1924.

Dillon, J., Stevenson, R. B. & Wals, A.E.J. (2016). Special section: Moving from citizen to civic science to address wicked conservation problems. *Conservation Biology* 30 (3): 450–455.

Dillon, J. & Wals, A.E.J. (2006). On the dangers of blurring methods, methodologies and ideologies in environmental education research. *Environmental Education Research* 12 (3-4): 549–558. doi:10.1080/13504620600799315

Dobson, A. (1980). *Green political thought, 4th Edition 2007*. London and New York: Routledge.

Dryzek, J. S. (1987). *Rational ecology: Environment and political economy*. Oxford: Blackwell.

Dryzek, J. S., Norgaard, R. & Schlosberg, D. (2013). *Climate challenged society*. Oxford: Oxford University Press.

Eckersley, R. (1992). *Environmentalism and political theory: Toward an ecocentric approach*. New York: State University of New York Press.

Engeström, Y. (1999). 'Innovative learning in work teams: Analyzing cycles of knowledge creation in practice', in Engeström, Y., Miettinen, R. & Punamaki, R. (Eds.) *Perspectives on activity theory*. Cambridge: Cambridge University Press. pp. 377–404.

Festinger, L. (1957). *A theory of cognitive dissonance*. New York: Harper & Row.

Florida, A. (2013). Participatory Democracy versus Deliberative Democracy: Elements for a Possible Theoretical Genealogy. Two Histories, Some Intersections. *Paper presented at 7th ECPR General Conference Sciences Po, Bordeaux*, 4–7 September 2013. Available at: https://ecpr.eu/Filestore/PaperProposal/71d7f83c-3fe4-4b11-82a2-c151cd3769f4.pdf

Forsyth, T. (2014). Deliberative democracy and climate change. *Public Administration* 92 (4): 1115–1123.

Funtowicz, S. & Ravetz, J. (1993). Science for the post-normal age. *Futures* 25: 739–755.

Gibbons, M., Limoges, C., Nowotny, H., Schwartzman, S., Scott, P. & Trow, M. (1994). *The new production of knowledge: The dynamics of science and research in contemporary societies*. London: Sage.

Griggs, D., Stafford-Smith, M., Gaffney, O., Rockström, J., Öhman, M. C., Shyamsundar, P., ... Noble, I. (2013). Sustainable development goals for people and planet. *Nature* 495: 305–307.

Guattari, F. (2000, orig. 1989). *The three ecologies.* Translated by Ian Pindar and Paul Sutton. London and New York: The Athlone Press.

Gudynas, E. (2011). Buen Vivir: todays tomorrow. *Development*, 54 (4): 441–47.

Habermas, J. (1984). *Theory of Communicative Action, Volume One: Reason and the Rationalization of Society.* Translated by Thomas A. McCarthy. Boston, MA: Beacon Press.

Habermas, J. (1987). *Theory of Communicative Action, Volume Two: Lifeworld and System: A Critique of Functionalist Reason.* Translated by Thomas A. McCarthy. Boston, MA: Beacon Press.

Haklay, M. (2015). *Citizen science and policy: A European perspective.* (Case Study Series No. Vol. 4). Commons Lab. Washington, DC: Woodrow Wilson International Center for Scholars.

Hawkes, J. (2001). *The fourth pillar of sustainability: Culture's essential role in public planning.* Melbourne: Common Ground Publishing Pty Ltd.

Hellberg, S. & Knutsson B. (2016): Sustaining the life-chance divide? Education for sustainable development and the global biopolitical regime, *Critical Studies in Education*, DOI:10 .1080/17508487.2016.1176064

Huckle, J. & Wals, A.E.J. (2015). The UN decade of education for sustainable development: Business as usual in the end. *Environmental Education Research* 21 (3): 491–505.

IOU (2015). Fourth World Conference of Speakers of Parliament, New York, 31 August–2 September 2015.

Jickling, B. & Sterling, S. (Eds.) (2017). *Post-sustainability and environmental education.* Cham, Switzerland: Palgrave.

Jickling, B. & Wals, A.E.J. (2012). Debating education for sustainable development 20 years after Rio: A conversation between Bob Jickling and Arjen Wals. *Journal of Education for Sustainable Development* 6 (1): 49–57.

Kerosuo, H. (2004). Examining boundaries in health care: Outline of a method for studying organizational boundaries in interaction. *Critical Practice Studies* 6 (1): 35–60.

Kopnina, H. (2016). The victims of unsustainability: a challenge to sustainable development goals. *International Journal of Sustainable Development & World Ecology*, 23 (2): 113–121, DOI: 10.1080/13504509.2015.1111269

Koppen, van K., Lijmbach, S., Margadant, M. & Wals, A.E.J. (2002). Your nature is not mine! Developing pluralism in the classroom. *Environmental Education Research* 8 (2): 121–135. doi:10.1080/13504620220128202

Leonard, L. and Barry, J. (eds). (2009). *The Transition to Sustainable Living and Practice*, (Vol. 4). Bingley, UK, Emerald Group Publishing.

Lotz-Sisitka, H., Wals, A.E.J., Kronlid, D., McGarry, D. 2015. Transformative, transgressive social learning: rethinking higher education pedagogy in times of systemic global dysfunction. *Current Opinion in Environmental Sustainability*, 16: 73–80.

Lovelock, James (2009). *The Vanishing Face of Gaia: A Final Warning: Enjoy It While You Can.* London: Allen Lane.

Martens, P., Roorda, N. & Cörvers, R. (2010). Sustainability, science, and higher education – The need for new paradigms. *Sustainability* 3 (5): 294–303.

McKibben, B. (1990). *The end of nature.* London: Penguin.

Merchant, C. (1980). *The death of nature: Women, ecology and the scientific revolution.* San Francisco: Harper & Row.

Nussbaum, M. (2010). *Not for profit: Why democracy needs the humanities.* Princeton, NJ: Princeton University Press.

Peters, M. A. (2011). *Neoliberalism and after? Education, social policy and the crisis of capitalism.* New York: Peter Lang.

Peters, M. A. (2016). Dissident thought: Systems of repression, networks of hope. *Contemporary Readings in Law and Social Justice* 8 (1): 20–36.

Peters, M. A. & Marshall, J. D. (1996). *Individualism and community: Education and social policy in the postmodern condition*. London: Falmer Press.

Peters, S. & Wals, A.E.J. (2013). 'Learning and knowing in pursuit of sustainability: Concepts and tools for trans-disciplinary environmental research', in Krasny, M. & Dillon, J. (Eds.) *Trading zones in environmental education: Creating trans-disciplinary dialogue*. New York: Peter Lang. pp. 79–104.

Piaget, J. (1964). Development and learning. *Journal of Research in Science Teaching* 2: 176–186.

Plumwood, V. (1993). *Feminism and the mastery of nature*. London: Routledge.

Ravetz, J. (2004). The post-normal science of precaution. *Futures* 36: 347–357.

Regeer, B. J., Hoes, A. C., van Amstel-van Saane, M., Caron-Flinterman, F. F. & Bunders, J. F. (2009). Six guiding principles for evaluating mode-2 strategies for sustainable development. *American Journal of Evaluation* 30: 515–537.

Shellenberger, M. & Nordhaus, T. (2005). *The death of environmentalism: Global warming politics in a post-environmental world*. Retrieved from http://grist.org/article/doe-reprint/ (Accessed 27/01/2005)

Stehr, N. (2016). Exceptional circumstances: Does climate change Trump democracy? *Issues in Science and Technology* Winter: 30–39.

Stevenson, H. & Dryzek, J. S. (2014). *Democratizing global climate governance*. Cambridge: Cambridge University Press.

Tassone, V. & Eppink, H. (2016). *The EnRRICH tool for educators: (Re-) designing curricula in higher education from a 'responsible research and innovation' perspective*. Deliverable Report 2.3 from the EnRRICH project. Retrieved from www.livingknowledge.org/fileadmin/Dateien-Living-Knowledge/Dokumente_Dateien/EnRRICH/D2.3_The_EnRRICH_Tool_for_Educators.pdf (Accessed 14/03/2017)

The Guardian (2017). US federal department is censoring use of term 'climate change', emails reveal. *The Guardian*, August 7th, available at: https://www.theguardian.com/environment/2017/aug/07/usda-climate-change-language-censorship-emails

The New York Times (2017). Trump Signs Executive Order Unwinding Obama Climate Policies. *The New York Times*, March 28th, available at: https://www.nytimes.com/2017/03/28/climate/trump-executive-order-climate-change.html

Turner, J. (2015). Being young in the age of globalization: A look at recent literature on neoliberalism's effects on youth. *Social Justice* 41 (4): 8–22.

UNESCO. (2016). *Education for people and planet: Creating sustainable futures for all: Global education monitor report 2016*. Paris: UNESCO.

UNESCO. (2017). *Education for sustainable development goals: Learning objectives*. Paris: UNESCO.

van Gorder, C. (2008). Educating for social justice among the world's privileged: Re-inventing education for social justice. *Journal of Pedagogy, Pluralism and Practice* 7 (1). Retrieved from www.lesley.edu/journal-pedagogy-pluralism-practice/chris-van-gorder/pedagogy-children-oppressors/

van Mierlo, B., Regeer, B., van Amstel, M., Arkesteijn, M., Beekman, V., Bunders, J., . . . Leeuwis, C. (2010). *Reflexive monitoring in action: A guide for monitoring system innovation projects*. Oisterwijk: Boxpress. Retrieved from www.falw.vu.nl/en/Images/Reflexive_monitoring_in_Action_B_van_Mierlo_and_B_Regeer_2010_tcm246-399363.pdf

Wals, A.E.J. (Ed.) (2007). *Social learning towards a sustainable world*. Wageningen: Wageningen Academic Publishers.

Wals, A.E.J. & Schwarzin, L. (2012). Fostering organizational sustainability through dialogical interaction. *The Learning Organization* 19 (1): 11–27.

Young, J. (1990). *Post environmentalism*. London: Belhaven.

Part I

Embracing complexity and alternative futures

Conceptual tools and methods

3   Systems approaches for transforming social practice

Design requirements

Ariane König

> Our whole education tends to only draw simple logical conclusions and defining obvious cause and effect relationships [which] have no existence in reality. In reality all is indirect effects, networks, connections and time delays. Our civilisation will succeed [...] only if it acquires a far greater knowledge of systemic connections in complex systems.
>
> Frederic Vester (2007)

Challenges to designing systems approaches for societal transformation

Accelerating and interconnected change in the spheres of technology, economy, environment, and society raises existential questions about future reliable access to land, water, food, and energy in a growing number of geographic areas. In light of these considerations, perhaps one of the greatest challenges society faces lies in the way of how we conceive of and organize knowledge and learning in research, education, business, and governance. For example, water, food, and energy systems, and in particular the human, technological, and ecological elements of these systems, are usually considered separately by experts dedicated to specific knowledge fields. Disciplinary silos direct attention at a specific system, its spatial and time scales, and build on assumptions that usually reduce complexity, uncertainties, and aggregating contradictory perspectives. Science relies on peer review and career reward systems that often suppress specific kinds of questions, divergence, and contradictions, and to make matters worse, science education usually conveys specific sets of beliefs and worldviews as set frames for perceiving reality (Kuhn, 1962). Researchers in the natural and social science not only rely on different theories and methods, but also on disparate language and sets of values and worldviews that shape how they order and represent knowledge; they often don't understand each other, even when they genuinely try to engage in joint research. Developing some mutual understanding and shared language and representations can take years. Moreover, disciplines mostly drive towards abstract and generalisable knowledge rather than considering situated knowledge and locally specific interdependencies, experiences, and preferences. These may be essential for finding shared approaches to act on potential threats, for example, of resource depletion and pollution.

This chapter compares diverse analytic frameworks developed by researchers to understand ourselves and the environments we live in as complex and dynamic systems, in which our conceptions, actions, and environments change interdependently.

The frameworks we discuss in this chapter are designed to help us conceive of the environments we relate to as different interacting spheres (or sub-systems) of humans and society, technology, economy, and environment (or eco-systems). The frameworks help us to understand how the overall behaviour of each of these sub-systems and the entire set may depend on how each sub-system is connected to and depends on changes in the other sub-systems. The different frameworks we present on how humans relate to and depend on their environments have emerged from different fields of knowledge, including environmental science, economics and management, and human ecology.

There are several challenges to the design of such concepts and tools to guide collaborative sustainability science on dynamic complex systems, which we address in this chapter. First, insights from cognitive psychology suggest that cognitive pitfalls in trying to embrace such complexity in science, policy, and practice are many-fold (see Chapter 4 by Sonnleitner); our minds start to struggle when confronted with analysing relationships between just six or seven semi-interdependent variables. In practice and policy making, we are all caught up in the usual linear thinking we all have been trained for at school. We also usually lack the ability to embrace uncertainty and ignorance. These circumstances usually lead to policies and action that are ineffective or have side effects unintended from solutions (Forrester, 1969). Such pitfalls include incorrect descriptions of goals due to a focus on solving individual problems, such as energy security in the face of variable oil prices. The resulting one-dimensional analysis then does not consider feedbacks between systems and instead irreversibly foregrounds one possible solution for further analysis rather than trying to grasp a big systemic picture and first suggesting solutions that take account of interactions (Dörner, 1975). For example, in building the Aswan Dam for hydro-energy in Egypt, the huge and real impacts on food yields from a reduction in the annual flooding in the Nile valley, which had sustained Egyptian civilisation for millennia, was apparently entirely neglected. Further cognitive pitfalls also evident in this example include authoritarian behaviour from leaders and experts in technocracies in the face of uncertainty and ignorance. This leads to a reduced field of vision from consideration only of perspectives of those in power. Our current education systems, with a focus on identifying linear cause–effect relationships with disciplined knowledge fields and analytic practices, are particularly problematic. Society must now unlearn these assumptions and relearn different approaches to the understanding of complex problems by engaging in different processes for sense-making, knowing, learning, and creating (Boyden, 1981; Senge, 1990; Schön, 1993; Vester, 2002).

Radically new concepts of knowledge and learning, and associated social processes and spaces, are required (Wegerif, 2007, building on the sociologist Castells; see also the introduction of this book). Moreover, these new practices must emerge from and be fit for the networked society of the future, rather than an industrial society of the past. A networked society holds many untapped opportunities for scientific knowledge co-creation processes. Knowledge gaps and contradictions may become more easily apparent and less troubling to embrace, and these can present spaces for creativity (Wegerif, 2007). A serious challenge here is the requirement for new types of quality criteria for legitimating the emerging new shared knowledge.

In order to explore the emergence of different frameworks to address these challenges, this chapter first provides a short overview on the foundations of systems thinking that have been specifically developed to better understand human–environment interactions in the 1970s and 1980s, which are considered of significant influence to the selection of more recent frameworks presented in this chapter. Subsequently we develop a set of criteria to assess whether conceptual frameworks designed to describe human–environment interactions are suitable for structuring transformative sustainability sciences as a collaborative research process with diverse stakeholders (as described in Chapter 1 of this book). The subsequent section then singles out different systems approaches that are complementary in that they direct attention to different dynamics in human–environment interactions as they have emerged from different fields of knowledge, including environmental science, economics and management, and human ecology, and briefly discusses their merits and limitations. The frameworks presented in this chapter are selected on the basis that they have proven influential in science, policy, and practice. The conclusion highlights their significance and the urgent need to start thinking in terms of systemic interactions. It also stresses the need to use such frameworks as a basis for new approaches to combining research, learning, policy making, and social practice toward more sustainable governance.

Foundations of systems thinking: 1970s and 1980s

The exploding scientific literature with attempts to conceptualise and develop solutions to improve human–environment interactions was already of concern in the 1970s (Ehrlich & Holdren, 1971, 1972). Ecologist Paul Ehrlich and nuclear physicist John Holdren cautioned that theoretical solutions to problems are often not operational and that sometimes they are not solutions at all. A simple framework for realistic analysis as a basis for developing practical solutions was deemed necessary. The resulting 'IPAT equation' (Figure 3.1) conceives of environmental *Impacts* as a function of the size of the human *Population*, multiplied by the *Affluence* that affects material flows from our societal metabolism in the form of consumption and production, the impacts of which are in turn modulated (both positively and negatively) by the *Technologies* that are produced and in use.

This simple framework to structure discussions must be used with care, as there is no explicit warning that the variables on the right side of the equation are not independent: institutions, values, and beliefs can influence all of the variables and the nature of their interactions among them, and their interactions vary in space and over time. Further detailing of the main variables in italics and brackets is derived from Holdren et al. (1995).

This equation was developed to draw attention to what the authors deem are the major factors that affect the human–environment relationship. The framework clearly points out that at a sustainability limit there will be trade-offs between population, resource through-put, and thus economic activity per person, strongly suggesting that simply associating societal progress with economic growth no longer makes sense. Technology was considered an important factor for independent

Environmental **I**mpact = **P**opulation

    x **A**ffluence *(economic activity per person x resource-use per economic activity)*

    x **T**echnology *(stress on the environment per resource-use x damage per stress as measured by environmental susceptibility)*

Figure 3.1 The IPAT equation

consideration in the mediation of impacts of a growing human population on environments in which it is embedded.

Basic principles: stocks, flows, constraints, feedbacks, and experience

Stocks and flows: At the core of the problem for closer analysis – that of survival of a growing human population on a bounded planet – from an economist's point of view stand finite *stocks* of essential resources on earth that may be depleted or degraded by human activities more quickly by *outflows* than they can be renewed or replaced by *inflows* (Daly, 1973, 1977, 1991; Daly et al., 1989). Representations of stock–flow models as in Figure 3.2 were developed in early systematic analysis of the relation of economy and a planet with finite resources. Accordingly, flows replenishing and depleting key stocks should be in equilibrium over time. Stocks can decouple inflows from outflows, allowing for these to be out of balance for a limited amount of time, depending on their flow rates.

   Constraints. These examples of stocks and flows of non-renewable sources and renewable sources also clearly demonstrate the principle of 'constraints on system

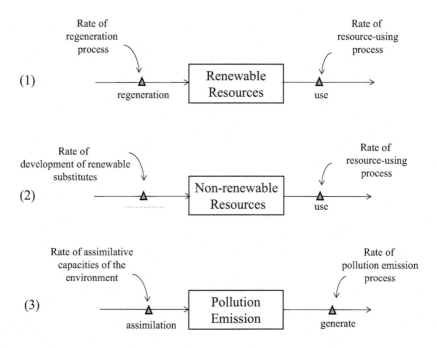

Figure 3.2 Stock–flow models of resource use on a finite planet (Daly, 1991)

This stock–flow model is based on Daly's work (1973, 1977, 1991; Daly et al., 1989) highlighting requisites for resource use on a finite planet

1 Rates of use of renewable resources do not exceed regeneration rates

2 Rates of use of non-renewables resources do not exceed rates of development of renewables substitutes

3 Rates of pollution emission do not exceed assimilative capacities of the environment

Figure also adapted from Dyball and Newell (2015, p. 95)

behaviour'. Systemic *constraints* are determined by the *size of the stock of a non-renewable resource* and/or *the rate of the flow for renewable resources* upon which the functioning of the system depends.

Feedbacks. Any self-organised system, including any life form on earth, depends on mechanisms to change and adapt flows in order to regulate the maintenance of its stocks. The vital mechanisms that regulate the inflow and outflow of these stocks in response to internal or external changes are called *feedbacks*. For example, any living system at any scale (unicellular, multi-cellular organisms, ecosystem) depends on the ability to regulate its activities in response to changes in its environment, including the availability of nutrients, water, temperature, or other elements it may depend upon or that may threaten its survival (Figure 3.3). A *feedback loop* can be defined as "a closed chain of causal connections from a stock, through a set of pre-programmed responses that are dependent of the level of the stock, to have a direct impact on a flow in order to change the stock" (Meadows, 2008). Multi-cellular organisms can thus be seen as a set of stocks and flows (in animals, for example, stocks of blood, muscles, bones, brainpower, experience, and cunning) that are interconnected and regulated to function in a coordinated manner by the presence of *feedback loops*, based on signalling pathways between its cells, including protein or peptide based (e.g. hormonal) and electrical impulses.

Ambient air temperature rises in summer can trigger adapting balancing responses in humans to counteract undesirable levels of bodily heat gain. Responses include automatic sweating or a conscious decision to jump into a cold pool – both responses enhance heat loss through skin. Similarly, if it gets colder in winter our body's metabolic rate may increase automatically, burning more fat and sugars to generate heat to keep warm, or we may intentionally dress warmer or turn our indoor heating up.

Figure 3.3 depicts feedback loops that are balancing or stabilizing in their overall effect on the stock level. Balancing loops are sources of resistance to change in a system. Reinforcing feedback loops, on the other hand, can jeopardise the stable maintenance of stock levels by reinforcing a particular dynamic trend relating to a stock level. For example, considering the stock of fertile top soil, soil erosion represents a depleting flow, which occurs in exposed areas where plant cover and root systems holding soil in place have decreased. With reduced fertile top soil there is in turn further reduced plant growth, resulting in a feedback loop that self-reinforces the decrease in top soil stock levels.

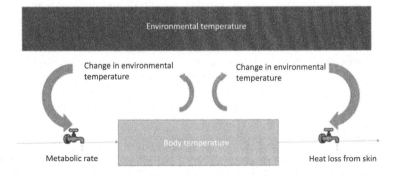

Figure 3.3 Balancing feedback loops for maintenance of a constant body temperature

Experience in self-organised systems. Evolutionary processes that allow species to persist over time by helping them to adapt to a changing environment rely on feedback mechanisms that allow organisms to transfer, test, and store experience on how to adapt to changes in the environment genetically. Similarly, ecosystems can be conceived as transferring, testing, and storing experience by forming patterns of behaviours and interactions that are self-organised and can repeat themselves. Ecologists have conceptualised such patterns and developed and tested convincing theories that the patterns are formed and refined by a set of interacting variables that function over specific scale ranges and form a mutually reinforcing core of relationships (Gunderson & Holling, 2002). Similarly, policies evolving in response to societal developments judged to have undesirable impacts can also be considered feedback loops of environmental and societal changes.

Complexity, systems models, and their purpose

A system of interest is usually developed as a conceptual tool and as such can be designed for a particular purpose. Users of such 'systems approaches' can specify boundaries and core elements of their systems of interest according to the goal of their analysis. The systems that are built in such dialogues consist of 'variables' and 'relationships'. Variables represent conceptual or concrete unities that have the ability to increase or decrease, for instance, in size, frequency, or impact; the desire for sustainability in a population or in a particular government may increase and decrease, just as the number of yearly built fossil fuel–fired power plants.

On systems diagrams, feedback arrows are denoted with a polarity that indicates the relationship of the direction of change in the affecting variable and the direction of change of the affected variable. A positive polarity (+) indicates that the direction of change in the affecting variable is the same as the resulting direction of change in the affected variable: an increase (or decrease) in the affecting variable will cause the affected variable to increase (or decrease) (all else being equal). A negative polarity (–) indicates that the direction of change in the affecting variable will cause a change in the opposite direction of the affected variable: an increase/decrease in the affecting variable will cause the affected variable to decrease/increase (all else being equal). The essential point is that a given polarity indicates the nature of the interaction between a single pair of variables. In order to assign the polarities correctly, you must always ask: What happens to the affected variable when the affecting variable is increased? The following is an example of a negative relationship. The affecting variable describes 'desire for sustainability' and the affected variable describes 'the number of fossil fuel–fired power plants built'; an increase of the desire for sustainability in a democratic nation will ideally lead to a decrease in the number of fossil fuel–fired power plants and vice versa, if the desire (particular by those in power) decreases the number of such power plants built in the next years will increase. Moreover, there is a time factor here, as such plants will usually remain operational for at least thirty years to recoup sunk investments (this particular systemic relationship may now be evidenced in the United States after the election of President Trump).

If a pair of variables is mutually affecting each other, they can form feedback loops. For example, if one's money in the bank account goes up, the interest gained goes up as well. In turn, the amount of money on that bank account increases, which then increases the interest gained again. This is a reinforcing or positive feedback loop. These loops cause runaway exponential growth or decline effects. On the other hand, if one's money in the bank account increases, the amount of money lost through taxes increases too, which in turn decreases the money in the bank account. This is a balancing or negative feedback loop. Balancing loops have a stabilizing effect on the system, which can either be beneficial or the cause of rigidity. So both effects of both types of loops, depending on the situation, can be desirable or uncalled for. If these loops are combined, you have a network of loops (see also Chapter 5).

Similarly, in human systems, learning can cause feedback loops. For example, policy indicators can serve as feedback mechanisms to stop undesirable trends or patterns of behaviour in governance systems. Monitoring the state of the environment in terms of its air, water, or soil quality may reveal pollution that progressively erodes a particular ecosystem service we rely upon and thus jeopardise the common good (e.g. undesirable farm practices polluting drinking water reservoirs). Such indicators can then suggest policies and measures to reduce such practices. However, system behaviour will then become particularly sensitive to such intentionally created, indicator-based feedback loops that were developed with particular goals in mind. However, evidence for policies in the form of statistics and indicators (regardless of which level of governance) often tends to assess just the level of a stock rather than trying to understand flows and feedback in terms of rates of replenishment or degradation of a resource and interdependencies between various sub-systems (Meadows, 2008). Chapter 4 by Sonnleitner outlines biological and evolutionary reasons for how our brain functions that explain why in practice we naturally are challenged and tend to avoid to embrace complexity and uncertainty.

Computer based modelling of complex systems: The first computer-based systems models of human–environment interactions were developed in the 1970s with the main purpose of enhancing our understanding of likely systemic behaviours in order to develop more sophisticated sets of indicators for improved policy making (Meadows et al., 1972). The boundaries and main constitutive elements of such system models are conceived depending on the purpose of the model. Hence, a complex nested system model is conceived to consist of a number of distinct parts that co-exist in a specific dynamic arrangement in order to investigate how relationships between parts give rise to the collective behaviours of a system as a whole and to better understand how the system interacts and forms relationships with its environment. Any conception or model of a system is artificially bounded; as in the real world all things are interdependent and interact, sometimes with significant time delays. The study of isolated parts of a complex system cannot be taken as indicative of overall behaviour of a system. Interdependencies between drivers of change, be it in nature or anthropogenic, are of interest for the identification of appropriate leverage points to change the overall behaviour of the system. The modelling of such systems can also help to reveal certain recurring patterns of behaviour in systems that can then be more easily discerned – these are called systems archetypes (Senge, 1990) or systems traps (Table 3.1, adapted from Meadows, 2008).

Table 3.1 Three examples of system traps and ways of escape

Archetype	Pattern of behaviour	Escape route
Tragedy of the commons	Shared public resources such as groundwater systems, clean air, surface water, or fish stocks in the open sea bring direct benefits to each individual user. Costs of abuse (pollution, depletion) are often differently distributed and sometimes subject to time delays, such that a balancing feedback loop is too weak to stop abusive behaviour.	Monitoring, communication, early warning systems about stock depletion; see Ostrom, 2009 and the second section of this chapter on the foundations of systems thinking
Success to the successful	This trap describes a situation in which winners are rewarded with means to improve their chance to win the next time. One example is the critique by Marx that capital breeds more capital, which will lead to a reinforcing feedback loop of inequitable distribution in a society in which certain types of capital stocks remain fairly level. For example, successful families generating high levels of income invest this into capital and will thus be able to invest more into the education of their children, which in turn may well improve their chances to engage in further successful capital accumulation on top of their inheritance. In agriculture, farm sizes have massively increased, and the number of farmers has dropped tremendously due to this reinforcing feedback loop.	Diversification of possibilities to engage in (For example, create policies for an economy in which initiatives/enterprises working with different logics such as market competition, redistribution, reciprocity can co-exist. Place strict limits to the fraction of the pie one winner may win, for example with anti-trust regulation).
Seeking the wrong goal	Feedback loops generated by policy indicators that are developed based on goals that are defined inaccurately or incompletely may cause undesirable results. Adoption of renewable energy sources and switches to biofuel production with consequences on land-use and cover change in the Amazon and impacts on food prices are an example here.	Specify indicators and goals that reflect carefully co-created conceptions of welfare of the system (states of ecosystems, economy, society, and human well-being).

Source: Adapted from Meadows, 2008

The main three criticisms to mathematical modelling of all global models, including the World 3 model developed at the Massachusetts Institute of Technology, might be summarised as:

- *Data introduced* into the model may be biased – beware of the GIGO (Garbage In = Garbage Out) effect. All data available for the World 3 model was produced based on prevailing logics and values determining resource allocation in present market-based economies. Many relations between sub-systems must be quantified, even when it is impossible to measure or verify such relationships empirically.

• *Data aggregation* can be tricky, and modelling in many cases will demand assumptions that allow us to aggregate apples and pears.

• *Computer models may not have taken into account feedback loops from human learning* and communication effects (in particular, with the limited computing capacity in the last century). For example it was not possible to reflect in the World 3 impacts of policy measures taken in response to scary messages about finite and rapidly depleting and degrading planetary resources (for an overview on the World 3 model please see Box 3.1).

Box 3.1   Classic study: the World 3 model as the basis for the book *Limits to Growth*

Approach: One of the first and most influential computer-based systems models of human activities on earth was 'World 3'. The model focuses on five variables: world population, industrialisation, pollution, food production, and resources depletion (Meadows et al., 1972). The underlying core assumption is that all these variables grow exponentially. Largely in line with assumptions underlying the IPAT equation, impacts of technological innovation are assumed to result only in linear increases in the availability of resources. The model serves to explore diverse feedback patterns to effect these growth trends in different manners, which then result in diverse overall system behaviours, which in turn are synthesised to three different scenarios.

Purpose: The authors emphasised that this model did not allow for making precise predictions, but served to better understand system dynamics and gain some broad indications of the system's behavioural tendencies in diverse circumstances.

Merits and limitations – a statement by Donella Meadows: 'We have great confidence in the basic qualitative assumptions and conclusions about the instability of the current global socioeconomic system and the general kinds of changes that will and will not lead to stability. We have relatively great confidence in the feedback-loop structure of the model, with some exceptions which I list below. We have a mixed degree of confidence in the numerical parameters of the model; some are well-known physical or biological constants that are unlikely to change, some are statistically derived social indices quite likely to change, and some are pure guesses that are perhaps only of the right order of magnitude. The structural assumptions in World 3 that I consider most dubious and also sensitive enough to be of concern are: (i) the constant capital-output ratio (which assumes no diminishing returns to capital), (ii) the residual nature of the investment function, (iii) the generally ineffective labour contribution to output'.

Main insights gained: The simple and robust insights gained from these modelling experiences and over three decades of probing and sensitivity analysis tweaking the different variables and scenarios are summarised in the book *Thinking in Systems: A Primer* (Meadows, 2008). These insights include some recurrent patterns of systemic behaviour in human–environment interactions in current governance systems in our Western culture; such patterns have also been called system archetypes or systems traps, three of which are summarised in Table 3.1.

Significance: This research and its clear and authoritative communication to policy makers and the public at large have played a significant role in directing attention to the limits of the biophysical-carrying capacity of the earth, and the likely strains on societal metabolism from human production and consumption and land use activities is presented (Meadows et al., 1972, 1992). The World 3 model can be seen to have inspired and allowed us to draw lessons from its critique of many other global systems computer models that have been developed since. Because the model itself was strongly criticised on scientific grounds and its conclusions can easily be accepted without an understanding of the model, it can be considered historically as a great consciousness-raising exercise rather than a conventional scientific production.

In sum, whether a model is really useful depends on its performance in a range of simulations of changing circumstances and whether the model's responses represent a realistic and informative pattern of behaviour. Specific model outputs are less constrained by empirical verifiability than is often supposed. Other types of quality criteria to assess their suitability to structure deliberation and the development of actionable knowledge for transformative sustainability science are therefore critical.

Collaborative systems approaches that are advocated for broader use in this book, such as those mentioned in this chapter's section on human ecology later and presented in more detail in Chapter 5, largely serve to bring people together from different backgrounds and enhance mutual understanding based on developing joint diagrams that describe relations between different components that matter in systems. These methods help to structure dialogues across differences in interests and expertise and build a shared and enhanced understanding of a complex dynamic system from diverse perspectives in a constructive manner. The dialogue to construct such networks helps diverse actors gain a shared understanding of what matters to each and why. The resulting diagrams that can combine qualitative relationships between variables, quantitative information, and descriptive language with pictures of webs of relations then further can be used for more detailed joint analysis to structure analytic conversations in diverse groups about a complex system. This in turn helps to jointly search and identify strategic leverage or intervention points for policy making and changes in practice. As a consequence, future collaboration across different interests and perspectives can become easier. Often, the 'reality' of the representation of the diagram in the end is not the goal, the real goal is enhanced mutual understanding and the start of a shared understanding of what may matter most to participants in the process (Newell & Siri, 2016).

Donella Meadows has described 'leverage points for systems change' (Figure 3.4), which are places to intervene in a system that have been listed in the order of increasing likelihood to affect a deeper transformation of the system. In many situations, substantial energy is used on leverage points on the left side of the diagram, which have limited influence. The highest leverage points are often the most

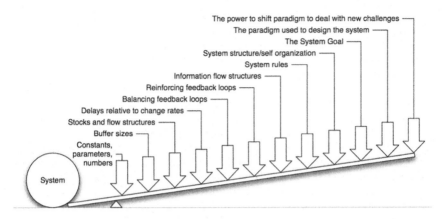

Figure 3.4 Leverage points for systems change

Source: Meadows, 2008

difficult to change (whether in individuals or at the population level), but hold the potential to bring the most far-reaching changes.

Criteria to assess systems approaches for the practice of transformative sustainability science

In order to provide an analytic framework for the subsequent sections of this chapter that present diverse approaches to describe social-ecological systems that are influential today, in this section we develop a set of criteria in order to assess the suitability of these approaches for the practice of transformative sustainability science (see Box 3.2). At the base are more general criteria for frameworks to characterise human–environment interactions developed by Holdren and colleagues (Holdren & Smith, 2000). These include that the emergent integrated understanding should match accepted quality criteria for knowledge and can gain legitimacy beyond a particular community. Moreover, such frameworks should enhance instrumental judgment and facilitate recognition of fundamental principles relating to the interactions between variables of different kinds. In particular in the normative endeavour of the practice of sustainability science a further criterion is that knowledge derived from the framework is suitable for use in working with futures (scenarios and visioning), as well as back-casting exercises and the development of indicators to assess social, technological, and environmental changes as a basis for transformative learning and concerted action in diverse groups of stakeholders and scientists.

The point of interaction between variables of different kinds can't be sufficiently emphasised. Exactly how the analytic framework enables the description of the relationship between human and environmental systems usually reflects the field of knowledge from which a particular systems approach is derived (Binder et al., 2013; O'Brien, 2012, 2015). Are these relations described in only one direction or are mutual interactions characterised? Does the conceptual framework take an anthropocentric or eco-centric perspective? Was the framework designed to inform action or analysis? Going one step further to solutions-oriented research, Wiek points out ways to think of impacts from research on such systems in terms of adoption in practice and transferability and scalability of the results (Wiek et al., 2012).

More fundamentally, diverse approaches to understand systemic interactions in social-ecological systems can be distinguished based on their fundamental assumptions, in particular relating to the understanding of 'cause and effects' and human agency and the role they play in planetary changes (O'Brien, 2012, 2015). These assumptions are usually taken from the field of knowledge or discipline of the lead scientists in the respective projects and influence the framing of questions by structuring what windows open on complex dynamic systems, which system boundaries are drawn, what is at stake, and what main elements (or sub-systems with variables) and which interactions are considered and how they are ordered and represented – by what evaluation criteria is their relative importance in the system determined? Hidden sets of values will often influence the power relations between the participants in collaborative interdisciplinary processes. In other words, they result in very different representations of our social and natural worlds and often shape what possible remedial actions for system improvement are revealed, suggested, distorted, or concealed.

The competence to develop shared creative new conceptions of complex dynamic systems with collaborators that defend very different worldviews is thus central to systemic transformations. Associated competences include the ability of engaging in diverse groups and embracing diverse perspectives and experiences (Newell & Siri, 2016). Such new shared conceptions can then influence what we perceive, how we understand and embody experience, how we relate to each other and our environments, and how we act. Accordingly, the following set of maxims of practice was proposed to judge the power of frameworks to understand social-ecological systems (Holling, 2001):

- Be as simple as possible but no simpler than is required for understanding and communication.
- Be dynamic and prescriptive, not static and descriptive. Monitoring of the present and past is static unless it connects to policies and actions and to the evaluation of different futures.
- Embrace uncertainty and unpredictability. Surprise and structural change are inevitable in systems of people and nature.

Another suggestion from research in ecology highlights the need to engage *diverse* segments of society (citizens, businesses, and government officials) in change processes with the goal of societal transformation and to foster a diversity of approaches to engage and address such challenges. Resilience to cope with rapidly changing environments is built in diverse sets of social practices and logics that can be readily and flexibly recombined for innovation.

Any systems approach we deem appropriate for transformative sustainability science as conceived of in this book has to make explicit and visible that knowledge, science, technology, and social practices and values through which we relate to each other and our environment are co-created. Sets of values serve as ordering principles that help in directing attention and resources to certain solutions over others and help to either cement or break up existing patterns of power distribution in society. Systemic transformation can emerge from transformative learning from social interactions that are designed to change the way we perceive, experience, conceive of, expect from, and through what social practices we relate to our surroundings. Emergent changes in social practice, if they respond to new widely shared expectations, may be stabilised through concomitantly evolving social structures such as prevailing values, norms, culture, rules and institutions, and associated changes in our technological and natural environment.

Our conception of transformative sustainability science in Chapter 1 suggests that a framework has to be able to make explicit and take account of four types of knowledge on complexity and systems; values, worldviews, and contradictions emerging therefrom; and futures and dynamics of social, technological, and environmental change resulting in 'actionable knowledge'. The wish to use emergent new knowledge in social learning processes also requires that representations are applicable across different scales of social organisation (individual, group, and system) and environments.

Box 3.2   Design criteria for systems approaches for social transformation

1   **Complexity of Systems Representations:** Resulting systems representations should:

- Allow for drawing on social and natural science to gain an integrated understanding of systems with attention to interactions and feedbacks between sub-systems, not focussing on identifying linear cause–effect relationships.
- Provide a basis for organizing information about different aspects of the culture–nature system under consideration that is open to both state of the biosphere and the actual life experience of human beings.
- Consider changes over time, rates of change, and scale of human impacts.

Whilst remaining as simple as possible to allow to structured sense-making and communication on the spot and avoid intransparent algorithms:

2   **Acknowledgement of uncertainty:** The process to develop systems diagrams should allow participants to build capacity for embracing and becoming comfortable in the face of uncertainty, unpredictability, unknowns, and ignorance. The resulting systems diagrams should direct attention at knowledge gaps and contradictions to foster emergence of creative solutions.

3   **A dialogic space to explore contingency and contradictions:** The framework and associated social process for collaboration should help to facilitate dialogue in which participants identify with the problem space rather than competing for legitimacy of their own perspectives and identities or entering into non-critical groupthink. It must be flexible; encourage creativity and learning; and embrace multiple and likely contradictory worldviews, values, and experiences that are important for the development of shared conceptual representations of systems and associated powerful metaphors for communication and provide a

structure for analysing, visualizing, and communicating about interactions between the different aspects of human situations. And finally it needs to allow accounting for what counts and not only for what is countable.

4   **A future orientation:** The process should allow us to structure conversations on the evaluation of different futures to offer a new and relatively safe space for creative and dynamic solutions to emerge.

5   **Social robustness of emergent recommendations/solutions:** Emergent integrated understanding matches accepted quality criteria for knowledge and can gain legitimacy beyond a particular community-enhanced instrumental judgment. It can facilitate recognition of fundamental principles relating to the interactions between variables of different kinds. It fosters amenability of development and representation of future scenarios, a desirable vision, back-casting exercises, and development of indicators to assess social, technological, and environmental changes to track change. This is highly desirable as a basis for transformative learning and action. It is amenable to communication through metaphors and indicators and pictures that can mediate shared experiences and expectations.

Accordingly, the main question we ask is whether such frameworks help to structure group processes and foster an understanding, in diverse groups, of which elements in a set of systems will contribute most in that situation to effect change. What are the interdependencies, feedbacks, reinforcing, or balancing loops, and what may be good leverage points to effect change? Only shared representations that are accepted and understandable from diverse viewpoints will be suitable to serve as a basis for concerted action. The diversity of participation and recognition of contradictions and disparate perspectives and approaches to effect change is also important to be maintained for the sake of resilience.

In sum, we ask whether the framework (i) enables us to capture a sophisticated scientific understanding of earth system processes, as well as of social and individual worlds and their interactions; (ii) helps to recognise and represent uncertainties and ignorance in the process to leave room for surprise and creativity; (iii) enables participation of diverse stakeholder interests and perspectives to get a shared understanding from a social knowledge-making process, but also to understand crucial differences in perspectives, interests, and values; (iv) enables us to recognise alternative futures and develop shared desirable visions and development pathways; and (v) captures the attention of decision makers and policy makers with power to effect change at all levels. The possibilities of pinpointing feedback effects (especially cross-sector) and the depth of transdisciplinary understanding between team members (inter-subjective understanding) are critical.

Influential systems approaches today

In this section we provide an overview on diverse systems approaches emerging from the environmental and earth system sciences and ecology, from economics and management, and from the field of human ecology. For each we discuss the purpose, approach, main assumptions, and main merits and limitations. They were selected based on the fact that all these approaches direct attention at different facets of the problem and suggest different solutions. All approaches are also deemed fairly influential in practice as they are not only published in high-impact journals, but are also referred to in national and intergovernmental policy processes.

Environmental sciences, earth systems models, and ecology

Planetary boundaries: Leading researchers in the environmental and earth system sciences are collaborating to develop a set of planetary and regional boundaries to inform policy making on constraints from the biophysical-carrying capacity of the earth system on human activities (Rockström et al., 2009; Steffen et al., 2015). The research community points to the need for a paradigm shift that integrates efforts for further development of human societies and preserving the resilience of the earth system, offering a science-based risk analysis to inform appropriate new approaches to policy making. Emphasis is placed on highlighting the limits of the earth system's carrying capacity with reference to, for example, anthropogenic land use and land-cover change and human interference with the flow of compounds in natural bio-geo-chemical cycles (see Figure 3.5). The functioning of the earth system depends upon these material flows associated with our societal/industrial metabolism. Particular attention is paid to anthropogenic carbon emissions, nitrogen fixation and phosphate used for the production of agricultural fertilizers, fresh water use, and the emissions of pollutants that further jeopardise biosphere integrity and the provision of vital ecosystem services. The main purpose and motivation of this research is communicating to decision makers the risks of non-linear change in the functioning of the earth system, possible tipping points, and the risks of potential irreversible earth system changes; advocating precaution in the face of uncertainty; and developing the normative concept of a 'safe operating space' as a measure for consideration in policy making.

The term earth system in this research refers to a set of interdependent physical, chemical, and biological global-scale cycles and energy fluxes that provide the support system for life at the surface of the planet. The earth system is composed of five main compartments or sub-systems: the atmosphere, the litosphere, the oceans, the biosphere, and the cryosphere. These compartments are linked through interdependent material and energy fluxes (water cycle, carbon cycle, etc.). Within this system over the last 10,000 years of biospherically stable conditions the human race could flourish. This is what we call the Holocene. Two planetary boundaries have now been defined as 'core boundaries': the boundaries relating to the climate system, which determines and is determined by the amount, distribution, and net balance of energy at the earth's surface, and the biosphere that regulates material and energy flows in the entire earth system and through its diversity increases resilience to changes in the earth system's functioning (e.g. through photosynthesis and transevaporation by plants or other metabolic processes in terrestrial and aquatic microbes and higher organisms). Forcings and feedbacks within the earth system are deemed as important as external drivers of change, such as the flux of energy from the sun.

Research progresses to determine the safe operating space as well as risk and high-risk zones by comparing the order of magnitude of anthropogenic flows to flows expected in a naturally functioning planetary system without purported human

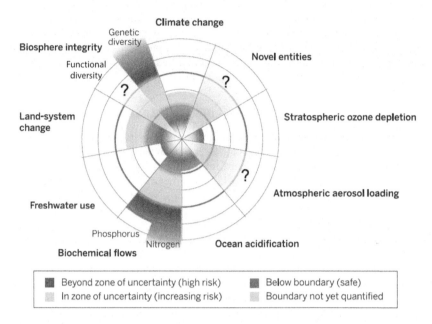

Figure 3.5 Current status of control variables for seven planetary boundaries
Source: Steffen et al., 2015

influence (Steffen et al., 2015), see Figure 3.5. Each boundary is associated with 'control variables' that suggest spaces for remedial action and associated measures, such as atmospheric CO_2 concentration for climate change or the maintenance of genetic and functional diversity of species in ecosystems for biosphere integrity. The boundaries are situated between the safe and increasing risk zones of quantified human impacts.

Integrated analysis of climate change, land use, energy, and water strategies (CLEWS): A second model that at the time of writing receives much attention at UN-level working groups, and that was developed in the context of implementation efforts of Agenda 21 for local action after the Rio Earth Summit in 1992, is the CLEWS systems model. The stated purpose of CLEWS is producing a scalable planning tool for the integrated assessment of land, energy, and water resource systems, drawing on existing well-tested methodologies and models by combining them into modules (Howells et al., 2013). The main motivation for this research was the systematic omission of cross-sectoral impacts in sectoral modelling–based decisions. Prime examples cited are policies to promote biofuels adopted by many governments that are however often dangerously short-sighted with respect to their real impacts on land use changes, food market prices, and net resulting greenhouse gas (GHG) emissions. The chosen approach here was to use existing sectoral models of land use, energy, and water systems. In seeking to deploy CLEWS at any scale, the first step is to identify points where the resource systems interact. Second, system elements of significance are identified, and existing data are compiled and entered into the relevant sectoral models insofar as they were not already in use in that geographic area. The framework integrates the Long-Range Energy Alternatives Planning (LEAP) tool and the Water Evaluation and Planning (WEP) tool developed by the Stockholm Environment Institute, the Agro-Ecological Zoning Approach developed by the Institute of Advanced Systems Applications (IASA), and the UN Food and Agricultural Organisation (UN FAO) with climate scenarios. These are not computer models in themselves but tools that can be used to create computer models of different systems of interest, each with its different data structure. LEAP, for example, can be used to assess energy use, production, and resource extraction across all sectors of an economy and can identify GHG emission sources and sinks. The third step involves the establishment of appropriate data exchanges between modules in an iterative fashion. Outputs from climate models such as changes in rainfall are then used to calibrate the model. The model allows exploration between interdependencies and relationships between the resource systems, and trade-offs and co-benefits between diverse mitigation and adaptation strategies.

A case study of a situation on the island of Mauritius illustrates successful deployment of the model to avert the adoption of energy policies that may have more detrimental side effects on other aspects essential for the viability of key ecosystems on the island (Howells et al., 2013). Stated policy priorities include making the island nation less dependent on energy imports and to reduce GHG emissions. This priority, in combination with the fact that the European Union's trade agreement with African-Caribbean Parties that safeguarded high margins on sugar cane exports to the EU was abolished, suggested the promotion of a local biofuel industry as an interesting option towards more independence from ever more volatile international markets. Deployment of the CLEWS model, however, suggested that predictions of reduced rainfall due to climate change in some areas of the island that had been earmarked for increased sugar cane production as they were previously fully rain-fed would then potentially lead to higher withdrawals of surface and ground water, which in turn is energy intensive and suggests other trade-offs in terms of fresh water provision and food production.

Resilience in complex social-ecological systems: perspectives from ecology: Traps of poverty and rigidity in systems, which can lead to overall systems collapse if sub-systems and elements are not diverse and adaptive enough, was also a central concern led by two ecologists in Sweden (Gunderson & Holling, 2002). The main objective of this research was to identify critical features in a system that affect or trigger reorganization and transformation in response to crisis, and as such determine the resilience of a system. Resilience can be defined as the ability of a substance or object to spring back into shape – elasticity – or the capacity of a life form to recover from difficulties (Oxford English Dictionary, 2015). Building on the critique of World 3 that the model did not sufficiently account for foresight, intentionality, and the ability of people to learn, an alternative computer-based systems approach was developed with the underlying assumption that the complexity of living systems of people and nature emerges not from a random association of a large number of interacting factors, but from a smaller number of critical controlling processes that create and maintain this self-organization (Holling, 2001; Gunderson & Holling, 2002). Higher-level resilience then comes from feedback loops that can learn, create, design, and evolve ever more complex restorative structures.

Core organizing principles for resulting systems models are foresight and learning, communication, and technology. Human foresight, intentionality, and learning can generate negative feedbacks in a system that effectively counteracts destructive, self-reinforcing positive forces. For example, in a market economy, doomsday scenarios about resource depletion can affect prices and thereby contribute to the conservation of a scarce resource. Human communication is seen as unique in its ability to allow testing and transfer of ideas in addition to storing and transferring experience. Successful ideas can become incorporated into levels of the panarchy that change much more slowly and are more stable, such as cultural myths, and social norms sedimented in legal constitutions and laws. In the information age, television, films, and the World Wide Web allow for global connections, storage, and transfer, contributing to a transformation of sedimented social practices such as culture, beliefs, and politics at the global scale. Technology allows us to scale human actions and technological innovation has progressively accelerated, changing the rules and context of the earth system.

The resulting integrating theory of panarchy describes a healthy socio-ecological system that is organised in several levels, with each developing at a different pace. The system can thus invent and experiment on several levels, scaling up successful experiments and benefiting from inventions that create opportunity, while it is kept safe from those that destabilise the system. The system of interconnected levels undergoes innovation cycles, but operates in a way that is both creative and conserving. In such a panarchy, the four processes of release, re-organisation, remembrance, and revolt are a set of the critical processes that allow us to balance change and sustainability (Figure 3.6). Some development cycles favour the creation and maintenance of resilience and opportunity through processes of 'release' and 'reorganization'; these processes represent connections from higher levels of organisation that change more slowly and lower levels of organisation that change more quickly. In subsequent cycles persistence and evolvability can be maintained through processes of 'remembrance' and 'revolt'. Thus the interactions between cycles in a panarchy combine learning with continuity. There are, however, also phases of conservation which promote only incremental change and resistance to change and rigidity. These can result in complete overhaul of a system in a sudden onset of rapid changes a conserving system can no longer cope with. (Figure 3.6 bears evidence that this research has also been one basis to build on for development of the Dutch multi-level perspective on socio-technical transitions referred to in Chapters 12 and 13 on energy transitions.)

In contrast to ecological hierarchies, this hierarchy is structured along dimensions of the number of people involved in rule sets and approximate turnover times.

Main merits and limitations: These three systems approaches emerging from the environmental and earth system sciences direct attention to consequences of human action. They do not further direct attention to or offer analysis of meaning-making and experiences motivating particular courses of human action. Detailed analyses are offered regarding what types and levels of human activities may jeopardise the viability of ecosystems and the provision of ecosystem services. Human agency, in this approach, is considered separate from ecosystems and the earth system, not considering that our conception and cognition of these systems, what we know about them, is dynamically merging from our interactions with them. Human learning is targeted, but is not represented as a

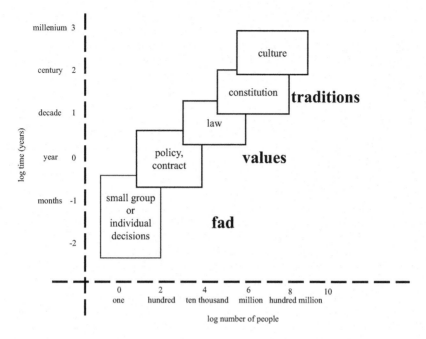

Figure 3.6 Institutional hierarchy of rule sets

Source: Adapted from Holling, 2001

feedback loop within the systems approaches, except for in the ecological model of resilient systems by Holling. Prevailing practices of production and consumption are considered, but the systems representations themselves do not consider more fundamental reasons and experiences for the prevalence of these practices as leverage points for change.

Following the call for a change in consciousness and values by some social scientists would require a complete change from the stance from which the papers are written, positing insights gained from environmental and earth system sciences as a given truth, not as socially constructed knowledge and concepts (Castree et al., 2014), but rather as directing attention to the fact that societies know the knowledge that 'serves them best'.

Economics and management studies

Economics, market failure, and the self-organization of user groups: Stocks of natural resources, including forests, fisheries, and fresh water, are often jeopardised from overuse and pollution. Assumptions of mainstream economics about the behaviour of humans is guided by bounded rationality suggesting that each individual will decide to optimize their actions upon weighing utility in terms of costs and benefits they incur (decisions may not be perfectly rational, as individuals may suffer from imperfect access to information). Given these assumptions, in situations of common pool resources, where ownership and liabilities are not clearly defined, and where there may be time delays between actions of individual users and impacts on stock with adverse consequences on all users, government regulation is required to maintain a balanced stock. Government policies developed based on a logic of prediction, command, and control in such circumstances have, however, also been observed to be prone to failure. Canadian fishing quotas, for example, just resulted in the malpractice of sorting catch for quality and throwing the remainder back into the sea in order to return to harbour with just the quota.

The concept of ecosystem services has been developed in an effort to foster the internalisation in economic pricing of externalities such as adverse environmental impacts (Costanza et al., 1997.) The uptake of this concept in policy making, including at the EU level, proves its effectiveness in structuring deliberations of stakeholders with competing interests in a setting in which principles of neoclassic economics prevail. (See also Chapter 9 by Dendoncker on ecosystem services.)

Another systems approach to characterizing such complex social-ecological systems was developed in order to better understand circumstances and motives for behaviours of user groups to successfully build self-organised systems for the maintenance of fragile shared stocks (Ostrom, 2009). From research spanning three decades on different cases across the world, a general framework for analysing sustainability of social ecological systems emerged that allowed us to pinpoint ten variables that affect users' likelihood to successfully self-organise to protect resources stocks. Figure 3.7 depicts that multi-level nested framework, which draws attention to the four first-level core sub-systems that help to organise this systems approach: resources systems (water, fish stocks, or other) and associated resource units, governance system and users. Ten variables that are suggested to play a crucial role in decisions of user groups to self-organise, include the size and productivity of the resource system, the

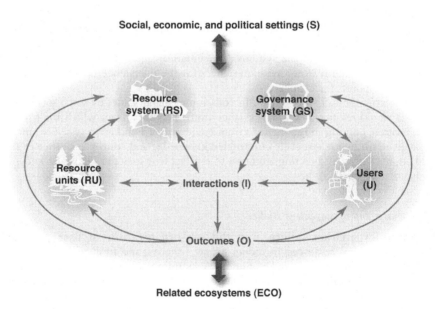

Figure 3.7 Core sub-systems in a framework for social ecological systems

Source: © Science Magazine

predictability of system dynamics, the number of users, and leadership, as well as norms and social capital, and knowledge of the complex social ecological system.

Self-organisation in common pool resource situations has been characterised as social learning about the circumstances under which these stock variables can produce maximal resource flows (Dietz et al., 2003). Such self-organised systems can be established not only to manage fishing grounds, groundwater basins, fragile Alpine land, or forests, but also systems in a sharing economy such as car sharing systems, parking, and mainframe computers. For effective governance in the face of declining resources, the concept of social learning has been proposed to present a mechanism to more effectively translate salient theory and empirical research findings for developing coping mechanisms in reality whilst also allowing room for some trial and error (Dietz et al., 2003).

Organising transformation from individual to collective action is usually an uncertain and complex undertaking. Reflexive and well-documented learning from trial and error is required. Design principles of local common resource management include high levels of face-to-face communication and monitoring and clear information on stocks, flows, and use of scarce resources and changes therein. Information should also address uncertainties in assessments; changes in behaviour; values of individuals and groups; platforms for analytic deliberation to deal with value conflicts; and physical, technological and social infrastructures to support institutionalisation of self-governance processes. Community monitoring projects are seen as central requisites for success of natural resource management (see also Chapter 15). Participatory monitoring involves multiple participants with diverse interests and forms of expertise in design and implementation of monitoring.

Merits and limitations: This analytic framework challenges the traditional stance of science located as objective knowledge outside of society. Learning might be represented by variables such as 'information exchange among users' – or values of natural resources economic value and investment activities.

However, the suggestions of these variables still very much reflect assumptions from the main domain of economics and a market economy. Diverse meanings and experiences and conceptual worlds are not so central to the analysis. This research did address challenges of overcoming disciplinary and conceptual boundaries (Poteete et al., 2010), but the categories with which this was operationalised, such as tracking conflicts among users, leave questions regarding at what depth these barriers to joint meaning-making were actually considered. The main aim of the framework in the end is the integration and accumulation of different knowledges rather than detailing and embracing their diversity.

Biocybernetics: Biocybernetics takes the functioning of viable organisms and ecosystems in the form of the eight principles to measure to assess and identify ways to improve the viability of social-ecological systems (Box 3.3). One main underlying question that drove the development of this approach is how we define and measure progress. Citing Vester, "It should not be any longer measured in material or technological terms, but at a new level of thought better suited to the altered state of our densely populated planet". Interconnected thinking is taken as a requisite for new approaches to planning and action that are oriented to principles of control and regulation in viable systems. Vester's sensitivity model offers an easy-to-grasp participatory planning tool to develop and assess solutions in this logic. (For more detail refer to Vester, 2007.)

Box 3.3   Eight principles of biocybernetics

Negative feedback must dominate positive feedback. Positive feedback sets things in motion through self-reinforcement. Negative feedback then ensures stability against disruptions and excess.

The functioning of the system must be independent of quantitative growth. The throughput of energy and matter is constant over the long term.

The system must operate in a function-oriented, not product-oriented manner. The system survives even when demand changes. Cf. circular economy.

Existing forces are used as in jiu-jitsu. External energy is used where possible while internal energy serves mainly as control energy.

Multiple uses of products, functions and organisational structures. This reduces throughput enhances inter-connectedness and cuts expenditure of energy and materials and information.

Recycling: using circular processes to keep refuse and sewage in the loop.

Symbiosis: reciprocal use of differences in kind through links ups and exchange.

Biological design of products, processes and forms of organisations through feedback and planning.

Source: Frederic Vester, 2007

Merits and limitations: The resulting model for system analysis and testing, 'The Sensitivity Model', presents a set of tools and procedures that were developed for this purpose that is applicable across different planning scales, and in participatory processes it highlights leverage points for action to balance vital stocks and flows. Viability is assessed largely on terms of self-regulation, flexibility, and controllability. That enables avoiding deterministic projections, vast accumulations of data that can no longer be interpreted using common sense, and closed simulations models with hidden and hard-to-understand algorithms. Feedback loops from instituting organizational change and learning mechanisms are deemed key leverage points to change overall system behaviour.

Changes over time, rates of change, and scale of human impacts are the focus. The model does focus attention on the state of the biosphere, but it neglects the representation of actual life experiences of human beings. Notions of human development and social change are not sufficiently problematized in the model at all (Castree, 2015; O'Brien, 2015). The intended use of the model for 'nudging away' from current development trajectories does little to challenge social orders, power relationships, or beliefs that are perpetuating current pathways (O'Brien, 2015), nor calling into questions prevailing values as organizing principles of society, power distributions, etc.

Human ecology

In human ecology, according to Dyball and Newell (2015), the building of sustainable communities can be seen as a learning process that pays attention to stocks and flows and feedbacks within complex social-ecological systems that are of a social and biophysical nature. One basic assumption of human ecology is that observation and learning processes about the state of ecosystems or human wellbeing or the effectiveness of institutions *can* create a desire to modify paradigms, which in turn leads to changes in institutional settings and the behaviours that they control. The goal is to create a core feedback loop that effects systemic transformation for sustainability. Such feedback loops can be developed at any level or scale of a system, at any level of abstraction. In human ecology, investigation of social ecological systems and their responses to possible policy interventions also requires great attention to cognitive factors and requirements for building shared understanding amongst constituents with divergent interests. Thinking together requires empathy and some notion of how others and oneself form mental models and predications they believe in. Collaborative mapping of complex social-ecological technological systems following the method developed by Newell and Proust (2012) (see also Chapter 5), allows us to identify particular problem spaces and collaboratively develop recommendations for policy and practice (for example, on how to promote sustainable farming practices in view of future challenges to the water–food–energy nexus) stemming from new insights on system dynamics gained from the collaborative mapping exercise.

In Chapter 10 Davila and Dyball employ a dynamic systems framework drawn from Dyball and Newell (2015) that explores the global food systems in terms of its environmental impacts and social illness and wellbeing issues, including hunger and obesity, as well as lock-in factors, including the dominance of powerful institutions that influence how food systems operate. This conceptual framework has been designed for the engagement of an extended peer community in the co-creation of knowledge. Dominant discourses in such systems are revealed in deliberations on the framework, and attention can be directed to issues of justice and sustainability.

Merits and limitations: Frameworks derived from human ecology provide a structure for analysing, visualising, and communicating about interactions between the different aspects of human situations and allow accounting for what counts and not only for what is countable. Moreover, the framework helps to facilitate dialogue in which participants identify with the problem space rather than competing for legitimacy of their own perspectives and identities or entering into non-critical groupthink. Resulting systems representations are flexible and encourage creativity and learning (see also Chapter 10).

Conclusion

New future-oriented systems approaches are urgently required in science, policy making, and practice for transforming our world towards sustainability. This is recognised in an increasing number of sustainability-related policies, both the more general and the more sectorally specific ones, such as the UN 2030 Agenda for Sustainable Development and the EU 2000 (European Commision, 2000) water framework directive.

Our difficulties in changing courses of action, be it at the individual, organisational, societal, or systemic level, in spite of the widest realisation that business as usual is no longer an option, suggest learning blockages in our society and its education system. This is not easy as we are in general organizing research, education, and governance in silos; masking uncertainties, ignorance, and complexity, conflicting expertise, interests and values; and often neglecting the question of how we get from knowledge to an improved repertoire of action where all of us can assume new responsibilities to effect change. Changing the way we create and organise new knowledge across disciplinary and expertise silos, as well as allowing for connecting research, policy making, and practice in a different manner in such knowledge co-creation processes is perhaps our greatest hurdle in our striving for more sustainable societies.

This chapter has presented a diversity of approaches that have been developed in particular circumstances for a particular purpose. Some overarching critiques often voiced by disciplined scientists faced with sustainability science applies to all these approaches. These include that:

- The science is interdisciplinary and acknowledges complexity, but there are as yet no clear quality criteria outside of established fields of knowledge.
- There is no adherence to basic scientific conventions to bring its point across.
- The uncertainties associated with producing such numbers or insights are enormous, as are the unknown unknowns. However, often these uncertainties are not well represented.
- It serves to establish a 'big' approximate picture of possibly interdependent circumstances rather than provable and reproducible facts.
- Resulting hypotheses/claims may not be falsifiable and may be situated rather than abstract, universally applicable knowledge.
- The science is often produced from an activist perspective, with a political mission (vs. the Mertonian norm of objectivity).

But all these approaches serve to build bridges between science policy and practice and across some of, if not all, the natural, social, and practice-based sciences and the humanities.

The development of conceptual frameworks as a basis for such new knowledge co-creation processes is critical. Quality or 'Design' criteria for such frameworks developed in this chapter include whether they (i) allow us to capture sophisticated scientific understanding of earth system processes as well as of social and individual worlds and their interactions; (ii) help to recognise and represent uncertainties and ignorance in the process to leave room for surprise and creativity; (iii) allow for participation of diverse stakeholder and interest and perspectives to get a shared understanding from a social knowledge-making process, but also to understand crucial differences in perspectives, interests, and values; (iv) allow to recognise alternative futures and develop shared desirable visions and development pathways; and (v) capture the attention of decision makers and policy makers with power to effect change at all levels. The possibility to pinpoint feedback effects (especially cross-sector) and the depth of transdisciplinary understanding between team members (inter-subjective understanding) are deemed critical (see Box 3.2, p. 67).

This chapter is written based on the assumption that, even if a complete identical representation of realities will never form in two minds and that pushing groups to develop sharp shared language-based definitions may result in more controversy than good, other ways of sharing perceptions, experiences, and expectations in social processes and spaces may offer new chances to foster mutual understanding amongst diverse parties. In an ideal case this mutual understanding can be represented in the form of an acceptable common representation of what may matter most in systemic interactions and what may need to be changed first for greater sustainability. This chapter thus places more emphasis on emerging new knowledge and approaches of collaborative sense-making processes than to advances in computing, as this is considered a requisite for concerted action towards systemic changes in the way we organize and frame our learning about impacts of economic activities and lifestyle choices across different scales of social organisation.

Developing approaches to collaborative monitoring, learning, and co-creating shared representations that reflect multiple diverse perspectives in a networked society is a requisite for societal transformation for sustainability (see also Chapter 14). The emphasis in this book is on collaboration across differences. The core question then is how can we get to and structure the kind of dialogue we need to train ourselves to engage in cognitive switching and to generate creative new solutions by keeping several contradictory perspectives in our minds at the same time, acknowledging uncertainty and ignorance, rather than focussing on what experts in single domains might emphasise as deep knowledge on specific cause–effect relationships within one system.

Questions for comprehension and reflection

1 Develop a table in which you draw systematic conclusions on how the quality criteria apply to each of the influential systems approaches presented in this chapter.
2 Select an issue of your choice with which you have familiarity to deploy one of the conceptual frameworks to characterise specific human–environment interactions that this framework directs attention to (drawing on the original research paper or book chapter that presents it). Alternatively, summarise Chapter 10 of this book in which Davila and Dyball elaborate on the application of a conceptual framework from human ecology to explore unsustainable trends in the prevailing global food system.

References

Binder, C. R., Hinkel, J., Bots, P.W.G. & Pahl-Wostl, C. (2013). Comparison of frameworks for analyzing social-ecological systems. *Ecology and Society* 18 (4): 26. doi:10.5751/ES-05551-180426

Boyden, S., Millar, S., Newcombe, K. & O'Neill, B. (1981). *Ecology of a city and its people: The case of Hong Kong.* Canberra: Australian National University Press.

Castree, N. (2015). Geography and global change science: Relationships necessary, absent, and possible. *Geographical Research* 53 (1): 1–15. doi:10.1111/1745-5871.12100

Castree, N., Adams, W. M., Barry, J., Brockington, D., Büscher, B., Corbera, E., . . . Wynne, B. (2014). Changing the intellectual climate. *Nature Climate Change* 4: 763–768. doi:10.1038/nclimate2339

Costanza, R., D'Arge, R., De Grott, R., Farber, S., Grasso, M., Hannon, B., Limburg, K., Naeem, S., O'Neill, R.V., Paruelo, J., Raskin, R.G., Sutton, P. & Van Den Belt, M. (1997). The value of the world's ecosystem services and natural capital. *Nature* 387: 253–260.

Daly, H. E. (Ed.) (1973). *Toward a steady-state economy.* San Francisco: W.H. Freeman.

Daly, H. E. (1977). *Steady-State economics: Second edition with new essays.* Washington, DC: Island Press.

Daly, H. E. (1991). 'Elements of environmental macroeconomics', in Costanza, R. (Ed.) *Ecological economics: The science and management of sustainability*. New York: Columbia University Press. pp. 32–46.

Daly, H. E. & Cobb, J. B. Jr. (1989). *For the common good: Redirecting the economy toward community, the environment, and a sustainable future*. Boston: Beacon Press.

Dietz, T., Ostrom, E. & Stern, P. (2003). The struggle to govern the commons. *Science* 302: 1907–1912. doi:10.1126/science.1091015

Dörner, D. (1975). Wie menschen eine welt verbessern wollten und sie dabei zerstören. *Bild der Wissenschaft* 12 (2): 48–53. (Populärwissenschaftlicher Aufsatz).

Dyball, R. & Newell, B. (2015). *Understanding human ecology: A systems approach to sustainability*. London: Routledge.

Ehrlich, P. R. & Holdren, J. P. (1971). Impact of population growth. *Science* 171 (3977): 1212–1217.

Ehrlich, P. R. & Holdren, J. P. (1972). One-dimensional ecology. *Science and Public Affairs: Bulletin of the Atomic Scientists* 28 (5): 16–27.

European Commision. (2000). Directive 2000/60/EC of the European Parliament and of the Council of 23 October 2000 establishing a framework for Community action in the field of water policy. Official Journal of the European Communities.

Forrester, J. W. (1969). *Urban dynamics*. Cambridge, MA: MIT Press.

Gunderson, L. H. & Holling, C. S. (Eds.) (2002). *Panarchy: Understanding transformations in human and natural systems*. Washington, DC: Island Press.

Holdren, J. P., Daily, G. C. & Ehrlich, P. R. (1995). 'The meaning of sustainability: Biogeophysical aspects', in Munasinghe, M. & Shearer, W. (Eds.) *Defining and measuring sustainability: The biogeophysical foundation*. Washington, DC: World Bank. pp. 3–18.

Holdren, J. P. & Smith, K. R. (2000). 'Energy, the environment and health', in Goldemberg, J. (Ed.) *World energy assessment: Energy and the challenge of sustainability*. Washington, DC: UNDP. pp. 61–110.

Holling, C. S. (2001). Understanding the complexity of economic, ecological, and social systems. *Ecosystems* 4 (5): 390–405.

Howells, M., Hermann, S., Welsch, M., Bazilian, M., Segerström, R., Alfstad, T., . . . Ramma, I. (2013). Integrated analysis of climate change, land-use, energy and water strategies. *Nature Climate Change* 3: 621–626. doi:10.1038/nclimate1789

Kuhn, T. S. (1962). *The structure of scientific revolutions*. Chicago: University of Chicago Press.

Meadows, D. H. (2008). *Thinking in systems: A primer*. Vermont: Chelsea Green Publishing.

Meadows, D. H. & Meadows, D. L. (2007). The history and conclusions of the limits to growth. *System Dynamics Review* 23 (2–3): 191–197.

Meadows, D. H., Meadows, D. L. & Randers, I. (1992). *Beyond the limits: A global collapse or a sustainable future*. London: Earthscan.

Meadows, D. H., Meadows, D. L., Randers, J. & Behrens, W. W. III. (1972). *The limits to growth*. New York: Universe Books.

Newell, B. (2012). Simple models, powerful ideas: Towards effective integrative practice. *Global Environmental Change* 22 (3): 776–783. doi:10.1016/j.gloenvcha.2012.03.006

Newell, B. & Proust, K. (2012). *Introduction to collaborative conceptual modelling*. Working paper, ANU open access research. Retrieved from https://digitalcollections.anu.edu.au/handle/1885/9386

Newell, B. & Siri, J. (2016). A role for low-order system dynamics models in urban health policy making. *Environment International* 95: 93–97.

O'Brien, K. (2012). Global environmental change II: From adaptation to deliberate transformation. *Progress in Human Geography* 36 (5): 667–676. doi:10.1177/0309132511425767

O'Brien, K. (2015). Political agency: The key to tackling climate change. *Science* 350 (6265): 1170–1171. doi:10.1126/science.aad0267

Ostrom, E. (2009). A general framework for analyzing sustainability of social-ecological systems. *Science* 325 (5939): 419–422. doi:10.1126/science.1172133

Oxford English Dictionary (2015). OED Online: The definitive record of the English language. www.oed.com/ Oxford: Oxford University Press.

Poteete, A. R., Janssen, M. A. & Ostrom, E. (2010). *Working together: Collective action, the commons, and multiple methods in practice*. Princeton: Princeton University Press.

Rockstrom, J., Steffen, W., Noone, K., Persson, A., Chapin, F. S., Lambin, E. F., . . . Foley, J. A. (2009). A safe operating space for humanity. *Nature* 461 (7263): 472–475. doi:10.1038/461472a

Schön, D. A. (Ed.) (1993). *The reflective turn: Case studies in and on educational practice*. New York: Teachers College Press.

Senge, P. (1990). *The fifth discipline: The art and science of the learning organization*. New York: Currency Doubleday.

Steffen, W., Richardon, K., Rockström, J., Cornell, S. E., Fetzer, I., Bennett, E. M., . . . Sörlin, S. (2015). Planetary boundaries: Guiding human development on a changing planet. *Science* 347 (6223): 736–746. doi:10.1126/science.1259855

Vester, F. (2002). *Unsere Welt – ein vernetztes System [Our world – A networked system]*. Munich: Deutscher Taschenbuch Verlag.

Vester, F. (2007). *The art of interconnected thinking: Ideas and tools for tackling with complexity*. Munich: MCB-Verlag.

Walker, B., Holling, C. S., Carpenter, S. R. & Kinzig, A. (2004). Resilience, adaptability and transformability in social – ecological systems. *Ecology and Society* 9 (2): 5. Retrieved from www.ecologyandsociety.org/vol9/iss2/art5/

Wegerif, R. (2007). *Dialogic education and technology: Expanding the space of learning*. New York: Springer.

Wiek, A., Ness, B., Schweizer-Ries, P., Brand, F.S. & Farioli, F. (2012). From complex systems analysis to transformational change: A comparative appraisal of sustainability science projects. *Sustainability Science* 7 (Supplement 1): 5–24. doi: 10.1007/s11625-011-0148-y

4 Cognitive pitfalls in dealing with sustainability

Philipp Sonnleitner

The challenge

The human mind developed to make sense of the world, to structure it, and hence, control it as far as possible to maximize a human's chance of survival (Geary, 2005). The way it works was optimized for the environment in which the genus *Homo* was evolutionarily shaped, a time period roughly ranging from 2.6 million years ago until 12,000 years ago, also called the Paleolithic (Wood, 2005). This so-called *environment of evolutionary adaptedness*, however, is foundationally different from the world of today. Instead of being overwhelmed by easily available food choices in the local supermarket, early humans were struggling to secure food at all. Instead of dealing with long-term risks of instant pleasures like a decent meal or smoking a cigarette, they dealt with immediate threats that could only be solved by quick, physical decisions like running for their life. Instead of living in a complex, inter-connected world, in which every action has (often unintended and unforeseeable) side and long-term effects on a local as well as on a global level, small social clans were formed in which actions had immediate consequences and problems were just locally relevant and comparatively straightforward. In short, large parts of the brain were optimized for a world that was completely different from today's.

This chapter discusses how this cognitive heritage is influencing human decision making, especially in the context of sustainability. First, the theoretical framework for this endeavour is set by introducing dual-process models, the current approach in cognitive psychology to describe human thinking. On basis of these models, typical errors in human decision making, such as framing effects, temporal discounting, or problems in dealing with complex systems, are explored and linked to the environmental context. This chapter closes by reinterpreting these systematic errors as part of the solution and discussing how existing knowledge about the way human thinking works could be leveraged for solving the numerous challenges today's society faces.

Human thinking through the lens of cognitive psychology

Phenomena of human decision making are best studied and described by using a (cognitive) model that formalizes and at the same time structures our conception of how the brain works. To this end, this chapter adopts the theoretical framework of cognitive psychology which offers a wealth of studies enlightening the reasons for cognitive fallacies in everyday decision making and complex problem solving. Mainly drawing on an experimental approach, cognitive psychology tries to understand and adequately describe how people perceive information, process it, and hence act upon it (e.g. Solso, 2001). Although resulting theoretical models can best be understood as approximations of human thinking processes, they enable a glimpse into what is going on in the mind and allow for structuring and predicting complex cognitive performances, such as problem solving, reasoning, or reading.

Current approaches to describe human thinking are called dual-process theories and mainly distinguish between two cognitive systems (e.g. Evans & Over, 1996; Sloman, 1996; Stanovich & West, 2000). The first one encompasses so-called "cognitive modules" that evolved relatively early and enabled individuals to make faster and beneficial choices given the ecological context they lived in (Geary, 2005). Preferring high caloric, sweet, or fat food that delivers more energy compared to other types of nutrition would be an example of such a "cognitive module" (e.g. Beidler, 1982). Because functioning of these modules is largely unconscious and automatic, they are often summarized using the umbrella term Autonomous Set of Systems, Implicit System, or System 1, which is optimized to respond to immediate cues or threats (Evans, 2003; Kahneman, 2011; Stanovich, 2009). The second system, usually called Analytic System, Explicit System, or System 2, in contrast, evolved relatively late; first indications of it are found in cave drawings from the Upper Paleolithic about 60,000 to 30,000 years ago, demonstrating that some form of abstract thinking was possible, allowing for symbolizing a bear by an abstract image painted on a cave wall. System 2 encompasses conscious thinking processes and allows humans to reason and to plan by extending the time perspective and mentally simulating cause-and-effect chains. These complex performances, however, are limited by working memory and are therefore slower than cognitive modules of System 1. System 2 is therefore more prone to errors and causes a higher metabolism of the brain, which often leads to actions based on "intuitive" suggestions of System 1 instead of carefully planned outcomes (Evans, 2003; Geary, 2005; Kahneman, 2011; Stanovich, 2009).

Depending on the situation, the two cognitive systems are differentially suited (Geary, 2005). Choosing the most appropriate credit rate is something that should not be decided by your gut feelings, which are a product of System 1. Social situations, in turn, might quickly become awkward when your vis-à-vis' body posture, facial expression, and voice intonation is overthought too much by System 2. Problems in decision making start when situations have cues that trigger the automatic System 1 but should better be approached by the conscious System 2, leading to irrational and suboptimal solutions.

Merits and limitations of the chosen theoretical approach

Especially when talking about psychological phenomena, it has to be kept in mind that no model is identical with what it aims to describe, but is more like a simplified image of it. The same applies for cognitive models of human thinking, which often use a highly mechanistic account and strongly shape the way we think and do research about certain topics. Further, a short glance at the history of cognitive psychology shows that such models mostly are temporary aids to describe certain phenomena of thinking and are modified or replaced after some decades (e.g. Solso, 2001). Whether dual-process theories stand the test of time remains to be seen, but they face several criticisms, including convincing alternative approaches, unifying all cognitive processes within one coherent system (cf. Evans & Stanovich, 2013 or Osman, 2004, 2014).

Interestingly, one reason for the simplicity of existing models to describe the human mind can be seen in the mind itself. As will be discussed later in the chapter, people have considerable problems in dealing with complexity, and interpreting or using cognitive models to explain human behaviour is no exception. However, one reason for the development of System 2 was the need to explain what is going on in the world by building abstract models that are used to explain and hence control the environment (Geary, 2005). Thus, applying somewhat inaccurate but useful working metaphors to guide actions and make predictions is typically human.

Note that research on cognitive biases, fallacies, or pitfalls has already had tremendous impact by disenchanting the myth of humans as economically rational actors (Kahneman, 2003, 2011). If society strives for a sustainable development, however, it has to be even more aware of the implications of these cognitive biases on everyday decision making. Not only do they explain why people behave in an irrational and unsustainable way in the first place, but knowledge of them also offers possible solutions for avoiding or counteracting them in the future. In the following, it will be demonstrated how this "cognitive architecture" of the brain leads to a variety of fallacies when people are confronted with decisions.

Cognitive fallacies in decision making

Prospect theory and framing effects

Depending on whether the focus is set on gains or losses, people opt for different choices, an (irrational) behaviour first explained by Kahneman and Tversky's (1979) prospect theory. The credit card industry, for example, made sure that any surcharge for paying with a card was relabelled as a discount for paying in cash. Losses are usually perceived to weigh heavier than gains; therefore, paying a surcharge for using a credit card would be much less attractive than simply forgoing the benefit of paying in cash (see Thaler, 1980). Thus, the way a problem is framed, that is how a certain issue is introduced and presented, heavily influences how people see, understand, and decide on that problem (Kahneman, 2003). Several studies consistently demonstrated that people seek risk when the focus of a problem is set on sure losses and avoid risk when the focus is set on gains (see Box 4.1 for a typical example). Crucially, this pattern not only affects lay people, but also professionals in their domain of expertise. In the context of deciding between cancer therapies, McNeil et al. (1982) showed that even trained physicians' choice heavily depended on the framing of the problem, avoiding risks when presented with sure gains and preferring risky options when confronted with unavoidable losses.

This discrepancy arises from two characteristics of System 1 (e.g. Stanovich, 2004). First, System 1 is biased to accept given information and the context or "frame" as is. Reformulating the initial propositions would be mentally effortful and System 2 would be needed. Second, the described utility function of preferring secure gains over more promising but risky options must have been advantageous in the *environment of evolutionary adaptedness*, true to the motto "a bird in the hand is worth two in the bush". Such behaviour absolutely makes sense in an insecure environment in which resources are scarce. Because System 1 acts automatically and faster than System 2, when reading such scenarios as in Box 4.1, our gut feelings thus immediately give us a preference for a certain option. Note that when participants were told to directly compare both scenarios, in other words, when System 2 was activated, all choice preferences between the scenarios vanished. Consequently, it's not that people weren't able to figure out that the scenarios were in fact identical; they just gave the answer that first came to their mind.

Box 4.1   Engaged learning activity: framing effects

Read through Scenario 1 and decide which option you would take. Note your decision.

Scenario 1

Imagine an African village with 600 inhabitants. You would like to ensure a power supply for this village, but given the specific circumstances (financial situation, local climate), there are only two alternatives:

A) If a diesel generator is purchased, the supply for 200 inhabitants is guaranteed.
B) If a hydroelectric power production is installed, there is a one-third probability that 600 inhabitants can be provided with energy and a two-thirds probability that due to a drought year, nobody has power.

Now read through Scenario 2 and again note down your decision.

Scenario 2

Imagine an African village with 600 inhabitants. You would like to ensure a power supply for this village, but given the specific circumstances (financial situation, local climate), there are only two alternatives:

A′) If a diesel generator is purchased, 400 inhabitants will be without power supply.
B′) If a hydroelectric power production is installed, there is a one-third probability that everybody will have power supply and a two-thirds probability that due to a drought year, 600 inhabitants will be without power supply.

This example is a typical problem used in research studies on framing effects and identical concerning the numbers to the example used in Kahneman (2003). Essentially, both scenarios are exactly the same – only in Scenario 1 focus is set on sure gains which render the uncertain, risky decision B less attractive, whereas in Scenario 2, the focus is set on sure losses, making the riskier decision B′ more attractive to choose. In fact, this prediction by prospect theory has been confirmed in a huge amount of studies, reporting a majority of people choosing A in Scenario 1 and B′ in Scenario 2.

Framing in the environmental context

It has been frequently shown that framing effects also play a central role within the context of sustainability. For example, Amelung and Funke (2015) investigated opinion formation concerning climate engineering techniques, such as cloud whitening or injecting aerosols in the stratosphere to reduce solar radiation reaching the earth's surface, as an alternative to mitigating CO_2 emissions when fighting global warming. They could show that participants focusing on the risks of current political efforts to mitigate CO_2 emissions and doubting their success were more likely to adopt climate engineering strategies than participants reporting more faith in more conservative approaches to the problem. Results clearly show that the framing of climate engineering techniques as a plan B or backup strategy that is usually employed in related research proposals and media coverage strongly shapes public opinion on the topic and henceforth steers the debate in a more benevolent direction towards these techniques (Amelung & Funke, 2015; Bellamy et al., 2012). Thus, risk framings, directly targeting certain features of System 1, can create powerful momentum to steer public and hence political opinion. A prominent example in this context can also be seen in the German nuclear phase-out in June 2011 that was a direct response to the Fukushima Daiichi nuclear disaster that happened three months earlier in Japan. Initially, the then ruling government led by Chancellor Angela Merkel had decided to significantly extend the operating time for German nuclear power plants only to decide on immediate shut-downs and restrictions of operating time after the incident (Appunn, 2015).

Substantial framing effects were also found concerning the environmental discourse in the United States that is known to be highly polarized between liberals and conservatives (Feinberg & Willer, 2013). Liberals' stronger pro-environmental attitudes (e.g. McCright & Dunlap, 2011) could largely be explained by the current framing of the environmental debate in terms of harm and care principles – moral concepts that were found to be more strongly embraced by liberals than by conservatives. Strikingly, if environmental messages were reformulated corresponding to moral values of purity and sanctity, differences in pro-environmental attitude between liberals and conservatives completely vanished. One suggested reason for this shift was that the purity/sanctity framing referred to moral principles typically advocated by conservatives, thus triggering System 1 of the conservative participants to detect familiarity and, hence, to put greater trust in the presented message. These findings not only demonstrate the strong impact of certain framings of environmental information and messages on public opinion, but also highlight that framing effects can be general but also differential. Because messages or choices cannot be "unframed", it is especially important to critically reflect the frame that is used.

Temporal discounting and climate change

Risk or uncertainty is also crucial when thinking about the future. Given the challenging and unpredictable setting in which the genus *Homo* mainly developed, it comes as no surprise that System 1 is optimized to prefer instant gains over long-term rewards, a pattern also known as temporal discounting. Within the climate change context, the problem is even more aggravated by the fact that potential but uncertain benefits of immediate costs (e.g. significant reduction of resource consumption) do not only lie in the far future; they only become perceivable within the next generation(s) at best. Consequently, Jacquet and colleagues (2013) could impressively show that a common climate protection goal (the financing of a related newspaper ad) was reached in 70% of the participating groups when there was an instant (monetary) reward, in 36% of the groups if there was a seven-week delay of their endowment, and in none of the groups if the common benefit was invested in planting trees, a proxy for an intergenerational advantage. Especially in the last condition, selfish behaviour of group participants dominated through-out the experiment. According to the authors, international climate change nego-tiations won't succeed until powerful short-term incentives, such as punishment, reward, or reputation, will be introduced to the debate, thus somewhat cancelling the effects of temporal discounting.

Compared to stable and secure conditions, predictions for the future become even more uncertain when the environment is harsh. How this factor addition-ally influences decision making was shown in a recent experiment by Laran and Salerno (2013) in which participants were primed with either neutral cues or cues of environmental harshness while having to choose between high-caloric or low-caloric food alternatives. In the environmental harshness condition, trigger-ing primordial reactions of System 1, participants significantly chose high-caloric food over low-caloric alternatives, indicating that they (subconsciously) prepared for tough times by securing high-energy food. Crucially, this preference vanished when neutral cues were presented, strongly suggesting that people adopt different strategies depending on the perceived environmental conditions. This relation has also been demonstrated in other studies, showing that the harsher and the more uncertain people perceive their personal future, the more likely they are to adopt a so-called fast life-history strategy that is associated with higher consumption, gambling, or a lower age at giving the first birth (Nettle, 2010; Griskevicius et al., 2012). In the light of these findings, however, it seems highly doubtful that current warnings of climate change's consequences, a looming scarcity of resources, and a constant environmental alarmism will have their envisaged effects. Instead, such messages might even increase non-sustainable behaviour in that they trigger cogni-tive systems that are sensitive to such threats and urge people to think about their own advantage first.

Dealing with problems that are complex

Most problems in the context of sustainability are not isolated but arise within sys-tems and networks and were thus described as complex problems (see also Chap-ter 3). Such problems typically incorporate five characteristics that define them as complex (e.g. Funke, 2001; Dörner, 1989): They (a) consist of a large number of elements that (b) are constantly influencing each other. The underlying relations (c) include dynamics, for example, side effects or variables that, depending on their state, autoregressively change on their own and are (d) not transparent to the prob-lem solver. In addition, most scenarios (e) require the problem solver to achieve several, partly contradictory goals. How these characteristics can be found in the (complex) problem of applying climate engineering is demonstrated in Box 4.2.

Basically, interacting with or trying to control complex systems can be seen as a constant stream of decisions. Because cognitive capacities are limited, the first decision already starts with choosing where to draw attention within the system. If a problem is identified, appropriate strategies have to be selected and applied in order to gather knowledge on this issue, for example, how a certain variable (e.g.

the application of climate engineering) influences others (e.g. local temperatures, public opinion on this topic). The impact of interventions has to be evaluated though, and the decision on the right indicators (e.g. solar radiation, frequency of extreme weather phenomena) and the right time frame (e.g. months vs. years) has to be made. If finally, it was decided which information is transferred into knowledge, this understanding has to be purposefully applied, and choices concerning the right targets have to be made. However, because full knowledge of the system's underlying connections is mostly impossible, as is the definition of goals that are optimal for all involved stakeholders, complex problem solving can be modelled as a sequence of decisions that are made under uncertainty (e.g. Osman, 2014). This, however, also implies that all phenomena discussed earlier, such as framing effects or prospect theory, come into play.

Box 4.2   Climate engineering as a complex problem?

In an excellent paper on the uncertainties of climate engineering, Amelung and Funke (2013) illustrate how the decision on injecting aerosols in the stratosphere to reduce solar radiation and thus counter extreme weather events can be described using criteria of the complex problem-solving research paradigm:

(a) *Complexity*: A large number of variables have to be considered as informing this decision. Besides obvious physical and biological parameters, such as the expected change in local temperatures and weather conditions and its effects on the prevailing fauna and flora, psychological, social, economic, legal, and, hence, political aspects also have to be considered. The application of climate engineering techniques might be refused by the public, but needs sufficient support of political decision makers who have to negotiate this supranational issue also with neighbouring states. A successful application might have varying effects on different economic branches (agricultural sector vs. tourism), and benefits may not outweigh deployment costs and related risks.

(b) *Connectivity:* The involved variables are part of a system and in a constant feedback loop with each other. The prospect of a solution to climate change might lead to a less responsible consumer behavior neutralising some of the positive effects. In addition, the technique itself, injecting aerosols in the stratosphere, would have several effects, indirect and probably, due to a lack of case studies, also unknown ones on variables that are not yet considered.

(c) *Dynamics*: Even if the technique would successfully reduce solar radiation, the impact on temperature and local weather phenomena might not be linear in nature and, thus, be delayed. This, however, may lead to a disappointed public putting increased pressure on policy makers who could then feel forced to intensify their efforts with even more drastic interventions.

(d) *Intransparency*: Although some climate parameters can approximately be reproduced in computer simulations, understanding the whole (itself complex) climate system is still beyond reach. Together with unpredictable social impacts, these unknown processes add to the intransparency of the decision's consequences.

(e) *Polytely*. The manifold and various effects of climate engineering would benefit all involved stakeholders differently. Consequently, multiple and partly contradicting goals have to be considered and weighted for this decision. Given the large number of involved factions with contradicting regional interests, this might render an optimal solution even impossible.

Typical errors when dealing with complex systems

Human performance in solving complex problems has been addressed in various disciplines, ranging from economics to cognitive psychology, using a broad variety of computer-based simulations that mirror dynamic and non-linear real-word scenarios (for a thorough overview see Osman, 2010). In general, by interacting with such scenarios over a certain period, people could learn to control these systems reasonably well, with knowledge on the underlying connections being the key aspect (e.g. Beckmann & Goode, 2010; Sonnleitner et al., 2013; Osman, 2014). But research also revealed a significant amount of typical errors that are made.

If you know nothing about a system, the first step is to explore its underlying mechanics in order to build a mental model of how the variables are linked to each other. But where to turn first and how to find out? Most studies have revealed that people don't apply a systematic way to gather knowledge, probably because no hypotheses are formulated and hence tested (Dörner, 1989). Partly, this could be explained with the lack of clear goal setting during this initial exploration phase. With the diffuse and uncertain aim of finding out how the scenario works, the mind looks for familiar anchors, and System 1 "intuitively" suggests topics or problems that fit this criteria. Beckmann and Goode (2014) found out, however, that such "false familiarity" – vague but in no way perfect knowledge that is triggered by the semantic meaning of a certain system variable – has detrimental effects on knowledge acquisition. Existing assumptions are taken as correct and are not systematically tested, thus leading to a faulty or incomplete mental model of the scenario. The missing of a deliberately planned exploration strategy and the subsequent reliance on familiarity also leads to so-called "thematic vagabonding" and the focus on single goals instead of a necessary holistic problem-solving approach (Brehmer, 2005). This fragmentary system exploration is aggravated by the tendency to overestimate the status quo of a system – in other words, the information that is available right now – and overlook feedback delays or future side effects and tipping points of their interventions. Again, in a relatively uncertain environment, System 1's predisposition of a myopic time perspective may be advantageous, but not in the long run.

A recent study by Sonnleitner, König, and Sikharulidze (2017) showed that students within a course on sustainable development and social innovation had problems interpreting information created while exploring a problem correctly. Effects of an invisible system variable were either completely attributed to other visually represented system elements or accounted for twice by simply not differentiating them from other visible effects. Reasons for these errors were seen in a complex interaction of overemphasizing visually represented information (a feature of System 1) and the high cognitive workload to disentangle two sources of an effect (limitations of System 2). These findings also point to the high impact of reasoning ability on complex problem-solving performance. Studies showed that even full information about the underlying connections does not guarantee perfect achievement of envisaged targets; it depends on what you make out of this knowledge (e.g. Goode & Beckmann, 2010; Sonnleitner et al., 2013). Other mental pitfalls that were found in this phase of knowledge generation are mistaking correlations with causation, the illusion of being able to control variables that are beyond the influence of the problem solver, and the preference to look for evidence that supports the preconceived model instead of disproving it, which would be mentally more demanding (Brehmer, 2005; Dörner, 1989). Taken together, people were found to apply suboptimal strategies to generate information on unknown systems and, in addition, build wrong mental models by drawing wrong conclusions about this information, either because of mental limitations in understanding the generated information or because of mental shortcuts provided by System 1. Thus, the basis for systemic interventions in most cases is already flawed.

Because problem solvers are in a constant feedback loop with the systems they interact with, they can learn from their mistakes, adapt their faulty mental models, and reasonably reach their envisaged targets (Brehmer, 2005; Osman, 2014). Frequently, however, this feedback loop ends in a downward spiral, especially when due to wrong knowledge interventions have unexpected outcomes, and hence, increase uncertainty and emotional pressure to solve the situation. Typical reactions to reduce these negative emotions range from rash, pointless actions to extensive and paralyzing information seeking to fill in the existing knowledge gaps (Dörner, 1989). Even if meaningful interventions are made, their impact could be delayed due to an exponential change rate or initially neutralizing side effects. Most people, however, expect and act upon linear changes and therefore increase their activities in the absence of estimated effects. The focus on present states and the neglect of processes underlying a certain system, finally, leads to an oscillation of actions causing a constant overshooting and undershooting of targeted values.

Leveraging insights from cognitive psychology

Informing and "nudging" the public

In the light of the reviewed examples, it seems evident that cognitive biases play a crucial role in the context of sustainability. But what can be done to lessen their impact or prevent them in the first place? One important step is raising awareness of cognitive biases in the broad public (for an excellent example see Marshall, 2014) and disciplines other than (cognitive) psychology. With the introduction of behavioural economics to many curricula focusing on finance and economy, for example, this slowly gains momentum. Because environmental issues can only be tackled by a multidisciplinary effort, knowledge on human decision making should also be part of other disciplines that train students to act within complex systems (e.g. engineering, politics, or spatial planning).

Another route is the explicit consideration of cognitive biases in policy making and the design of so-called decision architectures, which has already been done in some contexts (Thaler & Sunstein, 2008). In fact, for several years governments have been doing exactly this by applying a soft paternalism that "nudges" people in a certain direction when confronted with decisions (for a detailed discussion of this account and examples see Thaler & Sunstein, 2008). However, despite being promising, this approach has been empirically challenged and is far away from being the key to all sustainability problems (Osman, 2014, 2015). It has to be complemented by a stronger inclusion of the wider public through a social dialogue; otherwise, such "nudging" will provoke reactance in people, the unwillingness to comply with actually beneficial measures, just because behavioural freedom is seemingly threatened through their introduction (e.g. Miller et al., 2007). Instead of patronizing the public, governments should start initiatives that inform about existing mental pitfalls, and that invite people to conjointly develop ways to face uncertainty by avoiding such fallacies. Devolving responsibility downwards so decisions are taken nearer to the people they will affect, may not only be more politically mature but also more likely to succeed (Osman, 2014). An idea that Ravetz (2006) conceived as post-normal science.

Adapting educational curricula

One consequence of studying human decision making in complex environments was the request for adapting educational curricula by explicit training of systems thinking using computer-based scenarios (Dörner, 1989; Vester, 2012). People should not only be informed about cognitive pitfalls in dealing with complex problems, but should make their own experiences in order to learn from them. The question arises, however, how the training simulations should be designed in order to maximize the learning outcome. Should they reflect real, specific scenarios or be formulated in a more abstract, general way? Funke (2006) highlighted the discrepancy in this context that the more detailed scenarios are, the more situation specific and therefore useless for other scenarios the acquired knowledge will be. On the other hand, if very general and abstract scenarios are used for training, domain-general principles might be demonstrated, but their usefulness for domain-specific situations might be questionable. In addition, Beckmann and Goode (2014) pointed out that scenarios which use meaningful labels for the included variables, thus being "semantically embedded", might hamper learning about the underlying connections and dynamics because learners might have false prefixed associations with the scenario's variables that are not tested systematically (see earlier). They consequently suggest that novel or abstract contexts might therefore be advantageous for learning general skills like systematically exploring relations between variables. If semantically embedded scenarios have to be used in order to teach specific knowledge, essential steps of the training would be the explication and, hence, the consequent and systematic test of learners' assumptions about the system. Beckmann and Goode (2014) also highlight that if concrete goals are given in the phase of exploring a system, people aim to reach them simultaneously by manipulating several input variables at the same time. Needless to say, such a mix of interventions makes it impossible to draw clear-cut conclusions about cause-and-effect chains and, thus, significantly reduces the gathered knowledge. The interventions might have been successful but the learner does not know why; hence he or she does not develop a secure foundation of understanding the system and finally develops (unjustified) overconfidence in his or her problem-solving skills. From this point of view, existing computer-based simulations aiming to teach systems thinking, such as Vester's Ecopolicy (2011), that asks the learner to solve several problems of a fictitious state (e.g. poor educational level and health conditions of the population, high environmental pollution, or a faltering economy) within a few steps might not reach their goals.

Studies on the training of systems thinking and complex problem solving are still scarce and produced mixed results. Kretzschmar and Süß (2015) showed that university students were able to extract general principles of system exploration by just interacting with several heterogeneous scenarios over time. Although they outperformed peers without training in gathering knowledge about a new scenario, they could not use this advantage to reach specific target values. Micheli (2016) used another approach to teach complex problem solving by discussing characteristics of complex problems and its implications for interacting with them within a one-hour lecture and afterwards assessing students' problem-solving skills. However, the intervention showed almost no effect. A successful training approach was reported by Akcaoglu, Gutierrez, Hodges, and Sonnleitner (2016). They found that students enrolled in a course on how to design and basically program computer games significantly improved their performance interacting with a complex problem-solving scenario, thus pointing to the possibility that also indirect training methods might be efficient. An approach in the context of sustainability education was recently presented by Sonnleitner et al. (2017), who confronted learners with a gamified problem-solving scenario of reduced complexity. While working on the problem, students took notes about experienced emotions which were then discussed within a lecture focusing on cognitive pitfalls in dealing with complexity. Data revealed that typical problems and reactions observed in more complex and larger scenarios (as discussed earlier) could be replicated even in this simpler simulation, and hence used to increase the awareness and sensitivity of learners to their own mistakes and limitations. Exercises to reflect, for instance, on framing effects, such as the one presented in Box 4.1 could additionally help to make these phenomena more tangible.

The need for sensitizing students to cognitive biases and for training them in systems thinking on a larger scale, even by integrating these aspects into educational curricula, is beyond doubt, given its importance (outlined in Chapter 3) and by the frequent, severe mistakes people make when learning and applying it. Recent efforts to include skills relevant for systems thinking in educational curricula (e.g. Micheli, 2016; OECD, 2014) and curricula on sustainable development (Sonnleitner et al.) paint a promising picture. But a clear and convincing concept of how to teach systems thinking and which tools to use for it is still missing. For the moment, informing about and demonstrating typical cognitive biases through experience using simulations in this context seems to be a solid foundation for further developments. Only if decision makers face their own limitations in the light of uncertainty can they appreciate the complex nature of interacting with systems and avoid some of the mistakes mentioned earlier.

Conclusion

The present chapter set out to highlight the influence of humans' cognitive heritage on today's intricate decisions, especially in the context of sustainable development. Although in the reviewed examples, this heritage seems more like a burden than a rich inheritance, history has shown that overcoming barriers by creatively solving problems using newly invented tools and techniques is at the core of the human species. In this sense, utilizing insights of (the tool) cognitive psychology and orchestrating it with other techniques reviewed throughout this book might be a promising path to solve the problems of sustainability in the best tradition of mankind.

Questions for comprehension and reflection

1   Why are spontaneous preferences for certain decisions sometimes suboptimal?
2   What could be the unintended consequences of campaigns against global warming emphasizing that society is running out of time to make a change?
3   What characterizes complex problems, and what problems do people have in dealing with them?
4   Observe yourself the next time you go grocery shopping. How many of your purchase decisions do you make consciously? How many of them are based on spontaneous impulses?
5   Imagine your local supermarket reorganizes its shelves and at the place of your favourite chocolate bar, you find a low-calorie alternative. A bit irritated, you find your preferred one at the very bottom of the shelf that is hard to reach. How would you feel about this "nudging" towards healthier alternatives?
6   What could be done to better prepare executives to deal with complex problems, and what are potential reasons for resistance?

References

Core references are marked with two asterisks (**)

Akcaoglu, M., Gutierrez, A. P., Hodges, C. B. & Sonnleitner, P. (2016). 'Game design as a complex problem solving process', in Zheng, R. & Gardner, M. K. (Eds.) *Handbook of research on serious games for educational applications*. Hershey, PA: IGI Global. pp. 217–233.

Amelung, D. & Funke, J. (2013). Dealing with the uncertainties of climate engineering: Warnings from a psychological complex problem solving perspective. *Technology in Society* 35 (1): 32–40. doi:10.1016/j.techsoc.2013.03.001

Amelung, D. & Funke, J. (2015). Laypeople's risky decisions in the climate change context: Climate engineering as a risk-defusing strategy? *Human and Ecological Risk Assessment* 21 (2): 533–559. doi:10.1080/10807039.2014.932203

Appunn, K. (2015). The history behind Germany's nuclear phase-out. *Clean Energy Wire Factsheet*. Retrieved from www.cleanenergywire.org/factsheets/history-behind-germanys-nuclear-phase-out

Beckmann, J. F. & Goode, N. (2014). The benefit of being naïve and knowing it: The unfavourable impact of perceived context familiarity on learning in complex problem solving tasks. *Instructional Science* 42 (2): 271–290.

Beidler, L. M. (1982). 'Biological basis of food selection', in Barker, L. M. (Ed.) *The psychology of human food selection*. Chichester, UK: England Ellis Horwood Limited. pp. 3–15.

Bellamy, R., Chilvers, J., Vaughan, N. & Lenton, T. (2012). A review of climate geoengineering appraisals. *WIREs Climate Change* 3 (6): 597–615. doi:10.1002/wcc.197

**Brehmer, B. (2005). Micro-worlds and the circular relation between people and their environment. *Theoretical Issues in Ergonomics Science* 6 (1): 73–93. doi:10.1080/146392205 12331311580

Dörner, D. (1989). *Die logik des misslingens: Strategisches denken in komplexen situationen: The logic of failing: Strategic thinking in complex situations.* Hamburg: Rowohlt.

Evans, J. St. B. T. (2003). In two minds: Dual-process accounts of reasoning. *Trends in Cogntive Science* 7: 454–459.

Evans, J. St. B. T. & Over, D. E. (1996). Rationality in the selection task: Epistemic utility versus uncertainty reduction. *Psychological Review* 103 (2): 356–363. doi:10.1037/0033–295X.103.2.356

Evans, J. St. B. T. & Stanovich, K. E. (2013). Dual-process theories of higher cognition: Advancing the debate. *Perspectives on Psychological Science* 8 (3): 223–241.

Feinberg, M. & Willer, R. (2013). The moral roots of environmental attitudes. *Psychological Science* 24 (1): 56–62. doi:10.1177/0956797612449177

Funke, J. (2001). Dynamic systems as tools for analysing human judgement. *Thinking and Reasoning* 7 (1): 69–89. doi:10.1080/13546780042000046

Funke, J. (2006). 'Komplexes problemlösen (Complex problem solving)', in Funke, J. & Birbaumer, N. (Eds.) *Denken und Problemlösen: Enzyklopädie der Psychologie.* Göttingen: Hogrefe. pp. 375–446.

Geary, D. C. (2005). *The origin of mind: Evolution of brain, cognition, and general intelligence.* Washington, DC: American Psychological Association.

Goode, N. & Beckmann, J. F. (2010). You need to know: There is a causal relationship between structural knowledge and control performance in complex problem solving tasks. *Intelligence* 38 (3): 345–352. doi:10.1016/j.intell.2010.01.001

Griskevicius, V., Ackerman, J. M., Cantú, S. M., Delton, A. W., Robertson, T. E., Simpson, J. A., . . . Tybur, J. M. (2012). When the economy falters, do people spend or save? Responses to resource scarcity depend on childhood environments. *Psychological Science* 24 (2): 197–205. doi:10.1177/0956797612451471

Jacquet, J., Hagel, K., Hauert, C., Marotzke, J., Röhl, T. & Milinski, M. (2013). Intra- and intergenerational discounting in the climate game. *Nature Climate Change* 3: 1025–1028.

★★Kahneman, D. (2003). A perspective on judgment and choice: Mapping bounded rationality. *American Psychologist* 58 (9): 697–720. doi:10.1037/0003–066X.58.9.697

Kahneman, D. (2011). *Thinking, fast and slow.* New York: Farrar, Straus and Giroux.

Kahneman, D. & Tversky, A. (1979). Prospect theory: An analysis of decision under risk. *Econometrica* 47 (2): 263–291. doi:10.2307/1914185

Kretzschmar, A. & Süß, H. M. (2015). A study on the training of complex problem solving competence. *Journal of Dynamic Decision Making* 1 (4): 1–15. doi:10.11588/jddm.2015.1.15455

Laran, J. & Salerno, A. (2013). Life-history strategy, food choice, and caloric consumption. *Psychological Science* 24: 167–173. doi:10.1177/0956797612450033

★★Marshall, G. (2014). *Don't even think about it: Why our brains are wired to ignore climate change.* New York: Bloomsbury USA.

McCright, A. M. & Dunlap, R. E. (2011). The politicization of climate change: Political polarization in the American public's views of global warming. *The Sociological Quarterly* 52: 155–194.

McNeil, B. J., Pauker, S. G., Sox, H. C. Jr. & Tversky, A. (1982). On the elicitation of preferences for alternative therapies. *New England Journal of Medicine* 306 (21): 1259–1262. doi:10.1056/NEJM198205273062103

Micheli, E. G. (2016). *Complex problem solving and the theory of complexity in high school teaching.* Unpublished dissertation. Italy: University of Bergamo.

Miller, C. H., Lane, L. T., Deatrick, L. M., Young, A. M. & Potts, K. A. (2007). Psychological reactance and promotional health messages: The effects of controlling language, lexical concreteness, and the restoration of freedom. *Human Communication Research* 33 (2): 219–240. doi:10.1111/j.1468–2958.2007.00297.x

Nettle, D. (2010). Dying young and living fast: Variation in life history across English neighborhoods. *Behavioral Ecology* 21 (2): 387–395. doi:10.1093/beheco/arp202

OECD. (2014). *PISA 2012 results: Creative problem solving: Students' skills in tackling real-life problems (Volume V)*. Paris: PISA, OECD Publishing.

Osman, M. (2004). An evaluation of dual-process theories of reasoning. *Psychonomic Bulletin and Review* 11 (6): 988–1010. doi:10.3758/BF03196730

Osman, M. (2010). *Controlling uncertainty: Decision making and learning in complex worlds.* Chichester: Wiley-Blackwell.

Osman, M. (2014). *Future-minded: The psychology of agency and control.* Basingstoke: Palgrave MacMillen.

Osman, M. (2015). Does our unconscious rule? *Psychologist* 28 (2): 114–117.

Ravetz, J. R. (2006). Post-normal science and the complexity of transitions towards sustainability. *Ecological Complexity* 3 (4): 275–284. doi:10.1016/j.ecocom.2007.02.001

Sloman, S. A. (1996). The empirical case for two systems of reasoning. *Psychological Bulletin* 119 (1): 3–22. doi:10.1037/0033–2909.119.1.3

Solso, R. L. (2001). *Cognitive psychology.* Boston: Allyn and Bacon.

Sonnleitner, P., Keller, U., Martin, R. & Brunner, M. (2013). Students' complex problem-solving abilities: Their structure and relations to reasoning ability and educational success. *Intelligence* 41 (5): 289–305. doi:10.1016/j.intell.2013.05.002

Sonnleitner, P., König, A. & Sikharulidze, T. (2017). Learning to confront complexity: What roles can a computer-based problem-solving scenario play? Environmental Education Research. doi: 10.1080/13504622.2017.1378623.

Stanovich, K. E. (2004). *The robot's rebellion: Finding meaning the age of Darwin.* Chicago: University of Chicago Press.

Stanovich, K. E. (2009). *What intelligence tests miss: The psychology of rational thought.* Yale: University Press.

Stanovich, K. E. & West, R. F. (2000). Individual differences in reasoning: Implications for the rationality debate? *Behavioral and Brain Sciences* 23: 645–726.

Thaler, R. H. (1980). Toward a positive theory of consumer choice. *Journal of Economic Behavior and Organization* 1 (1): 39–60. doi:10.1016/0167–2681(80)90051–7

Thaler, R. H. & Sunstein, C. R. (2008). *Nudge: Improving decisions about health, wealth, and happiness.* New Haven and London: Yale University Press.

Vester, F. (2011). *Ecopolicy, das kybernetische Strategiespiel* [The cybernetic strategy game]. Munich, Germany: MCB Verlag.

★★Vester, F. (2012). *The art of interconnected thinking: Ideas and tools for tackling with complexity.* Munich: MCB-Verlag.

Wood, B. (2005). *Human evolution: A very short introduction.* New York: Oxford University Press.

5 Escaping the complexity dilemma

Barry Newell and Katrina Proust

The challenge

The characteristic behaviour of a social-ecological system (SES) emerges from feedback interactions between its parts. This means that its response to human activities cannot be predicted on the basis of studies of its individual parts taken separately. As expressed by Ackoff (1986), "A system is more than the sum of its parts; it is the product of their interactions. If taken apart, it simply disappears". But anyone who tries to look at an SES as a whole will be overwhelmed by its complexity – such a system has too many parts, interacting in too many different ways, at too many different scales, for it to be understood as a whole. This is the 'complexity dilemma'. On the one hand, practical approaches to policy development require the identification of relatively simple, understandable sub-systems. On the other hand, the reductive process of isolating a sub-system draws a boundary around a limited set of state variables and breaks causal links with key variables in other sub-systems. This process runs the risk of leading to unsustainable policies that sooner or later are besieged by unexpected outcomes – often outcomes that the new policies themselves have triggered.

The complexity dilemma presents a clear challenge to sustainability science: Are there practical ways to isolate relatively simple SES sub-systems that (a) remain linked to the dynamics of the wider system and (b) are relevant in efforts to build resilience and sustainability? If we can answer "Yes" in a given context, then we have ways to escape the complexity dilemma – at least in that context.

The identification of a viable sub-system is a highly context-specific task. What can be done, however, is to establish generic methods that can guide the process. Such methods can help the members of a social-learning group to develop a coherent approach while leaving them free to design detailed operational procedures tailored to their specific context. In this chapter we outline a high-level process that can support efforts to isolate and study the internal dynamics of significant SES sub-systems while, at the same time, taking account of the main feedback interactions that operate between these sub-systems. This approach, which we call *Collaborative Conceptual Modelling* (CCM), has grown out of some 30 years of theoretical studies and practical collaborative work with a wide range of community, student, academic, and professional groups (Newell & Proust, 2012). CCM is designed to provide practical guidance for the collaborative development of understandings that have a combination of breadth and depth that is typically beyond the reach of isolated individuals.

The approach – Collaborative Conceptual Modelling (CCM)[1]

The name 'Collaborative Conceptual Modelling' was chosen to emphasise several fundamental ideas. First, we intend the term 'modelling' to encompass the development of cause–effect models that range from an individual's tacit mental models to various informal and formal models shared by the members of a social group or society. Cause–effect models strongly condition the way people perceive the world and so guide their decisions and actions. Second, the term 'conceptual' serves to pick out a particular subset of the possible cause–effect models. The aim of a CCM exercise is to articulate, extend, and blend the mental models of the members of an adaptive group, rather than attempt to produce definitive predictions of future behaviour. Accordingly, although detailed, high-order[2] models can be important in some endeavours, in CCM we focus on the iterative development of influence diagrams, causal-loop diagrams, and low-order stock-and-flow models (Proust & Newell, 2006; Ghaffarzadegan et al., 2011; Proust et al., 2012; Newell & Siri, 2016). Third, we use the term 'collaborative' to stress the necessity of social learning in any attempt to take a comprehensive approach to sustainability. It is not possible to build useful systemic understandings, which take account of feedback interactions that cross the boundaries between conventional sectors and disciplines, without meshing the mental models of people with a wide range of experiences, values, and allegiances (Newell, 2012)

Conceptual foundations and basic assumptions

CCM draws on concepts from applied history (Proust, 2004), complexity (Axelrod & Cohen, 1999), resilience thinking (Walker & Salt, 2006), system dynamics (Sterman, 2000; Meadows, 2009), and cognitive linguistics (Lakoff & Johnson, 1999, 2003; Reddy, 1993; Newell, 2012). Concepts from system dynamics (hereinafter SD) are particularly important in social-learning contexts because of their practicality and accessibility and their focus on feedback, endogenously generated behaviour, and collaboration (Senge, 1990; Vennix, 1996; Richardson, 2011). In Box 5.1 we present systems-thinking principles that summarise the perspective of many in the SD community.

Box 5.1   Collaborative Conceptual Modelling systems-thinking principles

1   *The Feedback Principle*: Feedback effects are dominant drivers of behaviour in any complex system.
2   *The Holistic Principle*: The behaviour of a complex system emerges from the feedback interactions between its parts, and therefore cannot be optimised by optimising the behaviour of its parts taken one by one.
3   *The Inertia Principle*: The filling and draining of accumulations (stocks) is a pervasive process in complex systems. The presence of accumulations causes delayed responses, thereby giving rise to system inertia.

---

4   *The Surprise Principle*: Any action taken in a complex system will have multiple outcomes, some expected and some unexpected. The expected outcomes *might* occur – unexpected outcomes will *always* occur. The unexpected outcomes are usually unwanted and delayed – the delays make it difficult to identify the triggering actions.

5   *The History Principle*: Knowledge of past activities and patterns of behaviour is essential in any attempt to understand how causation 'works' in a complex system.

6   *The Myopia Principle*: No one person can see the whole of a complex system.

7   *The Collaboration Principle*: The boundaries of a complex system cut across traditional disciplines, organisations, governance sectors, and subcultures. An effective systems approach therefore requires deep collaboration between people with different backgrounds, worldviews, and allegiances.

CCM is intended to provide coherent support to efforts to build sustainable communities. In seeking this coherence, it is necessary to identify the principal operations required and to order them according to their natural dependencies. In CCM we assume the following ordering (Figure 5.1): survival (sustainability) requires adaptation (which we take to include mitigation and innovation); successful adaptation requires holistic governance based on broad systemic worldviews; the development of holistic governance requires a transdisciplinary approach. These operations are nested. It is not possible to operate effectively at the higher levels without first operating effectively at the lower levels.

We further assume that there are at least three sets of 'essential interactions' that must be taken into account in any attempt to build a sustainable community (Table 5.1). CCM is intended to help a social group to understand and manage these interactions:

1   *Feedback interactions between the parts of the system of interest*. Feedback drives system behaviour at all scales, affecting families and national governments alike. As a result, the response of an SES to human intervention can seem counterintuitive to people who are unaware of feedback effects (Forrester, 1969; Sterman, 2000). Thus, efforts to understand such causal effects can help communities minimise policy failure. In particular, the realisation that cross-sector feedback plays an essential role in system behaviour, yet is often invisible to community members, makes it clear that ways to take account of, and manage, cross-sector feedback are a necessary part of any adaptation plan.

2   *Interactions between the past and the future*. Historical studies of the outcomes of human actions over a wide range of time scales are essential in the development of the cause–effect insights needed to understand the dynamics of an SES. Such studies can, therefore, provide the basis for the construction of scenarios that explore possible futures. Scenario development is a powerful tool for strategic conversations and social learning (Schwartz, 1991; de Geus, 1997; Swanson & Bhadwal, 2009; Kröger & Schäfer, 2016).

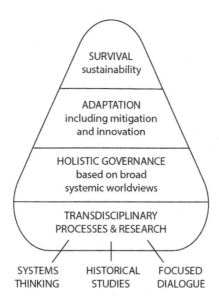

Figure 5.1 The Collaborative Conceptual Modelling (CCM) hierarchy of operations. This diagram summarises our assumptions concerning the principal processes required to develop sustainable societies. The three activities shown at the bottom of the diagram can help a social-learning team to generate insights into the 'essential interactions' listed in Table 5.1.

Table 5.1 The essential interactions

Interactions . . .	System principles (Box 5.1)	Key disciplines/methods
1. Between the parts of the system of interest	Feedback Principle, Holistic Principle	System Dynamics, Systems Thinking
2. Between the past and the future	History Principle, Inertia Principle, Surprise Principle	History, Dynamical Modelling, Scenario Development
3. Between community members	Myopia Principle, Collaboration Principle	Cognitive Linguistics, Focused Dialogue

3   *Interactions between the members of a social-learning team.* An isolated individual cannot build a satisfactory understanding of the dynamics of a complex system. If the perceptions and theories of many individuals can be meshed synergistically, there is the possibility that a more encompassing, more coherent understanding can emerge. This is not a trivial task, given the challenges of establishing mutual understanding between individuals with different backgrounds, experiences, worldviews, and aims (Reddy, 1993; Newell, 2012). Nevertheless, sustained dialogue, focused on the key challenges identified by group members, can be highly productive. Such interaction is crucial in any

attempt to develop the shared cause–effect frameworks and common language that are prerequisites for sustainable societies (Newell, 2012; Wals & Schwarzin, 2012).

Finally, we assume that it is worthwhile attempting to isolate specific sub-systems and build an understanding of their dynamics. We are encouraged in this assumption by the existence of ecosystems where multiple entities self-organise in novel and advantageous ways under environmental pressures. Such ecosystems tend to function as relatively isolated, slowly interacting sub-systems – that is, the interactions between the parts of the sub-systems are stronger than those between the sub-systems. Further, to the extent that the overall system is hierarchical, there is a tendency for the high-frequency internal dynamics of the sub-systems to be decoupled from the low-frequency dynamics of the interactions between the sub-systems (Simon, 1981, p. 217).

Of course, *social*-ecological systems differ from ecosystems in important ways. Central to these differences is the tendency of humans to conceive and implement behavioural policies that are not bound by natural laws. Nevertheless, the emergence of polycentric governance in complex urban systems suggests the operation of ecosystem-like evolution. Polycentric structures, where responsibility and authority are devolved to semi-autonomous decision making and management units operating at a range of scales, offer many practical advantages (Ostrom et al., 1961; Aligica & Tarko, 2012).[3] In particular, they can support the evolution of institutions and policies that are effective and adaptive because they are based on a deep understanding of changing local conditions and needs. Polycentric governance can also facilitate the development of a rich spectrum of creative and experimental management approaches that increase the adaptive capacity of the community. Because dynamic complexity (Sterman, 2000, p. 22) increases rapidly with the order of an SES, we believe that it is worthwhile exploring the possibility of isolating low-order sub-systems that are semi-autonomous. Provided that the group remains aware of the limitations of working with sub-systems, such an approach can play crucial educational and operational roles.

The CCM process

In Figure 5.2 we show the overall structure of the CCM approach. The six shaded areas represent 'co-evolving activities'. The activities are co-evolving in the sense that, although there is an overall need for a group to progress from Activity 1 through to Activity 6, it is usually necessary to loop back and revisit earlier activities as new understandings emerge. The activities are divided into two phases whose scope is indicated by the outer curved lines in the diagram. Phase I comprises Activities 1 to 3. These activities are designed to help a social-learning group generate a broad overview of the SES within which they live and work. The Phase I protocols support this process by fostering systems thinking, focused dialogue, and conceptual integration (Newell et al., 2005; Newell, 2012). Phase II comprises Activities 4 to 6. These activities are designed to support the group's efforts to develop a better understanding of the dominant dynamics of their specific system of interest and to apply their new understanding and models to construct scenarios that can guide community decision making (Swanson &

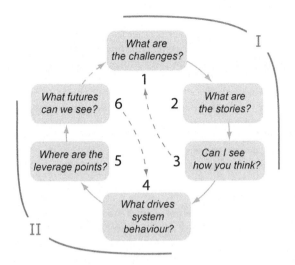

Figure 5.2 The iterative structure of the Collaborative Conceptual Modelling (CCM) approach. The numbered grey boxes represent the co-evolving activities, the solid arrows show the basic sequence of activities, and the dashed arrows indicate typical iterative pathways. The solid arcs, labelled 'I' and 'II', indicate the scope of CCM Phase I and Phase II activities.

Bhadwal, 2009). Phase II is more challenging than Phase I and requires a greater commitment of time. In some cases it will need the involvement of experienced modellers.

Each of the co-evolving activities shown in Figure 5.2 is labelled with a leading question that is designed to focus the initial discussion. The activities are described briefly next. More detail is presented in Newell and Proust (2012) and in Part II of Dyball and Newell (2015).

Activity 1: What are the challenges?

As indicated by the leading question, the emphasis early in a CCM project is on 'challenges' rather than '*the* problem'. This keeps the initial discussions wider than is often the case when the members of a group seek a tightly defined research problem on which to base their collaborative work. Hasty acceptance of a specific research focus can lead to premature convergence on a superficial problem or to a focus on symptoms instead of fundamental issues. It can also disenfranchise those who see the less obvious problems, and give a misleading sense of unity among group members who, in reality, do not yet understand each other's worldviews.

Activity 2: What are the stories?

A crucial step in building an awareness of a system's possible future behaviour is to examine its past behaviour – in particular, its response to human actions (Forrester,

1961, p. 352; Sterman, 2000; Proust, 2004). In CCM we assume that any attempt to understand change requires a base in historical data (Jordanova, 2000). At one end of the scale are simple 'cause-and-effect stories' and oral histories that reveal the human dimensions so often missing in traditional scientific and economic studies. At the other end of the scale, quantitative historical data (such as time series) can provide essential insights in formal studies of the dynamics of complex sub-systems. Historical studies can contribute information about the sources of dynamic complexity (delays and feedback effects), the multiple consequences of past actions, and the multiple drivers of current situations. They can help a group to build an understanding of historical contingency and path dependence (Arthur, 1990) and to define baseline conditions for tracking change.

Activity 3: Can I see how you think?

The inability of individuals 'to see' the whole system is one of the main impediments to the development of sustainable societies. The development of mutual understanding requires a *shared context for communication* (Reddy, 1993). The shared context comprises (a) sets of specific concepts, understood in particular ways and (b) an agreed set of words used to label these concepts. Nevertheless, we agree with the statement (usually attributed to George Bernard Shaw) that "the single biggest problem in communication is the illusion that it has taken place". In CCM workshops we use a protocol that we call 'pair-blending' (Box 5.2) to help a group develop a *visual* cause–effect language that allows them to overcome the communication illusion – that is, *to see* more accurately how each other thinks (Newell, 2012). Pair blending is designed to help a social-learning group work together to define useful sub-systems and develop a genuine shared understanding of the endogenous and exogenous interactions that drive their behaviour. As emphasised by Newell et al. (2005)

> such practical activities are essential in our attempts to forge the *similarities* in our worldviews into robust communication links – links that are strong enough to enable us to use the *differences* in our worldviews to create powerful new approaches to sustainability.

Box 5.2   Collaborative Conceptual Modelling pair blending

There are three steps in a CCM pair-blending activity:

1   Each person constructs an individual ID (Figure 5.3) that captures his or her mental model of the way that cause and effect operates in the system of interest. Participants are encouraged to regard their IDs as tentative dynamic hypotheses rather than 'true' descriptions of the structure of the system. They are, nevertheless, asked to adhere to a set of rules for the construction of their diagrams. The diagrams are built around a specific

focus variable, following a procedure similar to that recommended by Vennix (1996, p. 120). Considerable stress is laid on the importance of expressing variable names according to set rules of 'grammar' (Box 5.3). Participants are asked to minimise the number of variables used (preferably ≤10) and to attempt to identify possible feedback loops.

2   Group members work in pairs to combine their individual diagrams to form a single, blended diagram that incorporates the essential features of their two worldviews. They are again advised to minimise the number of variables in their joint diagram.

3   Each pair presents their blended diagram to the whole group for discussion and constructive criticism. Because (a) all pairs present IDs, (b) all group members understand the 'shared visual language' provided by the diagrams, and (c) the diagrams represent differing views of the same system of interest, these presentations tend to generate rich, focused dialogues (Newell, 2012). These dialogues help the group to move towards a holistic approach.

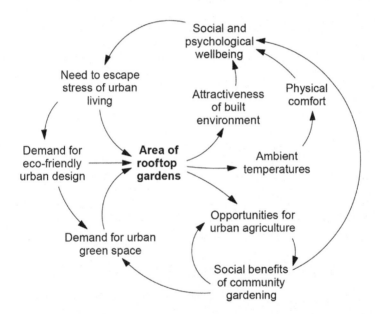

Figure 5.3 An influence diagram (ID). The blocks of text represent system state variables (stocks) and the arrows represent state-change processes (flows) that can change the values of the state variables. Group members are encouraged to use language carefully when they are naming the state variables. They are also encouraged 'to annotate' the arrows, using numerals (or other labels) as an aid to describing the corresponding state-change processes. Provided that clear 'rules of grammar' are followed (Box 5.3), the diagrams become part of a shared visual language that helps the group members communicate effectively about their individual views of causation.

Box 5.3   Selecting and naming system variables

In CCM, influence diagrams (IDs) are used as the basis of a shared visual language to support communication between the members of a social-learning group. The first step in using an ID is to think about the story that the diagram is intended to convey. How did things change in the past? How do you expect them to change in the future? The second step is to identify the variables that play a prominent role in your change-over-time story. This selection of variables sets the boundary of your system of interest – this is the cause–effect structure that you believe can explain the change-over-time story.

People with different worldviews will tell different stories. They will select different sets of variables and so will propose different system boundaries. This variety is essential, as it can lead to a much deeper understanding of the system. Nevertheless, taking advantage of differences in perception requires that each person's conceptual model of cause and effect is articulated very clearly. You need, therefore, to choose your variables carefully, selecting only those that you consider to be the most important. That is, you should aim to describe the simplest cause–effect structure that explains your story. One constraint that can help to maintain a focus on the dominant effects is to allow yourself to use no more than 10 variables in your diagram. Always remember that you are not trying to model the world – your aim is to produce a visual representation of your explanation (your theory) of a particular change-over-time story.

How you name your variables is very important. You need to use nouns or noun phrases that allow you to talk about *changes* in the *amount* (or *quantity*) of something. For example, each of the following variable names involves a quantifiable noun that indicates that an amount is involved:

Area of pasture	**Risk** of injury
Fraction of houses with verandas	**Level** of public engagement
Number of insect-breeding sites	**Demand** for clean water
Amount of energy	**Extent** of cycle paths

Avoid phrases that lock in a specific *direction* of change: For example, avoid labels like *Lack of rain* or *Increased rainfall* – instead use the direction-neutral term *Amount of rain* (mm). Also avoid phrases like *Travelling to work* – instead use *Distance travelled to work* (km) or *Commuting time* (minutes). Use words and phrases that are appropriate for the concept – when describing *discrete* items, use words like *Number of [people, houses, or dogs]*; for *continuous* quantities, use phrases like *Area of land* or *volume of water*. Abstract concepts such as *love*, *resilience*, and *awareness* can always be 'measured' on qualitative scales that run from *less* to *more* (or some equivalent). Some concepts, such as *diet*, need to be unpacked or explained in more extensive phrases – for example, *nutritional quality of diet*.

Activity 4: What drives system behaviour?

Activity 4 takes the group from systems thinking to system dynamics. The historical data, IDs, and shared understandings developed in CCM Phase I are used to identify feedback structures that have the potential to provide an *endogenous*[4] explanation of system behaviour (Richardson, 2011). Conceptual models, which express the group's dynamic hypotheses concerning the way that their system of interest operates, are then constructed by elaborating one or more of the candidate structures. Depending on the needs and capacity of the group, these conceptual models can be presented as stock-and-flow maps or low-order system-dynamics (LOSD) models (Newell & Siri, 2016).[5] The process of identifying feedback structures that are relatively simple, but that are dynamically dominant, is a reductive process. In CCM a protocol called Feedback-Guided Analysis is used to ensure that this reduction preserves key feedback links between sub-systems – see Newell (2015) for a detailed discussion.

Activity 5: Where are the leverage points?

The identification of leverage points, where a relatively small local change can produce major effects throughout the system, is a principal aim of CCM studies. CCM Activities 1 through 4 are designed to generate the insights required to identify potential leverage points. Very often these leverage points operate through relatively simple, but dominant, feedback structures. Meadows (2009) provides an insightful discussion of the nature of leverage points. In Table 5.2, which is adapted from Meadow's discussion, the system leverage points are listed in order of increasing effectiveness.

Activity 6: What futures can we see?

We all see the world through the lenses of our mental models – perception and decision making are both model dependent (Chalmers, 1976). If we want to see more clearly, if we want to make better decisions, then we need better models – more realistic and more reliable understandings of how the world works. These understandings can exist as private mental models that tacitly guide an individual, or they can be expressed as formal theoretical frameworks that provide a coherent approach for a social-learning team (Newell, 2015).

CCM Activity 6 typically involves the exploration of a range of possible futures. The development of 'systemic scenarios' is used to support this process. Scenarios are usually built following the approach developed by the Royal Dutch/Shell Group in the 1980s (Schwartz, 1991; van der Heijden, 1996; de Geus, 1997). The development of *systemic* scenarios follows these standard steps, but places more emphasis than usual on the dynamics of the group's system of interest. Such an approach can help group members to develop 'feedback eyes' and so anticipate unwanted system effects.

Scenario building is important in any dynamical study. It brings the underlying abstract concepts to life by embedding them in captivating stories. In addition, the development of multiple stories prevents the group from focusing on developing

Table 5.2 The Meadows leverage-point scale

Leverage point	Description
1. Numbers	Constants and parameters such as subsidies, taxes, and standards.
2. Buffers	The size of stabilising stocks and inventories relative to their flows.
3. Stock-and-flow structures	Physical and social structures and the way that they interact.
4. Delays	The length of time delays relative to the rates of system change.
5. Balancing feedback loops	The strength of stabilising loops relative to the strength of the changes that they oppose.
6. Reinforcing feedback loops	The strength (gain) of the driving loops.
7. Information flows	The structure of who does and who does not have access to information.
8. Rules	Policies and laws, including incentives, punishments, and constraints.
9. Self-organisation	The ability of the system to change its own structure.
10. Goals	The purpose or function of the system.
11. Paradigms	The mind-set out of which the system arises. This mind-set determines the system's goals, structures, rules, delays, and parameters.
12. Transcending paradigms	The ability to look at paradigms 'from the outside', to recognise that no one paradigm is 'true' and all encompassing.

'the' correct model and the production of accurate predictions. From the pedagogical point of view, the process works best when the whole group builds the scenarios. The educational approach called 'constructionism' rests on the basic principle that individuals learn best by 'making' – by tinkering, by doing something (Martinez & Stager, 2013). This idea resonates with the ancient Chinese proverb that is usually rendered into English as *I hear and I forget; I see and I remember; I do and I understand.* There is good modern evidence that this principle is correct (see, for example, Papert, 1980; Kolb, 1984; Martinez & Stager, 2013). It is supported by the demonstration that human conceptual systems are 'embodied' – that is, based on metaphorical elaboration of real-world, bodily experiences (Lakoff & Johnson, 2003; Newell, 2012).

Involvement in the construction of scenarios can help a community to develop 'memories of the future'. This term, which was coined by neurobiologist David Ingvar (1985), refers to the heightened sensitivity to significant variables and events that comes from the activity of seriously imagining a range of plausible futures. For this reason, scenario development can increase a social learning group's ability to detect the signals that new opportunities or dangers are emerging. Also, especially if the scenarios are based on low-order system dynamics models, the process can help community members to grasp the importance of cross-sector feedback effects in complex systems. It can alert them to the very real possibility that the effect of actions taken in one sector can loop around, through other sectors, to amplify or oppose the original actions.

Merits and limitations of CCM

Merits

An accessible approach to transdisciplinary engagement

In developing CCM we have blended concepts and methods drawn from a range of disciplines in the sciences, social sciences, and humanities. In so doing we have tried to identify relatively simple, generic protocols that can be accessed efficiently by a wide range of people.

A balance between firm guidance and flexibility in operation

CCM specifies a high-level process consisting of a set of six co-evolving activities that can be implemented using a variety of methods. Although we recommend specific protocols, such as pair-blending (Box 5.2), feedback-guided analysis (Newell, 2015), and scenario development, other protocols can be substituted within the CCM activities. This allows a social-learning group to tune the process to increase its resonance with their particular challenges and research methods. This means that the CCM process can be readily adapted to new contexts and challenges.

An iterative approach to learning

The six CCM activities can be carried out in sequence, thus providing a clear sense of direction in a social-learning process. But the activities can be revisited iteratively to incorporate the new insights generated as the group gains experience.

A practical way to develop a shared cause-effect language

In CCM workshops we use pair blending of IDs (Box 5.2) to help participants develop a shared visual language. This approach works very well in practice. After working alone to produce their individual diagrams, participants in a pair-blending exercise always welcome the arrival of a colleague (Essential Interactions 1 and 3). We have found that it is usually possible to find points of contact between individual's diagrams, even when these represent conflicting worldviews – we typically see significant increases in mutual understanding in a matter of hours. Pair blending offers a way to escape the communication illusion.

A practical way to blend history, system dynamics, and scenario development

The stories that people tell enliven a CCM discussion, particularly when it becomes apparent that different people tell different stories about the same events. Such differences reveal the basic subjectivity that dominates peoples' understanding of the behaviour of complex SESs and the consequent need to articulate and closely examine their own and their colleagues' worldviews (Essential Interaction 3). In CCM workshops we stress the need to use studies of the past to develop dynamical understandings that aid in anticipating future policy outcomes (Essential Interactions 1 and 2).

Limitations

The CCM process can lead to over-simplification

CCM is based on the assumption that complex systems are 'near decomposable' (Simon, 1981). That is, that they can be looked at validly as a set of inter-related sub-systems, where the variables within sub-systems interact more strongly with each other than with the variables in related sub-systems. Thus, CCM requires a form of 'holistic reduction' that focuses on identifying sub-systems whose behaviour can be explained largely in terms of endogenous forces operating within the slowly changing constraints of exogenous influences (Richardson, 2011). The identification of such semi-autonomous sub-systems is as much an art as a science – it depends on the collaborative efforts of experienced, thoughtful people. And there is always the danger of over-simplification. Even highly skilled people can produce simplistic conclusions in their efforts to escape the complexity dilemma.

The CCM process requires personal development

Although participation in a one-day CCM workshop can give participants a glimpse of the need to work together and take account of cross-sector feedback, this is obviously only the first step towards the insights and skills needed for the design of adaptive policy. The CCM process depends on participants' willingness to improve their personal mastery (Senge, 1990). For example, Bohm (1996, p. 20) has pointed out that effective dialogue requires people 'to suspend' their assumptions, that is, 'to hang them up' for collaborative examination and possible revision. Participants in a transdisciplinary program need to accept that their individual worldviews are severely limited – not something that comes easily to most people (Kuhn, 1996, p. 111). When taken seriously, worldview revision is a non-trivial, time-consuming process. As a result, the development of genuine shared understandings is rare, even in groups established to take a transdisciplinary approach (Newell, 2012). There is also the need for education in specialist skills. For example, CCM depends on SD concepts and tools that are unfamiliar to many people. Research has shown that even individuals with training in mathematics and science can have poor intuitions concerning the operation of basic causal mechanisms, such as feedback and accumulation, that drive the behaviour of complex systems (Sterman, 2008; Cronin et al., 2009). The members of a social-learning group must develop the required systems-thinking skills – another non-trivial, time-consuming task (Booth Sweeney & Sterman, 2000).

Ultimately, CCM cannot help a social-learning group escape the consequences of ignorance and uncertainty

In SESs, where human activity can cause unexpected and unwanted change, social-learning groups are always skating on thin ice. No matter how well educated, no matter how experienced and wise their members, they cannot have full knowledge of the system's dynamics and so cannot accurately predict the outcomes of their decisions. Although some management problems are relatively 'tame' and can be handled using conventional approaches, there is inevitably a wide spectrum of 'wicked' problems that "are complex, unpredictable, open ended, or intractable" (Head & Alford, 2015). Indeed, in the presence of uncertainty and ignorance, it is difficult to escape situations where well-intended policies exacerbate the problems that they are intended to solve, or create entirely new problems. That is why it is necessary to develop the ability to craft and test policies that can adapt to unforeseen conditions (Swanson & Bhadwal, 2009).

Conclusion

There is a growing awareness that the development of sustainable societies requires researchers, policy makers, and community members to take a transdisciplinary approach. CCM brings together, in a structured way, a set of linked activities designed to answer the question of *how* this might be done in practice. The central process that colours all six CCM activities involves (a) the articulation of an individual group member's perception of how the system of interest has behaved over time and why it has behaved this way and (b) the meshing of these individual perceptions to produce shared understandings.

Despite its limitations, CCM has proven to be an effective way to explore possible approaches to escaping the complexity dilemma. Above all, it focuses attention on the need for practical social-learning activities that can play a central role in efforts to develop adaptive policy for sustainable communities. As stated in Section 1, the complexity dilemma presents a clear challenge to sustainability science: Are there practical ways to isolate relatively simple SES sub-systems that (a) remain linked to the dynamics of the wider system and (b) are relevant in efforts to build resilience and sustainability? On the basis of the insights that we have gained in CCM workshops, drawing on the experience and wisdom of many inspiring people, our tentative answer to this question is "Yes".

Acknowledgements

We acknowledge the contributions to our thinking made by many workshop participants. They have helped us keep our feet on the ground. We have had valuable discussions with long-term critical colleagues Helen Brown, Chris Browne, Anthony Capon, Paul Compston, Robert Dyball, Nordin Hassan, Roderick Lawrence, Candice Lung, Craig Miller, David Newell, José Siri, and Robert Wasson.

Questions for comprehension and reflection

1   The following phrases are not well expressed as variable names for use in an influence diagram. Explain what is wrong with the language used and suggest better forms of expression. Refer to Box 5.3 for guidance.

Lack of community interest
Amount of dogs
Poor air quality
Playing tennis
High rainfall
Policy

2   Draw arrows to indicate cause–effect links between the following variables in a water management system:

Pressure from conservationists to increase environmental flows
Pressure from farmers to reduce environmental flows
Wetland health
Threat of flooding of agricultural lands
Volume of water released from dam

Add more variables if needed to clarify the processes or mechanisms represented by your cause–effect links.

3   The arrows in an influence diagram represent processes or mechanisms whereby changes in one variable cause changes in another variable. Consider the influence diagram about rooftop gardens shown in Figure 5.3. Describe plausible cause–effect relationships that can be associated with each arrow.

4   Freeways are often seen as the solution to traffic congestion problems in modern cities. But cities are dynamically complex. Although urban freeways can reduce congestion in the short term, they rarely provide a sustainable solution. In fact, freeway construction almost always makes traffic congestion worse in the long term. Discuss this situation, accounting for the unintended consequences associated with urban freeway construction.

5   Suggest alternative ways of addressing urban traffic congestion by considering the wider system. Select about 10 variables that together describe your alternative approach, and assemble them into an influence diagram that shows plausible interactions.

Notes

1 Parts of this section come from the CCM working paper by Newell and Proust (2012).
2 The 'order' of a system is the number of state variables (stocks) that it contains.
3 There is, however, a strong tendency for semi-autonomous governance units to develop into isolated management 'silos'. The effects of silo formation include a reduction in the ability of managers to see the cross-sector feedback forces that can drive policy failure (Tett, 2015).
4 *Endogenous* forces are those generated inside the sub-system. *Exogenous* forces are those imposed by (or on) variables that lie outside the sub-system.
5 Dynamics cannot be inferred reliably from influence diagrams or causal-loop diagrams. The limitations of causal-loop diagrams are discussed by Richardson (1986, 1997).

References

Ackoff, R. L. (1986). *Management in small doses*. Hoboken, NJ: Wiley.

Aligica, P. D. & Tarko, V. (2012). Polycentricity: From Polanyi to Ostrom, and beyond. *Governance: An International Journal of Policy, Administration, and Institutions* 25 (2): 237–262.

Arthur, W. B. (1990). Positive feedbacks in the economy. *Scientific American* February: 92–99.

Axelrod, R. & Cohen, M. D. (1999). *Harnessing complexity: Organizational implications of a scientific frontier*. New York: The Free Press.

Bohm, D. (1996). *On dialogue*. London: Routledge.

Booth Sweeney, L. & Sterman, J. D. (2000). Bathtub dynamics: Initial results of a systems thinking inventory. *System Dynamics Review* 16: 249–286.

Chalmers, A. F. (1976). *What is this thing called science?* St. Lucia: University of Queensland Press.

Cronin, M. A., Gonzalez, C. & Sterman, J. D. (2009). Why don't well-educated adults understand accumulation? A challenge to researchers, educators, and citizens. *Organizational Behavior and Human Decision Processes* 108: 116–130.

de Geus, A. (1997). *The living company: Growth, learning and longevity in business.* London: Nicholas Brealey.

Dyball, R. & Newell, B. (2015). *Understanding human ecology: A systems approach to sustainability.* London: Earthscan/Routledge.

Forrester, J. W. (1961). *Industrial dynamics.* Cambridge, MA: Productivity Press.

Forrester, J. W. (1969). *Urban dynamics.* Waltham, MA: Pegasus.

Ghaffarzadegan, N., Lyneis, J. & Richardson, G. P. (2011). How small system dynamics models can help the public policy process. *System Dynamics Review* 27 (1): 22–44.

Head, B. W. & Alford, J. (2015). Wicked problems: Implications for public policy and management. *Administration and Society* 47 (6): 711–739.

Ingvar, D. H. (1985). Memory of the future: An essay on the temporal organization of conscious awareness. *Human Neurobiology* 4 (3): 127–136.

Jordanova, L. (2000). *History in practice.* London: Arnold Publishers.

Kolb, D. A. (1984). *Experiential learning: Experience as the source of learning and development.* Englewood Cliffs, NJ: Prentice-Hall.

Kröger, M. & Schäfer, M. (2016). Scenario development as a tool for interdisciplinary Integration processes in sustainable land use research. *Futures* 84: 64–81.

Kuhn, T. S. (1996). *The structure of scientific revolutions, 3rd Edition.* Chicago: The University of Chicago Press.

Lakoff, G. & Johnson, M. (1999). *Philosophy in the flesh: The embodied mind and its challenge to western thought.* New York: Basic Books.

Lakoff, G. & Johnson, M. (2003). *Metaphors we live by.* Chicago: University of Chicago Press.

Martinez, L. M. & Stager, G. S. (2013). *Invent to learn: Making, tinkering and engineering in the classroom.* Torrence, CA: Constructing Modern Knowledge Press.

Meadows, D. (2009). *Thinking in systems: A primer.* London: Earthscan.

Newell, B. (2012). Simple models, powerful ideas: Towards effective integrative practice. *Global Environmental Change* 22 (3): 776–783. doi:10.1016/j.gloenvcha.2012.03.006

Newell, B. (2015). 'Towards a shared theoretical framework', in Dyball, R. & Newell, B. (Eds.) *Understanding human ecology: A systems approach to sustainability.* London: Earthscan/Routledge. pp. 114–134.

Newell, B., Crumley, C. L., Hassan, N., Lambin, E. F., Pahl-Wostl, C., Underdal, A. & Wasson, R. (2005). A conceptual template for integrative human-environment research. *Global Environmental Change* 15 (4): 299–307. doi:10.1016/j.gloenvcha.2005.06.003

Newell, B. & Proust, K. (2012). *Introduction to collaborative conceptual modelling.* Working paper, ANU Open Access Research. Retrieved from https://digitalcollections.anu.edu.au/handle/1885/9386 (Accessed 10/02/2017)

Newell, B. & Siri, J. (2016). A role for low-order system dynamics models in urban health policy making. *Environment International* 95: 93–97.

Ostrom, V., Tiebout, C. M. & Warren, R. (1961). The organization of government in metropolitan areas: A theoretical inquiry. *The American Political Science Review* 55 (4): 831–842.

Papert, S. (1980). *Mindstorms: Children, computers, and powerful ideas.* Brighton: Basic Books.

Proust, K. M. (2004). *Learning from the past for sustainability: Towards an integrated approach.* Ph.D. thesis. Canberra: The Australian National University. Retrieved from https://digitalcollections.anu.edu.au/handle/1885/48001 (Accessed 10/02/2017)

Proust, K. & Newell, B. (2006). *Catchment & community: Towards a management focused dynamical study of the act water system*. Final report, Actew project WF-30038. Retrieved from www.water.anu.edu.au/pdf/publications/2006/Proust_Newell06.pdf (Accessed 10/02/2017)

Proust, K., Newell, B., Brown, H., Capon, A., Browne, C., Burton, A., . . . Zarafu, M. (2012). Human health and climate change: Leverage points for adaptation in urban environments. *International Journal of Environmental Research and Public Health* 9 (6): 2134–2158.

Reddy, M. J. (1993). 'The conduit metaphor: A case of frame conflict in our language about language', in Ortony, A. (Ed.) *Metaphor and thought, 2nd Edition.* Cambridge: Cambridge University Press. pp. 164–201.

Richardson, G. P. (1986). Problems with causal loop diagrams. *System Dynamics Review* 2 (2): 158–170.

Richardson, G. P. (1997). Problems in causal loop diagrams revisited. *System Dynamics Review* 13 (3): 247–252.

Richardson, G. P. (2011). Reflections on the foundations of system dynamics. *System Dynamics Review* 27 (3): 219–243.

Schwartz, P. (1991). *The art of the long view: Planning for the future in an uncertain world.* New York: Currency, Doubleday.

Senge, P. M. (1990). *The fifth discipline: The art & practice of the learning organization.* Sydney: Random House.

Simon, H. A. (1981). *The sciences of the artificial.* Cambridge, MA: The MIT Press.

Sterman, J. D. (2000). *Business dynamics: Systems thinking and modeling for a complex world.* Boston: Irwin McGraw-Hill.

Sterman, J. D. (2008). Risk communication on climate: Mental models and mass balance. *Science* 322: 532–533.

Swanson, D. & Bhadwal, S. (Eds.) (2009). *Creating adaptive policies: A guide for policy-making in an uncertain world.* Los Angeles: Sage.

Tett, G. (2015). *The silo effect: Why putting everything in its place isn't such a bright idea.* London: Little, Brown and Company.

van der Heijden, K. (1996). *Scenarios: The art of strategic conversation.* Chichester: Wiley.

Vennix, J.A.M. (1996). *Group model building: Facilitation team learning using system dynamics.* Chichester: Wiley.

Walker, B. & Salt, D. (2006). *Resilience thinking: Sustaining ecosystems and people in a changing world.* Washington: Island Press.

Wals, A.E.J. & Schwarzin, L. (2012). Fostering organizational sustainability through dialogical interaction. *The Learning Organization* 19 (1): 11–27.

6 Exploring alternative futures with scenarios

Gerard Drenth, Shirin Elahi and Ariane König

The challenge: embracing complexity and uncertainty in times of accelerating change

Global change, including in the geo-political, environmental, economic, demographic, cultural and technological spheres, is accelerating and is transforming our worlds more rapidly than we can think about it. Moreover, changes within all these spheres are interconnected and interdependent. For example, the growing human population exceeding 7 billion people with its socio-industrial metabolism and land-use change affect environmental changes such that in an increasing number of world areas looming scarcities at the food–water–energy nexus are becoming apparent; these existential challenges in turn are changing worldviews and prevailing values expressed in technological choices and lifestyles. Whilst some of these changes may be fairly predictable, others are much more uncertain. What is certain, however, is that future impacts of these changes will be influenced by human agency and that the capacity to learn at the individual, organisational and systemic level is pivotal to enhancing a society's ability to change and to exert influence where possible.

There is no easy way to determine the 'best course of action' towards organising society, economy, organisations and individual lifestyles in a more sustainable manner. Such issues are complex and dynamic: there are a number of different interconnected scales, from local to global; many different stakeholders with competing interests; and the natural, social and technological environments in which they are operating are dynamic and rapidly changing. Accordingly, one main challenge in deliberations on sustainability in times of accelerating change is that over time criteria of success may change. Not only do changes in complex *systems* need to be better understood, but also the changes in the *criteria* that the actors are going to be using to make normative judgements on sustainability of alternative courses of actions (Kemp & van Lente, 2011). Added to this there is the problem that even at a particular point in time, different actors may have different criteria for 'success'. These challenges can be described as 'wicked problems': problems that are indeterminate in scope, impossible to solve and where solution attempts in turn can have unintended and possibly irreversible consequences, so you can't possibly experiment with one approach and see if it works (Rittel & Webber, 1973). Traditional static approaches to management, governance and science relying on prediction, regulation and control can only play a limited role in resolving such dynamic and complex problems. Wicked problems require a systemic and process-oriented approach that fosters iterative learning over time in diverse groups of affected actors, which is by nature adaptive, participatory and trans-disciplinary (Wals, 2015).

This chapter posits scenario planning as a method to structure such participatory deliberation processes. Scenario planning (see e.g. Ramirez & Wilkinson, 2016), however, is not to be confounded with traditional control- and prediction-based planning approaches that prevail in many business sectors. Scenario planning refers to the inquisitive, learner-centric, iterative and adaptive form of strategic planning that is researched and taught at the University of Oxford as the Oxford Scenario Planning Approach (OSPA). There are diverse groups engaging in scenario approaches (see also Vervoort, 2015) with common ground but also some differences, the analysis of which, however, is beyond the scope of this chapter. Scenario approaches can be designed to combine research, governance and learning in communities of public authorities, stakeholders and scientists and acknowledge the effects of human–environment interactions, with diverse actors representing diverse stakes, interests and values and worldviews. Future-oriented methods for 'world making' best illustrated in scenario approaches assume that individuals conceive of the world they live in in multiple ways; thus there are many overlapping worlds, all of which are co-created in interactions with others and our environment (Vervoort et al., 2015). Developing joint scenarios jointly can enhance a shared knowledge base for shared expectations and therefore facilitate jointly changing social practice at higher levels of social organisation than just individuals – a prerequisite for sustainability transitions. Scenarios for sustainability are coherent and plausible stories that combine narratives with quantitative representations, direct attention at elements in a system and their interactions and describe plausible interdependent development pathways. As such, these scenarios are uniquely suited to gain an enhanced understanding of human–environment interactions by scanning the future in a creative, rigorous and policy-relevant manner that reflects the normative character of sustainability and incorporates different perspectives (Swart et al., 2004). Combining scenario practice with systems approaches described by Newell and Proust (Chapter 5), König (Chapter 3) or Davila and Dyball (Chapter 10) facilitates the identification of path-dependent lock-in situations in unsustainable social practices. The collective nature of the scenario process is potentially more likely to direct attention to potential leverage points that might help to overcome barriers to change than deliberations based on advice from disciplined experts or by learning from the past.

This chapter first introduces the scenario method by juxtaposing it to practices of fore-casting and visioning, providing advice on the design of scenario processes and exploring what can be learnt from scenario practice. The application of scenario practice for identifying challenges at the food–water–energy nexus is then discussed as a basis to explore the merits and limitations of the approach. The conclusion highlights how the process of projecting different context-dependent forms of knowledge into the future, which is less contested, can reveal unrecognised contradictions and conflicts whilst at the same time enhancing our insights about uncertainties and what we don't know and possibly can't know. The resulting shared understanding will offer a more socially and temporally robust knowledge base for concerted action for sustainability.

Scenario approaches to structuring future-oriented social learning

Humans have the ability to imagine futures and thereby influence and contribute to changing the world, rather than to just endure a changing world. Scenario approaches as described in Vervoort (2015) are particularly well suited to 'world making' rather than just trying to adapt or becoming resilient to change, as imaginaries of the future developed in a participatory process with diverse stakeholders can become more widely shared and change expectations and social practice at higher levels of social organisation than just individuals. This section will first compare three methods to explore futures: forecasting, visioning and scenario planning. Subsequently, we will provide advice on the design of scenario planning and then discuss what can be learnt.

Comparing three approaches to exploring futures

There are many methods to explore futures, including visioning, forecasting, Delphi searches, environmental scanning, trend analysis and extrapolation, expert modelling and prediction, computer simulation, technology road mapping and historical analogy or also more esoteric practices such as astrology and horoscopes or reading tea leaves. Particularly common in policy making is *forecasting* using expert opinions combined with extrapolation, whereas citizens' initiatives for transition to sustainability increasingly rely on *visioning* techniques to create shared knowledge about possible futures for concerted action. This section will compare forecasting and visioning with scenario planning (see Table 6.1).

Forecasting focuses on certainties, as it largely relies on 'casting or extrapolating forward' representations of a particular past, with modelling techniques or trend analysis and extrapolation, usually led by disciplined experts of a particular field.

Table 6.1 A comparison of forecasts, visions and scenarios

	Forecast	*Vision*	*A set of scenarios*
Question of purpose	What is our best estimate, based on available and representative data, of what might happen?	Where do we want to be?	What are diverse views on what might happen and what we don't know?
A representation of . . .	A most likely state and derived states	A single most desirable future	A set of alternative open futures
The method	Mainly quantitative	Mainly qualitative with associated goals and targets	A mixture of qualitative and quantitative, with emphasis on stories
Process	Expert driven	Community driven	Trans-disciplinary
Relation to values in a pluralist society	Claims to be objective	Normative	Alternative worlds highlight disparate sets of values, worldviews and priorities
Risks and uncertainties	Hidden	Hidden	Revealed
Function: helps to analyse	What are the likely implications of 'business as usual'?	Where do we stand now in comparison, and how can we go about achieving this vision?	What are the implications and options for me and for us?

However, on the basis of prevailing forecasting methods, you can't forecast uncertainties. Thus, forecasting often hides uncertainties and conceals risks. Moreover, forecasts are also usually single-point projections, with sensitivity analysis around the base projection, which gives rise to alternative but closely related situations in the future, often described as 'best case' and 'worst case'. Such forecasting in practice obscures potential disruptive events, discontinuities and the existence of truly alternative futures.

Unfortunately, in practice there is much confusion between the terms 'forecasts' and 'scenarios' as these are often used interchangeably. Forecasts are easily distinguished from scenarios as they have a baseline that is usually a simple extrapolation of a particular trend presented as a 'business-as-usual case'; deviations from this baseline are generated with sensitivity analysis in which some variables range between expected maximum and minimum values; resulting variations are then often associated with a 'best case' and a 'worst case' scenario. (More elaborate, but based on the same principle, would be the use of Monte Carlo simulations around the baseline.) Examples of highly influential forecasts that are mistakenly called scenarios include 'energy scenarios' by the International Energy Agency and the U.S. Energy Information Administration and the 'climate scenarios' by the International Panel on Climate Change (IPCC) and the National Oceanic and Atmospheric Administration. In each case, these forecasts are the results of trend analysis and quantitative modelling and are very useful as analytic tools, but they do not serve the same purpose as the scenario approaches that are the subject of this chapter. The scenario approaches suitable for transformative sustainability science direct attention to uncertainties, unknowns, possible disruptive events and multiple worldviews and values; they result in sets of diverse and unrelated plausible alternative future worlds that are represented in the form of stories, often later complemented with quantitative elements.

Visions are representations of a desirable (aspect of a) future world; as such they are normative and are usually developed in order to give a direction to social change; they can also be developed as a foundation for a shared identity and in community building (Wiek & Iwaniec, 2014; Costanza, 2000). They are not predictions or forecasts, but they are usefully associated with specific goals and targets, qualitative and quantitative to make them actionable and tangible (Ravetz, 2000). The planetary boundaries advanced by Rockström and colleagues (2009) can be taken as such a vision. This method has also been adapted for use in organisations as a leadership tool that enables diverse teams to develop a shared understanding of the complex interactions of organisations and their environments combined with systems approaches (Senge, 1993). A vision gives people a sense of 'where we could take the future', and good visions take people along on the journey into that future – they are motivational. Again, there is potential for confusion as visions are sometimes called 'normative scenarios'. The authors of this chapter consider this misleading, as the purpose of a multifaceted scenario approach described in this chapter differs from the purpose of developing a unified vision.

Participant-driven back-casting exercises leading to the articulation of desirable futures have been proven useful for driving action on sustainability (Robinson et al., 2011). The desirable futures can be used as basis for developing pathways for how this future might be attained starting from the status quo. However, strongly normative vision and back-casting processes have been criticized for not leaving sufficient room for dissent and divergent view-points, in particular based on too reductionist computer model–based visualization of futures. Scenario approaches can usefully inform visioning and back-casting processes, as they, amongst other things, serve to identify uncertainties, risks, possible disruptive events and unknowns. Only after having explored these is it useful to narrow the focus and determine what one should *aspire* to.

Scanning the future with scenario analysis in combination with back-casting approaches in participatory processes helps in the understanding of conditions of complexity, uncertainty, potential disruptive changes and human choice and constraints (Robinson et al., 2011; Wiek & Lang, 2015). Scenarios are stories describing future worlds that illustrate alternative outcomes of developments. Exploratory scenario building engages research to better understand drivers of change, certain and uncertain, in a contextual environment that we cannot influence. The approach blends qualitative and quantitative analysis in order to explore alternative outcomes of global change and associated implications locally in a transactional environment, where some changes might be brought about if a critical mass of actors or stakeholders engages. Scenarios do this by working as a set of divergent stories of the future. Such a set of scenarios usually serves to highlight things we can or can't know about the future; uncertainties that matter but are rarely talked about and inter-dependencies in alternative future development paths, human choices and constraints; and differential power distribution in society. Sets of scenarios may also be designed to sketch *the interdependence of culture and values prevailing in society and how these are interdependent with technological choices*; this may also be related to experienced quality of life and environment and how distributional issues might play out in different futures. This dimension is, however, often neglected.

There are calls for further improvements of the methods for stakeholder interaction using scenario construction and systems analysis and the design of more interactive processes that allow for exploring highly complex and uncertain value-laden issues, bringing these processes closer to social learning and action in real life (Jerneck et al., 2011; Robinson et al., 2011; König, 2013, 2015; Wiek & Lang, 2015). For example, one risk brought up is that participants can be overwhelmed by the complexity of choices they are being asked to make.

All three approaches to explore futures are useful for different purposes, yet it would be unwise to view any of the futures they create as 'an end in itself'. Ultimately, each describes a future that has not yet taken place, something that is by its very nature uncertain. It is not about *having* the vision, forecast or scenarios; it is about the collaborative process of *developing* this. The value lies as much in the process as the product.

Designing participatory procedures for scenario practice

Participants in scenario processes often experience the initial phase as becoming ever more uncomfortable with the things they thought they knew and understood well – with the familiar. They start to realise that some of the assumptions they hold about the future may actually be quite dangerous or toxic. Or they thought they held similar views or criteria as co-workers only to find that they actually perceive the same reality with very different eyes. In some cases, there is a realisation that the dominant narrative explicitly excludes certain forms of knowledge or that thinking is done in disparate silos with scant acknowledgement for the insights from elsewhere. The scenario journey takes participants to a place of unfamiliarity and allows them to develop a level of comfort with ignorance. And this is the ultimate goal: the point of scenario planning is to give people the confidence and methodology to acknowledge and embrace uncertainty – as uncertainty is the overriding characteristic of the future.

Scenario approaches, if well designed, can empower participants in distinguishing which aspects of futures or specific impacts of future developments they may have influence on and which aspects they may have to adapt to. These insights can then form the basis whereby both the individual and the organisation can develop their own goals, together with targets and actions plans to achieve these, as well as monitoring plans for evaluation and iterative improvement. For this purpose, it is useful to distinguish between the contextual and transactional environments (van der Heijden, 2005). The key difference between the two is the ability to influence the environment, and the boundary between the two lies where the actor's influence reduces to nil (see Figure 6.1). Scenario research usually first focuses on identifying 'factors' (drivers of change and uncertainties) in the contextual environment (as, for example, global technological developments, demography and migration), especially those that are both not well understood and potentially impactful. Scenarios are then different plausible, challenging and relevant forms the transactional environment could take under the influence of these factors, shaped by actors, and including development pathways leading to them.

In terms of methodology, a scenario approach often starts with a 'framing or scoping workshop' in order to identify main project objectives, 'users' of the scenarios, purpose, salient research questions and stakeholders to be interviewed. The main stages in the scenario method may include conducting interviews, performing document research, developing and detailing a scenario set, systems thinking, affirmation and implication or strategy/policy workshops.

The main purpose of the interviews is usually to collect personal opinions, concerns and impressions from diverse points of view. Uncertainties voiced,

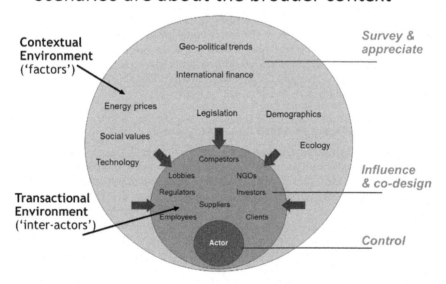

Figure 6.1 The contextual and transactional environments in scenario planning

Source: Used and adapted with author's permission from Ramírez & Wilkinson, 2016

contradictions and tensions are of interest. These interviews can serve to identify overarching themes and are often synthesised in a 'chorus of voices' reflecting
diverse viewpoints on each theme. These insights then can be used to structure
further research and participatory workshops (see also Box 6.1 for some structuring
questions to design a workshop series). Drivers of change and uncertainties that are
identified in interviews and collaborative workshops can be further characterized
by compiling relevant trend analyses and statistical forecasts.

Box 6.1   Questions useful for exploring in workshops

- Who are we? Values, identities, inclusive notions of agency and actors,
  including non-human actors, ideas etc. (Latour, 2004).
- Why did this world come about? Systems, relations between elements?
  Which beliefs and values have shaped it? Who has shaped it and according to what ends?
- Where are we? This question initiates spatial consideration about the
  scale and scope of the world. What is the function of place? Constitution
  of space – territorial? Networked?
- How do we experience the world? What does the world look, sound,
  feel and taste like? How are core beliefs and ideas about the world
  encountered and maintained?
- What are sources of discomfort and gaps in this world? In what situations in this world to do you need courage to act? Contradictory view
  points, disagreements, issues of uncertainty, ignorance and uncomfortable truths, when confronted within an alternative world, can offer spaces
  for creativity to seek novel approaches to act on problems old and new.

Questions about relationships between worlds

- How do new scenario worlds draw on available worlds? What is taken
  into account? What is left out? What is emphasized? Composition,
  ordering, weighting, deletion, supplementation and deformation are
  ways in which new worlds are created based on existing worlds. Which
  are useful? Which happen inadvertently?
- How are worlds interacting? Are there 'sub-worlds' within a scenario?
  What is the relationship between worlds of different scales, and how are
  they distributed? How might worlds previously unaware of each other
  cope with an encounter?
- What relevance does this new world have for present worlds? In what
  ways may a newly created scenario world threaten present worlds? What
  opportunities may it bring? What new ideas? How does it challenge the
  values associated with present worlds? Which elements could help fill
  knowledge gaps and blind spots?

Source: Adapted from Vervoort et al., 2015

Scenario planning is also a social process in the sense that it not just toler-ates, but actively *invites*, different points of view. Inviting 'outsiders' into scenario engagement workshops or as interviewees is critical if one intends to find known unknowns and even some *unknown* unknowns. So, much of it is about people disa-greeing, arguing even; and that is constructive because it is when people disagree that new insights are born.

Two logics often co-exist in the production of knowledge on sustainability chal-lenges: the first logic invites to seek '*accuracy* to guide decision making'. In this approach, as Wilkinson and Eidinow point out (Wilkinson & Eidinow, 2008), the output is the *product* of new learning. A second logic suggests that 'further learning can only be achieved through *co-production*'. Here the choice of participants (stake-holders in finding solutions) is critical.

A focus on finding accuracy relies on modelling, numbers and quantification. The IPCC scenarios (in our terminology these would actually be forecasts) are a good example of this. These forecasts are based on computer modelling of complex earth system processes and feedbacks that influence material and energy flows that then affect the constitution of the atmosphere, in order to predict, for example, the different levels of CO_2 in the atmosphere that one may find in different futures. The process is expert driven; the experts then disseminate the results to policy-makers and the public. Joint learning is less emphasised.

The second logic gives primacy to people to understand, learn and co-create new policies. The WBCSD (World Business Council for Sustainable Develop-ment) scenarios on the future of water are an example of this approach (WBCSD, 2009). Comparing the lists of participants for the IPCC forecasts and WBCSD scenarios illustrates the difference. At the IPCC you see *experts* in climate change and climate technology, and they collaborate doing the quantitative analysis. In the WBCSD scenarios one finds participants from every walk of life, because the purpose is on engaging all those who will participate in dealing with the global water challenge.

Combining both strands of research – expert driven and diversity mapping with systems approaches – is critical in scenario approaches (Seiffert & Loch, 2005). In scenarios, each story needs to 'hang together'; the way to ensure this is to apply causality analysis and system thinking. In system thinking one looks at the whole system, at what structures one can find that explain behaviour and events, identify-ing feedback loops in the system. In essence, system thinking is about seeing beyond events (Figure 6.2) by seeing them in the context of trends and patterns at a deeper level and systemic structures at the deepest level. The nuclear disaster at Fukushima was an event, an avian flu outbreak is an event, the collapse of the Lehman Broth-ers' firm was an event. But in systems thinking we attempt to go deeper and look for the answer to the question 'Why did this happen?' One analytical level below events (past or potentially disruptive future events) we find trends and patterns. But trends are dangerous to rely on. Scenario experts at Shell quote Alexander Cairn-cross: 'A trend is a trend is a trend. But the question is, will it bend? Will it alter its course through some unforeseen force and come to a premature end?' If one needs to understand how trends could possibly and plausibly be bent in practice (e.g. achieving reduced CO_2 emissions from energy efficiency measures by pre-empting rebound effects), systems analysis is crucial. Good scenarios are structured on visu-ally powerful and plausible system maps (see also Newell and Proust's Chapter 5 on how to develop these).

The role of system thinking in scenarios
The triangle of deep structure

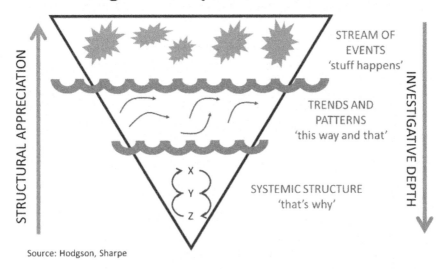

Source: Hodgson, Sharpe

Figure 6.2 The triangle of deep structure in past, present and future worlds
Source: Sharpe & van der Heijden, 2007, Figure 5.2
©Wiley Books. Copyright Clearance Center, Inc.

Box 6.2 Scenarios supporting the energy transition (WMB/CDP)

Purpose and context: We Mean Business (WMB) is a coalition of organisations working with thousands of businesses and investors. These businesses recognise that the transition to a low carbon economy is the only way to secure sustainable economic growth and prosperity for all. To accelerate this transition, WMB formed a platform to amplify the business voice, catalyse bold climate action by all and promote smart policy frameworks.[1] In 2015, two authors and their colleagues supported We Mean Business and the Carbon Disclosure Project (now CDP) in using a scenario process based foresight technique to break a historic logjam regarding carbon pricing and its role in tackling climate change. Stakeholders came from governments, business, science and the investment community. Each of these stakeholders may support the necessity of an 'energy transition' and the role carbon pricing could play in this, but each approach achieving this end in different ways and with different, often conflicting, agendas.

Process and content: Initially the output of the scenario engagement was a set of carbon pricing scenarios that could be put to immediate use to build momentum on the road to the 2015 COP21 meeting in Paris. But the scope widened as the engagement progressed. Over a period of four months, participants worked intensely to create a 'carbon pricing toolkit' consisting of scenarios, narratives, carbon pricing bands, carbon price trajectories and transformation pathways.[2]

Outcomes and their significance: The project was complex, with, as stated earlier, participants representing multiple stakeholders from the business, scientific and policymaking communities. The resulting toolkit has been widely communicated and forms a key component in the endeavour to stimulate productive dialogue about the future of carbon pricing in the pursuit of a sustainable, low carbon, economy.

What can we learn from engaging in scenario approaches?

Vickers's Appreciative System (Vickers, 1983) offers a theory and concepts for analysing organisational decision making. Vickers argued that when an organisation makes decisions it needs to first form judgements in three different domains. First there is the domain of 'what is going on?'; of *reality* judgement: What are the facts? What can we observe? What is observable? How do we frame reality? Next, judgements have to be made about *value*: What does it mean? Where do we want to go? What would be good and what would be bad? And, finally, decision making is also about setting direction and making *instrumental* judgements: now that we have analysed reality, and also analysed what values we use to determine what is right and what is wrong, we move into the domain of 'what shall we do?' Vickers argues that if there are unresolved issues in one of these three domains, one should not just ignore them and continue with making the decision, but instead address the deficiency in that domain.

Scenario approaches can contribute to each of the three judgement spaces Vickers identified (Burt & van der Heijden, 2008). First, scenarios support *making sense* of complexity: when things are really complex you cannot really capture them in a few simple variables. Complexity, ambiguity and uncertainty are easier to describe in a scenario than in a model. Scenarios actively explore uncertainty, aiming to understand it and give it a useful framing. Scenarios help in the surfacing of assumptions about the future that might be considered 'toxic'.

Scenario approaches, given certain design requisites for the process, can allow participants to gain a more sophisticated understanding of diverse *values* attached to alternative development pathways and trade-offs. Scenarios can help to create a 'common ground' in terms of a common understanding (Elahi, 2008).

Third, scenarios support *setting direction* and policy formulation. Each of the scenarios in the set represents a windtunnel through which a given policy or strategy can be tested or different options can be compared. The insights this generates enable strategies to be refined or redesigned or for new strategies and policies to be articulated. Using different plausible contexts stimulates creative and innovation processes aimed at finding new options, solutions and possible directions.

The purpose of scenarios can also be expressed in other terms (see also Wiek et al., 2006):

- Systematic understanding of critical variables in our context
- Appreciating different values and worldviews
- Holding strategic and courageous conversations
- Testing ('windtunnelling') strategies, options and policies
- Assessing risks
- Identifying new opportunities
- Generating new ideas and policies
- Communicating with stakeholders

Scenario practice has also been highlighted as a suitable approach to address some fundamental research challenges for sustainability science, and we indicate how scenario planning could contribute to these (Table 6.2) (Swart et al., 2004). These include spanning spatial scales benefits from taking scenarios and down-framing to lower system logics, whereby actors immerse themselves in a possible future context and work through possible courses of action and 're-action' by other actors (Ramírez & Wilkinson, 2016). 'Accounting for inertia and urgency' takes place when plausible scenario stories are told with explicit and impactful implications: this creates urgency. Scenario approaches encourage diversity of views and thus support recognition of a wide range of outlooks, norms and preferences. Scenarios clarify without simplifying, allowing complexity to be reflected. The cross-disciplinary nature of scenario processes promotes cross-theme and cross-issue thinking and acting. Uncertainties and surprises are, of course, key components of scenarios. Accounting for volition and actor influence is encouraged by scenario use workshops and in-scenario actor analysis. Combining qualitative and quantitative analysis is core to scenario approaches too, although here it is important to note that any quantification takes place after the qualitative definition of the scenarios. And finally, as inclusive social learning processes, scenario processes naturally engage stakeholders.

Table 6.2 The potential of scenarios in sustainability transition challenges

ST challenge	*Scenario contribution*
Spanning spatial scales	Global, focused, regional scenarios
Accounting for inertia and urgency	Connecting future to decisions taken now; monitoring and early warning systems
Recognising wide range of outlooks, norms and preferences	Encouraging diversity, differences of view and opinion – making disagreement an asset
Reflecting complexity and stresses	Clarifying not simplifying: system thinking: 'what if' analysis and implication sessions
Accounting for volition and actor influence	Encourages reflection on differing worldviews; 'use' sessions and actor analysis
Combining qualitative and quantitative analysis	Integral to scenario planning
Engaging stakeholders	Scenarios promote communication and testing and influencing human perceptions and goals

Source: After Swart et al., 2004

In sum, scenario planning offers an analytical and creative social process for assessing the factors and drivers of change in the contextual environment based on qualitative and quantitative information from very diverse knowledge fields and interpreting them to construct different plausible ways in which these could interact, thereby giving rise to different possible future transactional environments (Ogilvy, 2002; Ramírez et al., 2013). These then are called scenarios. Scenarios come in a small number and work as a set, which underlines that there is no one standalone version of the future. Scenarios are not projections or predictions. Scenarios are not about the actor themselves, but about the environment in which the actor might find itself. Scenarios are not 'truthful', 'accurate' or 'correct'; instead they are about what's plausible and useful to imagine. Plausibility and usefulness are important; the scenarios that will be told need to be believable and need to make people think or make people frame reality in a slightly different way. Reaching the right level of plausibility to resonate with but still challenge participants is essential for the art of scenario planning. If a scenario is too implausible, it will be rejected; but if it is too plausible it will not move scenario users out of their comfort zone.

Applying scenario approaches to gain an enhanced understanding and repertoire of action on possible future challenges at the water–food–energy nexus

One of the largest and most pressing challenges facing humanity this century is the water–food–energy nexus – the inextricable interlinkage between these basic resources upon which humans depend for survival. Without water, people die. Without energy, we cannot grow food; power our homes, schools and offices; or run the technological equipment that we rely on. Water and energy are necessary to grow and distribute the food on which we are dependent. A number of agricultural crops are converted into bioenergy. Vast quantities of water are used to generate electricity, and energy is necessary to clean water and move it proximate to the needs of the population. Not only are these resources becoming scarcer, but the interconnections and interdependencies between them have been unrecognised and underappreciated.

Currently, these resource systems are managed independently and in isolation, with little or no coordination between their use. There is scant awareness of where the critical system linkages and vulnerabilities lie, so that policies and measures that benefit one resource can lead to increased risks and detrimental effects for one of the other systems. Many observers argue that the world is facing a 'perfect storm' that is likely to result in acute shortages of water, food and energy.[3] The trajectories for resource demand for the necessities of life are all projected to increase sharply. Research projections (IRENA, 2015) are that by 2050 demand for energy will have risen by 80%, demand for food by 60% and demand for water by 55%. To make matters worse, climate change is likely to further exacerbate many of these inter-related pressures.

Power plants, both fossil fuel and nuclear ones, require vast quantities of water drawn from rivers and lakes for cooling. They are responsible for large percentages of total freshwater withdrawals: more than 43% of total in Europe, nearly 50% in the United States and more than 10% of China's water. Nuclear power plants use dramatically more water than gas or steam-powered plants and are therefore often situated alongside water sources. However, when water levels drop (or rise) this can cause the facility to become unusable. Without water, there is no energy security, and without reliable energy resources, our reliance on modern conveniences suddenly becomes a source of systemic risk. However, there are other stakeholder demands. Water used to produce energy (as in steam-driven turbines) is mainly not available for supply to residents, farms and industry, and although some of it condenses, it is with a different temperature and biological content. Some of the most critical interdependencies between water, energy and food are illustrated in the Figure 6.3.

When availability of basic resources comes under pressure, the trade-offs to be made are difficult and likely to be heavily contested by the many competing

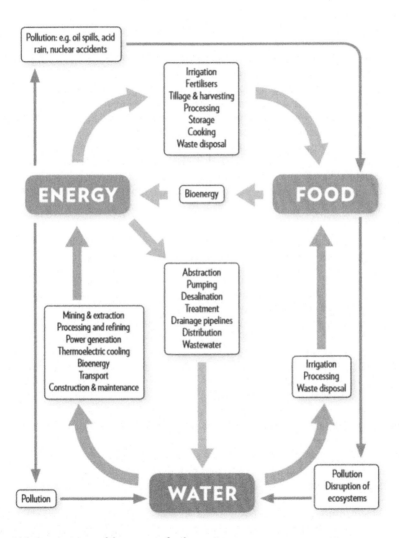

Figure 6.3 An overview of the energy–food–water nexus
Source: Wakeford et al., 2015, Figure 0–1

interests. This will require serious debates about short-term cost increases versus long-term savings; the balance between energy and water use; and compromises regarding transportation, market locations and raw material costs. Different water sources require very different amounts of energy: desalination of seawater can use ten times more energy than water from a lake or river. For example, using water to irrigate crops can promote food production, but also reduce availability of water and potential for hydropower, so what takes precedence?

Unlike many sustainability issues where the scale is mostly local, the water–food–energy nexus has a strong global dimension. Most of the world is industrialised to the extent that they have become embedded within the global trading system. This embeddedness implies dependency on imported goods, and with it vulnerability to resource constraints or shocks. Geopolitical tensions, resource nationalism or extreme weather events, such as droughts and floods, all have the potential to threaten security of supplies. If this were to occur, there would be surging oil and food prices, probably exacerbated by financial speculation. Even if there were no extreme shocks, there are interlinkages that will need to be better understood. So, for example, if one country exports a water-intensive product to another country, it exports water in virtual form. The 'virtual water' is the water necessary for the production of the particular item. So although the general perception is that trade of water is impossible due to the nature of water and the large distances between countries, water embodied in products implies virtual water trade. It would therefore be sensible for water-rich countries to produce water-intensive products, rather than water-poor countries, so that places with an abundance of water could export a critical resource. The production of water-intensive goods has implications for local water consumption and pollution relating to the export, whereas the consumption of water intensive goods leads to water saving, but also water dependency (Hoekstra, 2003).

Merits of scenario approaches to explore challenges at the nexus

The issue of the water–food–energy nexus is a textbook example of the 'wicked problem' discussed earlier in this chapter. Like almost every aspect of sustainability, it is a complex multifaceted issue that is impossible to define accurately. Here environmental, social, technical and political dilemmas coalesce and cut across boundaries of communities, organizations and nations. Countless stakeholders with competing interests need to be considered and reconciled. The demarcated boundaries between organisations, communities, departments and sectors have no place here, and any approaches to resolve dimensions affect the issue reflexively. Ultimately, approaches to resolve water–food–energy nexus issues are likely to be either adversarial or collaborative. If they are adversarial, everyone is likely to lose out. If there is any possibility of achieving a climate of cooperation when there are difficult potential losses to deal with, there will be a need to find the common ground to balance the conflicting interests of diverse stakeholders.

Everybody everywhere shares a common desire for a good future, and it is the focus on the future that will create the basis for cooperation, if it can be galvanised. Good scenario planning, drawing a diverse group of stakeholders together to search for shared future interests, offers a potential framework for building this common ground (Elahi, 2008). Such collaborative scenario-development work requires a disparate group of key stakeholders with many different and contradictory worldviews and interests. Depending on the design of the scenarios, these scenario builders can

form the 'extended peer community' that is necessary to create knowledge in situations of severe uncertainty, high stakes and disputed values (Funtowicz & Ravetz,
1993).

Their output, the resultant set of scenarios, has the potential to create a common
framework and language that can be utilised by a larger group. This cooperative
process of discovering the common ground has the potential to transform a wicked
problem into a shared problem, where different stakeholders and actors have a common ethical code or aligned values, or where at least the alternative point of view
is recognised. Ideally, this would increase the likelihood of acceptance of a solution
that will inevitably involve difficult trade-offs.

Box 6.3   A tale of three regimes

Context: The traditional agrarian regime is highly dependent on nature and
the environment. Agriculture is usually rain-fed, and local sources of energy
will be primarily biomass. Most of the population lives without access to
electricity, running water and any modern conveniences. In developed parts
of the world, people are reliant on an industrialised regime. Here, there is
a complex infrastructure supplying energy, water, food and other goods, all
interconnected into a global system that is both heavily dependent on fossil
fuels and resulting in widespread overexploitation of resources and degradation of the environment.

Cuba offers one of the few large-scale examples of an agroecological
alternative to the agrarian and industrial regimes. Once reliant on subsidised
Soviet fossil fuels, the collapse of the Soviet Union resulted in an abrupt ter-
mination of preferential trading in energy and food supplies. This shock to
the system resulted in a 35% decline in GDP between 1989 and 1993. During this period of economic turmoil, termed the 'special period', power cuts
in Havana were so widespread that residents called the few hours of available
electricity 'light-outs'.

Over the following two decades the country was forced to become largely
self-sufficient in terms of food and energy production. Urbanisation trends
were strictly controlled. Agroecological practices were extensively developed
to reduce agricultural reliance on energy, so that between 1990 and 2012,
total energy consumption halved and water consumption was also reduced.
The agroecological model was proven demonstrably viable, outperforming
yields of many agricultural products farmed more intensively in the industrial
model. However, it also exposed the shortcomings of agroecological farming
with regard to cereal, dairy and meat production, and Cuba has remained
dependent on foreign imports of food.

Over time, a dualistic agricultural sector evolved, investing in genetically
modified monocrops such as maize and soya. This development has been
accompanied by growing reliance on fossil fuel consumption, with most of
the imports coming from Venezuela. Yet once again, the consequences of
energy security have become apparent. In July 2016 Cuba's Minister of Economy announced that the country would have to cut its fuel consumption by

one-third due to the political unrest in Venezuela. According to President Raúl Castro: 'There is speculation and rumors of an imminent collapse of our economy and a return to the acute phase of the 'special period'. We don't deny that there may be ill effects, but we are in better conditions than we were then to face them' (Burnett, 2016). How Cuba responds this time remains to be seen.

Source: Wakeford et al., 2015

Limitations

When there is a crisis, it is usually too late to explore alternatives collaboratively. In such situations the response is likely to be a reactive one. The alternative to a reactive approach is a proactive one: where policymakers anticipate potential problems and explore how the systems can be better interconnected to remove inconsistencies, improve efficiencies and generate potential co-benefits across multiple sectors. This is resource intensive in terms of finances, time and intellect.

However, sustainability issues will invariably require shared context-specific knowledge. In terms of the water–food–energy nexus, each of these basic resources will need to be evaluated both dynamically and collectively. Issues that will require monitoring include stability of supply, affordability for those more vulnerable within the population, quality of water and food, and accessibility to everyone everywhere. Should serious constraints arise for any of these, very difficult trade-offs will need to be made if the limitations on the resource availability are fairly allocated across society.

Scenario planning is a participatory process embracing complexity, contingency, contradictions, uncertainty and ignorance. It can assist people in making sense of complex, interdependent and intractable 'wicked' problems. They have the potential to create a helpful framework, a common ground, that can reconcile differences and resolve trade-offs. Once the set of scenarios has been built, they can be used to pre-test alternative strategies and policies in contexts that are beyond the influence and control of the stakeholders and actors involved. Most importantly, scenarios allow, and actually encourage, multiple and sometimes irreconcilable viewpoints and opinions to be articulated and embedded into the findings. That does not mean that scenario building will remove conflicts. But the result, when successful, can enable fewer, better arguments to be presented and ideally encourage conflicts to be satisfactorily resolved (Tattersall, 2016).

However, many scenarios exercises fail to meet these expectations. So here, setting out and agreeing the purpose and expectations of the scenarios process up front becomes critical. All too often, there is knowledge that is unrecognised or unknown: the 'here be dragons'. This knowledge is not always unavailable; it is possible that it is unacceptable for some reason and has therefore been airbrushed out. Three key factors create this myopia, namely: the psychological desire for control, the demarcated boundaries of hierarchical bureaucracies and the rational scientific paradigm (Elahi, 2011). This omission can be critical when dealing with controversial and complex multifaceted issues and is then likely to result in sub-optimal decisions, unreliable trade-offs, erosion of institutional credibility and unattended gaps in the research.

Without a neutral unbiased forum, strong leadership and clear ground-rules participation will not be open and constructive. Without a strong focus on issues of fairness and a desire to build the common ground and trust between parties, the scenario-building process will simply be an expensive exercise and a waste of resources. Without an explicit acknowledgement of ignorance and uncertainty, no process can openly explore them.

Conclusion

In sum, scenario approaches are based on the premise that the future is not an extrapolation of the past. We see turbulence and uncertainty around us. Many models built yesterday fail to generate useful insights to understand today's realities. It no longer makes much sense to set out to predict the future in a globally interconnected, complex and dynamically changing world. Scenarios offer us sets of alternative futures to explore and better recognise risks and known unknowns and help us to realise there must be unknown unknowns and still formulate policies and action plans under deep uncertainty and ignorance.

The scenario process counters group thinking and encourage teamwork and intellectual and social generosity, as one cannot do it on one's own. In order to create different and plausible scenarios, you really need different people. Scenario methods foster feedback learning both 'within the scenario', during the development, and 'with the scenarios', when one goes out and uses the material. Scenario practice offers a process for bringing contradictions from different worldviews into focus and creating shared representations of these that make these discrepancies productive and useful for better understanding complexities of challenges of, for example, food or energy system transformations for sustainability. Such contradictions are often connected to rarely talked-about uncertainties and knowledge gaps and the trade-offs they might represent. Differences of viewpoints are expressed in many places, such as the agenda determined by the problem framing, the criteria for success of a process or project or what constitutes acceptable knowledge. If there is anything that destroys productive enquiry, it is the frustration felt by participants when they cannot agree on what would constitute success, on what a 'right' outcome would be or even on the criteria they should use to judge whether they are going in roughly the right direction. Such a process usually ends up with everybody standing still and arguing. Yet, as we have described, scenario planning offers such a way to bring seemingly contradictory viewpoints into a social learning process and turn disagreements into assets instead of liabilities.

Finally, as a participative and inclusive social learning process of communal creativity, the scenario approach can also help to generate a shared understanding of alternative futures, which then forms a basis for socially robust concerted action. With those who have participated in this interpretation and analysis of the future, it creates a wider context and more 'buy-in' into the policies and strategies that emerge from the work. By helping to capture and accessibly representing a diversity in understandings of multiple possible futures, scenario approaches can help to embrace contradictions, tensions and gaps as creative spaces, in which new ways of thinking and acting can be invented that combine different logics and sets of values for prioritising allocation of attention and resources for action in new ways, for societal transformation for sustainability.

Questions for comprehension and reflection

1   How can we speak meaningfully about the future, as it is by nature unknowable?
2   Describe a current science policy issue where scenarios might be useful.
3   Why do we need all three approaches: forecasting, visioning and scenario planning?

Notes

1 www.wemeanbusinesscoalition.org/about
2 www.cdp.net/CDPResults/carbon-pricing-pathways-2015.pdf
3 The UK Government Chief Scientist John Beddingdon was one of the first to warn of this. World faces 'perfect storm' of problems by 2030, chief scientist to warn. Guardian, 18 Mar 2009. www.theguardian.com/science/2009/mar/18/perfect-storm-john-beddington-energy-food-climate

References

Burnett, V. (2016). Amid grim economic forecasts, Cubans fear a return to darker times. *New York Times*, July 12, 2016.

Burt, G. & van der Heijden, K. (2008). Towards a framework to understand purpose in futures studies: The role of Vickers' appreciative system. *Technological Forecasting and Social Change* 75 (8): 1109–1127. doi:10.1016/j.techfore.2008.03.003

Costanza, R. (2000). Visions of alternative (unpredictable) futures and their use in policy analysis. *Conservation Ecology* 4 (1): 5.

Elahi, S. (2008). 'Conceptions of fairness and forming the common ground', in Ramírez, R., Selsky, J. W. & van der Heijden, K. (Eds.) *Business planning for turbulent times: New methods for applying scenarios*. London and New York: Earthscan. pp. 223–241.

Elahi, S. (2011). Here be dragons . . . Exploring the 'unknown unknowns'. *Futures* 43: 196–201. doi:10.1016/j.futures.2010.10.008

Funtowicz, S. O. & Ravetz, J. R. (1993). Science for the post-normal age. *Futures* 25 (7): 739–755.

Hoekstra, A. Y. (2013). *Virtual water trade*. Proceedings of the international expert meeting on virtual water trade. Value of Water Research Report Series No. 12, IHE Delft.

IRENA. (2015). *Renewable energy in the water, energy & food nexus*. www.irena.org/document-downloads/publications/irena_water_energy_food_nexus_2015.pdf

Jerneck, A., Olsson, L., Ness, B., Anderberg, S., Baier, M., Clark, E., . . . Persson, J. (2011). Structuring sustainability science. *Sustainability Science* 6 (1): 69–82. doi:10.1007/s11625-010-0117-x

Kemp, R. & van Lente, H. (2011). The dual challenge of sustainability transitions. *Environmental Innovation and Societal Transitions* 1 (1): 121–124. doi:10.1016/j.eist.2011.04.001

König, A. (Ed.) (2013). *Regenerative sustainable development of universities and cities: The role of living laboratories.* Cheltenham, UK: Edward Elgar.

Latour, B. (2004). *Politics of nature: How to bring the sciences into democracy.* Translated by Catherine Porter. Cambridge, MA: Harvard University Press.

Ogilvy, J.A (2002). *Creating better futures: Scenario planning as a tool for a better tomorrow.* Oxford: Oxford University Press.

Ramírez, R., Österman, R. & Grönquist, D. (2013). Scenarios and early warnings as dynamic capabilities to frame managerial attention. *Technological Forecasting and Social Change* 80 (4): 825–828. doi:10.1016/j.techfore.2012.10.029

Ramírez, R. & Wilkinson, A. (2016). *Strategic Reframing: The Oxford Scenario Planning Approach.* Oxford: Oxford University Press.

Ravetz, J. (2000). Integrated assessment for sustainability appraisal in cities and regions. *Environmental Impact Assessment Review* 20: 31–64.

Rittel, H.W.J. & Webber, M. M. (1973). Dilemmas in a general theory of planning. *Policy Sciences* 4: 155–169.

Robinson, J., Burch, S., Talwar, S., O'Shea, M. & Walsh, M. (2011). Envisioning sustainability: Recent progress in the use of participatory backcasting approaches for sustainability research. *Technological Forecasting and Social Change* 78 (5): 756–768.

Rockström, J., Steffen, W., Noone, K., Persson, A., Chapin, F. S. III, Lambin, E., . . . Foley, J. (2009). Planetary boundaries: Exploring the safe operating space for humanity. *Ecology and Society* 14 (2): 32. Retrieved from www.ecologyandsociety.org/vol14/iss2/art32/

Seiffert, M.E.B. & Loch, C. (2005). Systemic thinking in environmental management: Support for sustainable development. *Journal of Cleaner Production* 13 (2): 1197–1202. doi:10.1016/j.jclepro.2004.07.004

Senge, P. M. (1993). Transforming the practice of management. *Human Resource Development Quarterly* 4 (1): 5–32.

Sharpe, B. & van der Heijden, K. (Eds.) (2007). *Scenarios for success: Turning insights into action.* Chichester: John Wiley and Sons.

Swart, R. J., Raskin, P. & Robinson, J. (2004). The problem of the future: Sustainability science and scenario analysis. *Global Environmental Change* 14 (2): 137–146. doi:10.1016/j.gloenvcha.2003.10.002

Tattersall, P. (2016). Community-based auditing: A post-normal science methodology. *Nature and Culture* 11 (3): 322–336.

Van der Heijden, K. (2005). *Scenarios: The art of strategic conversation, 2nd Edition.* Chichester: Wiley.

Vervoort, J. M., Bendor, R., Kelliher, A., Strik, O. & Helfgott, A.E.R. (2015). Scenarios and the art of worldmaking. *Futures* 74: 62–70.

Vickers, G. (1983). *Human systems are different.* London: Harper & Row.

Wakeford, J., Kelly, C. & Lagrange S. M. (2015). *Mitigating risks and vulnerabilities in the energy-food-water nexus in developing countries: Summary for policymakers.* Stellenbosch, South Africa: Sustainability Institute.

Wals, A.E.J. (2015). *Beyond unreasonable doubt: Education and learning for socio-ecological sustainability in the anthropocene.* Second Inaugural address held on December 17th, 2015 upon accepting Personal Professorship in Transformative Learning for Socio-Ecological Sustainability. Wageningen: Wageningen University.

WBCSD. (2009). Business in the world of water. www.wbcsd.org/contentwbc/download/2308/29069

Wiek, A., Binder, C. & Scholz, R. W. (2006). Functions of scenarios in transition processes. *Futures* 38 (7): 740–766. doi:10.1016/j.futures.2005.12.003

Wiek, A. & Iwaniec. D. (2014). Quality criteria for visions and visioning in sustainability science. *Sustainability Science* 9: 497–512. doi:10.1007/s11625-013-0208-6

Wiek, A. & Lang, D. J. (2015). 'Transformational sustainability research methodology', in Heinrichs, H., Martens, P. & Michelsen, G. (Eds.) *Sustainability science – An introduction*. New York: Springer. pp. 1–12.

Wilkinson, A. & Eidinow, E. (2008). Evolving practices in environmental scenarios: A new scenario typology. *Environmental Research Letters* 3 (4): 1–11.

7 Social technology and Theory U

Co-creating actionable knowledge for leadership

Isabel Page

Experience is not what happens to a man; it is what a man does with what happens to him.

– Aldous Huxley, *Texts & Pretexts*

Rising to the challenge of sustainable change

Rare is the person who does not foster the ideal of living in a healthy life support system. Rare also is the person who does not reach a point at which ideals are sacrificed. As creative planetary citizens of the Anthropogenic Age, we seem to be co-creating unintended consequences and hostile, bleak and at times insurmountable outcomes. Given low stress levels and a healthy maternal bond, rapid self-directed learning is evident from the moment we are born. Transformative learning and leadership concepts prevail in literature, so why do we hit barriers, apparently unable to take consistent, integrated and sustainable actions for change according to our highest ideals?

Malcolm Gladwell defines a tipping point as "A place where the unexpected becomes expected, where radical change is more than possibility. It is – contrary to all our expectations – a certainty" (Gladwell, 2000). Social technology provides knowledge inspiring higher values and purpose, informing us daily about world conditions; particularly young people are deeply affected by Internet images and content on water, energy, food, waste and poverty. Tipping-point eco-perspective certainties may be supported through analysing such knowledge from diverse perspectives and across generations. Co-visioning and co-generating initiatives to create hypotheses of emergent futures may reduce uncertainties and convey a sense of shared imagery in common desirability, acceptability and plausibility. When uncertainties are made explicit, areas of ignorance may be uncovered and explored and potentially disruptive events identified. Knowledge for action is an outcome of this process and may be effected through early prototyping, iterating constantly through engaged stakeholder feedback to understand and dismantle barriers to change.

Developing the concepts and themes from the introductory chapters, we focus here on this struggle to achieve co-created challenge-driven actionable knowledge based on appreciative judgment of complex perceived realities, values, purpose, differences and sensing of the emerging future. We will examine Theory U, a collective capacity-building framework designed to see, appreciate and foster collaboration

across differences with new eyes, embracing sustainability science's diverse interpretations of meanings and purpose across research, economics, engineering and
urban planning. We will examine innovative content delivery and the impact and
value of social technologies through a massive open online course (MOOC) from
the Presencing Institute, part of the Massachusetts Institute of Technology (MIT) in
the United States. In the context of the certificate and wider interventions, critical
evaluation will be proposed based on how far this framework, its practices, processes
and delivery embrace quality requisites of transformative knowledge co-creation
processes expressed as the key properties of complexity, contingency, contradictions
and trade-offs, uncertainty and ignorance.

Social technology to facilitate collaboration across differences

What knowledge and learning for transformative change?

Knowledge of perceived realities concerning complex 'wicked' problems and the
messes of environmental degradation, economic and financial meltdowns and social
inequity requires an understanding of human–environment relations, and particularly stakeholder communication, as co-created complex dynamic systems (see
Figure 7.1). Systems thinking, advocated by Peter Senge, senior lecturer at the Sloan
School of Management, MIT (Senge, 2006), informs every systemic influence to
be both cause and effect. Prevailing educational systems with largely place-based
environments seem to foster cause-and-effect linear and transmissive perspectives.

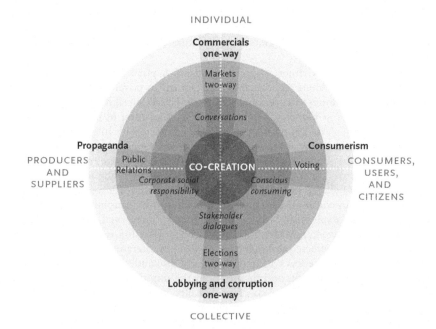

Figure 7.1 Four levels of stakeholder communication in economic systems

In today's fast-changing world, value- and purpose-based learning provides key skills to maintain focus, appreciate empathetically and comprehend multi-facets of challenges from the distinct perspectives of highly diverse experts and stakeholders. A shift is needed away from the unilateral and linear, low on inclusion and transparency and organized by an intention to serve the well-being of the few to co-creative and co-generative communication.

In organizational and social contexts, Senge stresses the anti-democratic nature of traditional leadership, advocating co-generative systems thinking as critical for Gladwell's proposition of context re-framing, transformative leadership development and crucial for adaptive social learning based on individual and collective values and purpose (cf. Box 7.1). For individuals, re-framing solutions to inner tensions through systems thinking may dissolve barriers to change by enabling empathetic shifts in mental models from isolated self-perception. This involves moving from seeing problems as created by those 'out there' towards self-responsibility for contributory actions.

Innovative frameworks, processes and skills, however, need to be integrated into learning content, advertised and disseminated, with social media providing channels to market. Since the turn of the century, combining educational platforms, both formal and informal, with social technologies such as Facebook and Twitter, has fostered communities of change through collective leadership, e.g. Occupy Wall Street, Brexit and the election of Donald Trump, to name but a few. Power and speed of movement connectivity and fast social learning seem to play a part in creating disconnect between established government and citizens who feel less privileged but nonetheless entitled. Ideas may be communicated effectively and at speed. Power over learning seems therefore to be disseminating beyond tertiary-level established educational institutions where quality is decided by an elite and only a privileged few may learn. Instead, individuals and/or communities of change are given fast virtual access to real-world skills for speedy change through, for example, coding, financial management, technology and skills to establish collective leadership.

Collaborating closely with Peter Senge on systems thinking, Otto Scharmer[1] and the team at the Presencing Institute, part of MIT in Boston, have developed Glasl's work on the U-Process at the Netherlands Pedagogical Institute (NPI). Scharmer, a former pupil of Glasl at Witten Herdecke University in the 1990s, has co-created Theory U into an innovative context-analysis collective leadership framework and personal change process that intends to foster improved conscious connections between the ecological, social and spiritual spheres. The Presencing Institute offers fee-paying face-to-face tuition at place-based centres around the globe (e.g. Boston, Berlin, Cape Town and Shanghai). In 2015 he co-founded the MITx U.Lab, a MOOC with innovative O2O (online-to-offline) learning architecture delivery to convene Innovation Labs for systemic change, create knowledge and tools for transformational change and build collective leadership capacity. For the September 2015 presentation, over 50,000 students registered from 185 countries, forming more than 1,000 self-organised virtual coaching circles and more than 560 self-organised local hubs, one of which is established at the University of Luxembourg as an antenna event to the certificate. Recognition of Scharmer for this innovation includes receiving the Jamieson Prize for Excellence in Teaching at MIT (2015) and the EU Leonardo Corporate Learning Award for the contributions of Theory U to the future of management and learning (2016), and being named one of the world's top thirty education professionals by globalgurus.org. Scharmer is co-founding and co-leading one of the biggest educational experiments of this century.

Box 7.1   Systems thinking for learning

The issue

W. H. Deming, the father of Total Quality Management, wrote to Peter Senge in 1999, 'Our prevailing system of management has destroyed our people. People are born with intrinsic motivation, self-respect, dignity, curiosity to learn, joy in learning. The forces of destruction begin with toddlers – a prize for best Halloween costume, grades in school, gold stars – and on up through the university'. Both Deming and Senge believe that educational system outcomes of thriving economic and social success depend on inner mental models and capacities of young students developed by educators (including parents and extended families) and supporting systems' structure.

Theory and methods used (disciplinary origins)

In his book *The Fifth Discipline* (1990, revised 2006), Senge, a Senior Lecturer at MIT Sloan School of Management, proposes an integrated perspective of the world, enabling people to expand patterns of thought, identify collective aspirations and learn continually together based on five component independent technologies, otherwise known as disciplines:

1   Personal mastery: continually clarifying and deepening our personal vision, focusing our energies, developing patience, and seeing reality objectively; (p. 7)
2   Understanding mental models: deeply ingrained assumptions, generalizations, or images that influence how we understand the world and take action; (p. 8)
3   Building shared vision: practicing shared vision through unearthing shared 'pictures of the future' fostering genuine commitment and enrollment (sic) rather than compliance; (p. 9)
4   Team learning: starts with dialogue, the capacity of team members of a team to suspend assumptions and genuinely think together; (p. 10) and
5   'Systems thinking: integrating all disciplines, fusing them into a coherent body of theory and practice'. (p. 11)

Main insights gained

Senge defines systems thinking as 'A way of thinking about, and a language for describing and understanding, the forces and interrelationships that shape the behaviour' (Senge, 2006). Crucial to social and organisational learning, systems thinking enables us to shift our mental model from being separate from to being part of the world, taking responsibility for our actions creating the problems we experience. 'Learning disabilities' actively prevent transformative learning: failure to self-question and recognise purpose, abdicating responsibility, viewing others as the enemy, blaming others, pseudo teams, reactionary leadership for quick fixes (the boiled frog parable), short-termism resulting in lack of appropriate and timely action and not experiencing the consequences of our actions. The withdrawal of politicians responsible for misleading strategies contributing to the UK vote to leave the European Union may be one such illustration.

Salience today

Personal mastery depends on innate and potential self-awareness, self-reflection and ability to change and act on our mental models. Competence in dialogic communication sets our ability for collective discussion. Our creative capacity fosters shared vision identified with collective big-picture and holistic understanding beyond individual and collective agency interests. Systems thinking needs the other four disciplines for integrated long-term sustainable commitment. Further, an education system reducing or eradicating Senge's learning disabilities might allow children to develop interconnected thinking and focus more widely beyond national boundaries to the great scope for innovation at a more integrated global level.

Theory U: co-creating the emerging future

Collaboration at MIT between Peter Senge and Otto Scharmer resulted in the book *Presence – Exploring Profound Change in People, Organizations and Society* (Senge et al., 2005). Inspired by William J. O'Brien, former CEO of the Hanover Insurance Group, the book relied on the following core principle: 'The primary determinant of the outcome of an intervention is the inner state of the intervenor'. One of many emerging collective leadership models (e.g. the Collective Leadership Framework for Ethical Leadership), Theory U claims to provide a critical and challenge-driven inquiry basis to facilitate learning across differences and diversity in groups. Sensing and actualizing an emerging future context based on values and purpose allows possible barriers to change to surface through individual and group dynamics and agendas. Through co-generation and co-creation, practices and processes emerge that are likely to overcome such barriers, resulting in actionable knowledge that may be prototyped and iterated.

As might be expected from conflict-resolution roots developed by Friedrich Glasl, and informing particularly the third level of integrated psychological leadership theory, Theory U's personal growth process reflects Senge's first four disciplines: personal mastery, understanding mental models, shared vision and team learning. Scharmer suggests that greater presence, improved attitude towards others and behavioural flexibility may be attained through raised self-awareness, relating to the core questions of 'Who is my Self?' (my higher Self in the wider context of eco perspective rather than myself from ego perspective) and 'What is my work?' (higher purpose). Understanding four different levels of conversation and inner dialogue constituting interference when engaging with others allows us to become aware of our own blind spots through suspension (opening the mind), redirecting (opening the heart) and letting go (opening the will) (Depraz et al., 2003). The last two connect to our deepest aspirations, allowing us to see cognitively and emotionally, cultivating a willingness to let go of our ego-driven agendas and predetermined goals. As social emergence through economies of creation, these gestures became incorporated in the six inflection points of the U: suspending, redirecting, letting go, letting come, enacting and embodying (Figure 7.2).

Senge's five learning disciplines transpose clearly into the U process: down the left side, co-sensing is claimed to transform perception and mental models, facilitate team learning through dialogue, practice fields and systems thinking. In presencing, on the cradle of the U lies personal vision and transformation of the Self and will to build shared vision. The process up the right side of the U suggests the possibility to crystallise systems thinking, team learning in prototyping and adjustment to co-realise and embed transformative actions.

However, presence and enlightenment are not the only factor in human psychology. It is suggested that presencing's antithesis, social pathology through economies of destruction, called absencing (see Figure 7.2), incorporates Senge's learning disabilities, resulting in a disengaged social pathology leading to three predictable leadership failures both in social, educational and economic organisational sub-systems. Strategy gaps cause failure to respond to external social or competitive challenges. Structural gaps cause failure to differentiate between innovation and expected

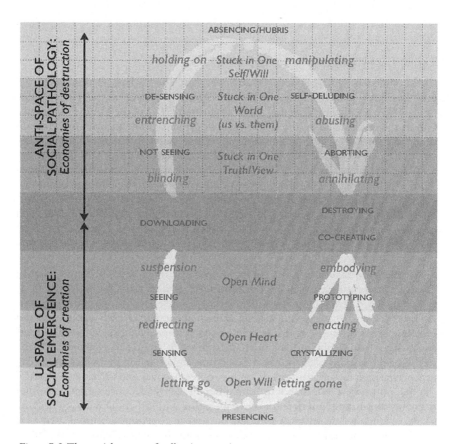

Figure 7.2 The social spaces of collective creation

service at the customer interface, whether the receiver of a service is a child experiencing social or educational poverty or a business client, an operations gap occurs when practices and processes result in outcomes nobody wants. Outcomes from childhood learning systems replicate largely socialised and transmissive communication for perceived positive social gain (being nice/liked/getting what we want).

It is suggested that groups genuinely motivated to reflect and generate shared insights may transcend learning difficulties by engaging in co-generative dialogue (Senge, 1994; Scharmer, 2009). Current participatory process norms based on dialogue and reflective inquiry on situational responsibility – for example, on climate change, poverty or exclusion – may not be sophisticated enough to achieve truly co-generative learning. It is suggested that diverse and distinct experts and stakeholders may achieve co-generative eco-perspective dialogue when sufficiently engaged in purpose and values to set aside personal agendas. Shifting identity from self to the collective through willingness to learn from each other may enable transformation beyond painful ego, and therefore strengthen systemic tipping-point certainty. Appreciating and comprehending multiple perspectives of challenges fostering co-generative learning may enable fast prototyping of ideas and action based on agreed and deeply shared desired outcomes (Scharmer, 2009). However, Scharmer and colleagues' assertion that the outcome of any process can only be as good as the interior condition of the intervenor brings us to individual responsibility.

Individual responsibility is developed through Theory U's personal change process designed to raise self-awareness of absencing in dialogue and offering the opportunity to develop co-generative dialogic skills. Scharmer offers twenty-one social field theory propositions defining four distinct social communication sources

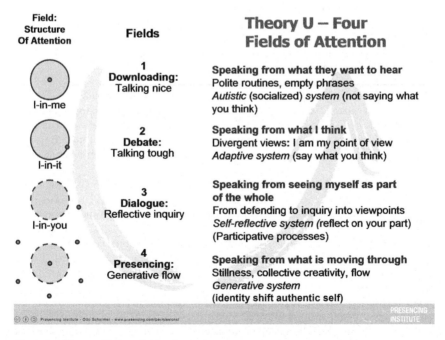

Figure 7.3 Theory U: four fields of attention

of attention, giving rise to four streams or fields of dialogic emergence (Scharmer, 2009, fig 15, p. 236).

Scharmer and Senge suggest our ability to move between attention fields depends on structural conflicts caused by forces external to people in dispute and often developed in childhood. Limited physical resources or authority, geographic constraints (proximity or distance), time (too little or too much) or organisational changes can make structural conflicts seem like a crisis, fostering tipping points causing inner self-interested interference resulting in absence. Internal dialogue is therefore often hijacked by three inner voices, the Voice of Judgment (VOJ), the Voice of Cynicism (VOC) and the Voice of Fear (VOF). These make themselves heard in phrases such as 'I can't do that, I'm too poor/fat/thin/old, etc.' – you can put your own parental script in there (VOJ), 'Well, she's from the north so she would say that'. (VOC), and 'Being at a networking event scares me, I don't know what to say to people'. (VOF). Structural conflict makes itself heard in interruptions when the listener's personal boundaries collapse due to tension: 'Why would you say that?' (projecting our own assumptions and values, often coming from our parents) or 'That's interesting (to me) tell me more', (remaining in the interest of the self rather than empathetically in the interest of the other). Overcoming structural conflict involves consequential recognition of resulting blind spots in behaviour and a disciplinary mental model shift.

Further forces act to dilute communication: depending on individual contexts, it is suggested that words count for around 7 per cent, tone of voice 38 per cent and unconscious body language 55 per cent (Mehrabian & Wiener, 1967; Mehrabian & Ferris, 1967). Theory U provides tools to raise awareness of our unconscious communication in ourselves and with others, amplified in the Social Presencing Theatre developed by Arawana Hayashi. Congruence of all three forces is necessary for positive impact on perception of authenticity to build relationship trust. Moving conversations between the fields of dialogic attention requires a high level of self-awareness, being able to recognising the partner's state and willingness to move between fields and a sophisticated understanding of speech patterns and toxic language.

Gartner defines social technology as

> [a]ny technology that facilitates social interactions and is enabled by a communications capability, such as the Internet or a mobile device. Examples are social software (e.g., wikis, blogs, social networks) and communication capabilities (e.g., Web conferencing) that are targeted at and enable social interactions.
>
> (Gartner Research website, 2016)

To deliver Theory U to a wider audience, in 2015 Scharmer and the team's U.School developed and launched U.Lab, a seven-week course on Theory U incorporating design thinking, into a virtual learning experience and action-based research. In 2017, the course will be extended to thirteen weeks to accommodate better the volume of material and give time for personal and practical skill development. Hosted as a MOOC on edX, the Presencing Institute's (PI) website (www.presencing.com) provides a platform for a virtual learning community through virtual coaching circles. The PI website, Facebook and Twitter provide social technology platforms for individual and community organisation and discussion. These platforms, frameworks and processes may be used by organisations seeking fast change based on deep strategic analysis, including those using business intelligence and analytics technologies for 'Big Data', Theory U providing practices and processes for accommodating social complexity, the non-linear nature of transformation and the need for human judgment (Korhonen & Page, 2015).

In addition to the self-organised virtual coaching circles hosted on the PI website, in an effort to mitigate the norm of the 85 per cent MOOC drop-out rate, value was added through off-line support (as far as possible for free). In 2016 this was provided by over 600 global hubs, one of which is situated as an antennae event of the Certificate in Sustainability and Social Innovation at the University of Luxembourg. Hubs are situated at the nexus of action-based learning (through tools such as sensing journeys, case clinics, and 3D and 4D mapping through social presencing theatre; innovation labs (through co-creating and prototyping); and capacity building (through the 0x and 1x programmes and networking on the Presencing Institute's website).

Hubs are self-organising and may choose how to organise and use the time themselves. At the University of Luxembourg, weekly sessions take the form of tutorials allowing students to catch up with and consolidate the week's learning and experience exercises designed to develop transformative and dialogic skills through deepening presencing and sensing practice (Wegerif, 2013) and using embodied presencing practice to experience collective unconscious processes. They also have the possibility, if they so wish, to have live coaching through the case clinics to raise their own awareness of the levels of listening, the language indicators for each level, and attuning themselves to recognising when the voice of judgment, cynicism and fear appear. The theory sounds like just what the doctor ordered, and substantial success claimed (cf. Theory U case study, Box 7.2) to bring about sustainable change in communities, and innovate education and learning processes, but does the theory stack up in practice?

Box 7.2   Theory U case study

Issue at cause

Healthcare systems in the developed world are increasingly in crisis. The baby boomer generation will only increase the burden of pressure on hospitals and regional healthcare providers both in the private and public sectors. Analysts in Germany warned that their "system will collapse under its own weight" without massive reform. In 1994, in Lahn-Dill, a region of 280,000 inhabitants north of Frankfurt, a hospital research report was published indicating 60 per cent of physicians surveyed felt "inwardly resigned" to job stress and 49 per cent had at least once thought about suicide. Dr Gert Schmidt from Giessen began forming a network of physicians, patients, government and other officials committed to large-scale change, starting in 1996 with a proposed new emergency care system.

Theory U

Between 1997 and 2000 Scharmer and a team worked with the group using Theory U change framework, practices and processes to build the new system. The networked system was helped to see itself through diverse actors 'sensing' collectively how they had created together a system that failed to meet aspirations. Core was the physicians' consideration that the weakest link in the system was their relationship with patients. Analysis led to four levels of care, each appropriate in differing circumstances: repair, therapy, reflection and self-transformation. The analysis and discussions led to developing a physician/patient platform for education and support.

Facilitated dialogue enabled the practitioners to resolve the financial and logistical difficulties of the new care system. Collaborative working relationships were facilitated among partners with previously little or no contact – hospitals, private practice and specialist clinics. Further initiatives were created including agreed sharing of specialized diagnostic equipment, new formats for information sharing between hospitals and external physicians, a care coordination office and specialist quality-improvement groups (e.g. management of heart disease). A second Theory U intervention in 2000 suggested that health and personal responsibility are tied. As patients became more responsible, for example patients developing local kitchens for teaching diabetic diet management, they challenged the dependencies in the traditional system, and physicians began creating infrastructures for reflection on individual patient care, for example differing appointment time length for acute and chronic patients, facilitated office meetings to reduce waiting times, documenting and sharing learning experiences in emergency care.

Source: Kaeufer et al., 2003

Theory U – a critical discussion

Critical discussion on knowledge of perceived purpose, values, the emerging future, action and results needs to focus on quality measures of process and measures of sustainable outcomes. Today, there is no common meaningful unit of measurement or weighting to gain an integrated understanding and repertoire of how knowledge contributes to action on situated problems. However, by taking account of information sourced from differing cognitive frameworks and complex realities, an attempt may be made to assess Theory U through how far it embraces four key requirements to transformative learning systems: (i) complexity, (ii) contingency, (iii) contradictions and trade-offs and (iv) uncertainty and ignorance (König, 2015).

Complexity

How far does adopting a systems perspective and engaging diverse disciplines, including critical and solution-oriented inquiry across the natural, social and engineering sciences, deal with dynamic complexity?

Beyond the boundaries of particular scientific disciplines, dynamic complexity involves value commitments in its diversity in culture, nationality, expectations and norms, gender and unequal social proof. Theory U's contextual framework was designed by Scharmer and Senge with the purpose of providing knowledge on perceived realities through a value-based systems perspective on the emerging future. Although value choices may invoke challenges of reliability and meaningfulness, key is the intention to provide a robust, though lengthy, process dealing with dynamic complexity to obviate quick fixes and provide for early iterative prototyping. When compared with other processes claiming to be human centred, for example, design thinking (DT), the U process seems more sophisticated. DT is used in the MOOC but, as it is more linear in process and assumptions, sits only in the crystallizing and prototyping phase of the right side of the U curve. In its results for the certificate, the Theory U process suggests potentially a more comprehensive and sophisticated process to breaking down systemic barriers, getting deeply into diverse stakeholder personality, revealing personal agendas, mandates and motivations. Policy preference for an inclusive stakeholder approach means success is highly dependent on the willingness to include stakeholders from systemic boundaries; listening deeply to concerns; and providing focus, time and resource attention.

Further, through coaching board members and executives across public, private and non-profit sectors, I have come to understand and share Senge's assertion that our neurophysiology ensures that focus is drawn consistently on day-to-day crisis, allowing creeping destructive change to go unnoticed in social structures and practices, technologies, research and learning approaches. Consequent disconnects and distortions, both at an individual and systemic level, tend to dilute personal power and foster feelings of helplessness. Recent events such as Brexit illustrate a context of disengaged political discourse, establishment leadership entrenched in low discourse and, as the recent U.S. elections indicate, out of touch with citizen ideology and the reality of social inequality.

Empathy emerging from diverse perspectives may turn to cynicism, particularly in young people around perceived adult unwillingness to engage deeply and continually to change behaviours and foster dynamic change. If young people cannot be certain that they can learn key skills with political and social policy support for actionable knowledge, unhelpful emotional disengagement may result.

Contingency

Connecting theory and practice and drawing on and producing place-based knowledge

It is suggested that the deep learning cycle of co-generated place-based actionable knowledge may be produced relevant to the individual group context, whether in the University of Luxembourg or rural Africa. One of the key values of certificate peer groups and U.Lab MOOC hubs is the commitment of learning spaces to define purpose and values, to use theory as pegs on which to hang hypotheses of emerging futures and to connect theory with practice. However, it's my experience that results are highly contingent on quality facilitation to ensure process structure and discipline necessary for context analysis. As MOOC hub facilitators receive no formal training, come from very different backgrounds and disciplines, and there is no requirement for experience in teaching or facilitating, quality and the off-line MOOC experience is inconsistent. With quality facilitation, trust and co-generative open attitude and behaviour in individuals may be fostered and nurtured to reveal willingness to go beyond the 'safe' to arrive at tipping points, or the 'crack', where group dynamics shift from individual to collective and co-generated solutions. This may also be arrived at without facilitation, but seems to take much longer for the group to find its way forward.

The purpose of online education derives from the value of massive democratizing of access to quality learning for actionable knowledge in their own communities. U.Lab's hosting platform edX, limited to universities such as MIT, Harvard, Stanford and University of California, Berkeley, bring the quality assurance of robust academic content, with efficient and effective delivery (right way, time place, online individual support and often for free). Although constantly evolving through iteration, current U.Lab MOOC delivery is rather clunky and time consuming, involving shifts between edX for learning content and the Presencing Institute's website for group hosting, plus Facebook, Twitter, etc., for messaging and crowd thought. Enrolment numbers are impressive; claims about completion rates are, however, obscure. For the neediest consumers, those in rural parts of developing countries without consistent high-speed Internet and access, the value commitment might be diluted, and a policy preference for emerging markets awash with technology may be emerging (e.g. China).

Contradiction and trade-offs

Taking an actor-oriented perspective

Knowledge on values claims to be at the heart of Theory U's actor orientation, based on the questions "Who is my Self?" and 'What is my work?' However, each individual sits in the larger context of society and work. Ensuring all relevant systemic actors or stakeholders, including those with decision power, are included and are heard is crucial to a robust outcome. Further, in dialogue the principle of equal time for each voice must be respected; exclusion will distort the quality of the outcome. This makes interventions time consuming and lengthy, but the trade-off may be policy and value commitment with a higher quality long-term outcome.

Theory U's personal process relies on raising actors' self-awareness and sensing of unconscious processes, particularly through Social Presencing Theatre, building trust through engaging in a deeply empathetic face-to-face interaction. Skype interventions go some way, but rely on reading body language and gestures rather than feeling energetic flow. The current introductory session in the certificate is only a sample; if facilitated well, weekly U.Lab hub meetings allow practice over thirteen weeks in an established group to begin to embed skills of non-toxic language, deep listening and understanding inner interference more consistently.

Realistically, learning and embedding skills to recognise our own blind spots takes a long time, and understanding a situation through anticipatory social learning, whether co-generative or not, does not necessarily mean that competences to change the situation are learned. Theory U assumes that actors are equal in the ability to get beyond personal mandates or beliefs and engage at a deep level with others. My experience of the certificate, MOOC and MBA students has shown that engagement is sometimes dependent on complex diversity norms – questioning a leader's perspective, command or control is heinous for some cultures. Further, changing our inner structural conflict is hard; the value commitment to reality truth, rather than truth perceived through the lens of inner structural conflicts, is painful, but suggests allowing generative creative tension.

Further, I observe a tendency for participants to conceal structural conflicts and establish 'pseudo'-communities where participants play the game of being present, but find themselves confused, lacking understanding and in degrees of absencing. Honest and open dialogue comes at the cost of having to give up saying what we think others want to hear to be nice, liked or get what we want. Revealing and allowing absencing is essential in interventions – if not, it is often expressed by one member of the community and springs onto the stage with surprising vehemence. Pseudo-communities are often revealed at the crystallizing and prototyping stage when actors actively try to reintroduce and impose their agendas, needing an iterative process to reconnect the group. Further, contradictions in our educational and organizational systems create both internal and external tensions, for example, in the expectation to problem-solve co-creatively and co-generatively with little individual attribution while having to compete individually for reward and promotion. Trade-offs therefore are made by leaders on a daily basis.

Conclusion

We framed this chapter with the question of why, when transformative learning and leadership concepts prevail in literature, do we hit barriers, apparently unable to take consistent, integrated and sustainable actions for change according to our highest ideals? This analysis of four requirements of transformative sustainable social learning systems through the lenses of Theory U highlights how fundamental are the transformations called for at the personal, organizational and systemic level. Social technology provides us with the means for global interconnectivity, and Theory U with a challenge-based collective leadership framework and process through which sustainable actionable knowledge may be co-generated. Yet ego-generated fear, judgment and cynicism are examples of factors that contribute barriers to transformative learning. Theory U provides us with a personal change process designed to raise self-awareness to help address those ego-based challenges, optimistically assuming all may benefit. However, cultural norms; the difficulty of overcoming childhood structural conflicts; and the lack of effective process facilitation, time and resources are some examples of barriers to the critical self-awareness of how we relate to ourselves, others and our surroundings. Research is needed to better assess and evaluate outcomes of learning and behavioural change through these programmes. Further intervention development may be useful to deepen awareness of and modify positively our speech patterns and toxic language and to foster deeper understanding to avoid creating pseudo communities. There is no magic bullet that will instantly propel us to higher levels of awareness, the essence of transformativity as we understand it, but, for a growing number, Theory U seems to be providing a compelling and effective means to that goal.

Questions for comprehension and reflection

1   What might your learning disabilities and inner structural conflicts look like, and how are they affecting your attitude, decisions, behaviour and actions towards others and our ecosystem?

2   Concerning the concept of absencing, how easy is it for you to hear the Voice of Judgment, Voice of Cynicism and Voice of Fear when they show up in your life? Journal daily examples.

3   Concerning the concept of being present, how aware are you of the inner interference that breaks your attention when listening to others? Are you tempted to interrupt before others finish speaking or fail to draw in quiet people present in meetings and discussions? How far do you consider your opinion as the last word? How far are you able to differentiate where you are operating in the four levels of attention? Journal daily examples.

Note

1 Senior Lecturer at Sloan School of Management, MIT, Thousand Talents Program Professor at Tsinghua University and co-founder of MIT's Presencing Institute, Otto Scharmer also chairs the MIT IDEAS program.

References

Depraz, N., Varela, J. & Vermersch, P. (Eds.) (2003). *On becoming aware: A pragmatics of experiencing*. Amsterdam and Philadelphia: John Benjamins Publishing Company.

Gladwell, M. (2000). *The tipping point: How little things can make a big difference*. Boston: Little, Brown & Company.

Glasl, F. & de la Houssaye, L. (1975a). *Organisatie-ontwikkeling in de praktijk*. Amsterdam and Brussel: Agon Elsevier.

Glasl, F. & de la Houssaye, L. (1975b). *Organisationsentwicklung: Das modell des instituts für organisationsentwicklung (npi) und seine praktische bewährung*. Bern and Stuttgart: Verlag Paul Haupt.

IT Glossary. *Gartner Research website*. (2016). Retrieved from www.gartner.com/it-glossary/social-technologies/ (Accessed 16/05/2016)

Kaeufer K., Scharmer, C. O. & Versteegen, U. (2003). Breathing life into a dying system: Recreating healthcare from within. *Reflections: The Sol Journal on Knowledge, Learning, and Change* 5 (3): 1–12.

König, A. (2015). Towards systemic change: On the co-creation and evaluation of a study programme in transformative sustainability science with stakeholders in Luxembourg. *Current Opinion in Environmental Sustainability* 16: 89–98.

Korhonen, J. J. & Page, I. (2015). *There is more to intelligent business than business intelligence.* IARIA 2015, Bustech 2015: The fifth International Conference on Business Intelligence and Technology.

Mehrabian, A. & Ferris, S. R. (1967). Inference of attitudes from nonverbal communication in two channels. *Journal of Consulting Psychology* 31 (3): 248–252. doi:10.1037/h0024648

Mehrabian, A. & Wiener, M. (1967). Decoding of inconsistent communications. *Journal of Personality and Social Psychology* 6: 109–114.

Scharmer, C. O. (2009). *Theory U: Leading from the future as it emerges*. San Francisco: Berrett-Koehler Publishers, Inc.

Senge, P. M. (1994). *The fifth discipline fieldbook: Strategies and tools for building a learning organization*. New York: The Crown Publishing Group.

Senge, P. M. (2006). *The fifth discipline: The art & practice of the learning organization*. New York: Doubleday (Random House).

Senge, P., Jaworski, J., Scharmer, C. O. & Flowers, B. S. (2005). *Presence: Exploring profound change in people, organizations and society*. London: Nicholas Brealey Publishing.

Wegerif, R. (2013). *Dialogic: Education for the internet age*. London: Routledge.

8 Staging design thinking for sustainability in practice

Guidance and watch-outs

Kilian Gericke, Boris Eisenbart and Gregor Waltersdorfer

Human-centred design for sustainability

Challenges of sustainability not only affect the environment, but in particular the global economies as well as societies (United Nations Department of Economic and Social Affairs, 2013). Poverty, inequality, food security, and energy transformation are among these challenges. They require a coherent dialogue between citizens, governments, industries, and science pertaining to the development of strategies and solutions that have the potential of changing the current situation in a lasting and sustainable manner. Technological innovation alone will not suffice to bring about the required changes to reconcile our societal metabolism with the planet's limited biophysical-carrying capacity without a fundamental change in social structures and practices.

A major goal, while addressing these challenges, is to replace existing technologies and associated structures and social practices with more sustainable ones that may very well be disruptive towards established ways and do not necessarily have to have evolved from the current structures and practices. In fact, in order to achieve the progress that is required, mankind will have to re-design and re-think many of the solutions that are currently in place in their entirety (Charter & Chick, 1997). The development of such sustainable solutions requires a different way of looking at the problems societies face today and opening up to potentially disrupting established ways of living. More specifically, the effects of problems and also the desired and undesired effects of (potential) solutions on the environment, global economies, and society along and beyond their lifecycles have to be understood holistically in order to develop adequate, sustainable solutions (McAloone & Andreasen, 2004; Tan, 2010).

One approach to analyse problems of sustainability and to develop appropriate solutions is to apply an explicit human-centred design approach with the purpose of changing apparent, ecologically destructive behavioural patterns. Human-centred design is an approach that particularly aims at addressing prevalent as well as sometimes less salient, subliminal needs and desires of human beings with solutions that are both technical feasible and that can lead to viable business models (Brown, 2008). By using these specific approaches, resulting solutions may not only be adequate in terms of meeting users' needs; they may at the same time ensure ecological sustainability. Therein, human-centred design unfolds its full potential by changing the way people behave and consume the limited resources of this planet.

In line with requisites for transformative social learning to enhance our understanding and repertoire of action on complex challenges with both social and technological/

environmental dimensions, human-centred design typically requires multidisciplinary teams, which have to develop an empathic understanding of user requirements and suitable solutions in an iterative manner by actively involving relevant users into their reasoning and deliberation processes (Maguire, 2001; Brown, 2008).

The underlying principles and methods of human-centred design are used in a variety of design disciplines, such as industrial design, software development, and business design, and are nowadays applied even in disciplines that would traditionally not necessarily be considered design disciplines, such as management and behavioural economics (Rumelt, 2011). The adaptation of human-centred design, as well as further strategies and methods from product development, to non-traditional disciplines is often referred to as design thinking.

Design thinking as a human-centred design approach offers people without formal training in design a mind-set and a toolkit for addressing problems and developing coherent solutions by applying processes, principles, and methods of designers (Martin, 2009; Beckman & Joyce, 2009; Brown & Katz, 2009; Plattner et al., 2009; IDEO, 2011b; Brown, 2008).

Taking such a designerly approach and applying a human-centred perspective in solution finding is expected to lead to new solutions and enhance re-thinking of existing ones.

A good example for the potential of human-centred design is the awarded project by ideo.org and the American Refugee Committee called "Asili". Initially starting out to improve childhood mortality rates in the Democratic Republic of the Congo, the designers, by taking a holistic human-centred design approach, ended up co-creating a community-owned system of social enterprises providing healthcare, clean water, farming cooperatives, and high-quality seeds and by that stiumlated the regional economy. Design decisions such as taking a systemic perspective, a membership pricing model, and transparency for all aspects of the enterprises contributed to the success of the project (IDEO, 2015, p. 129ff).

In the context of the educational program on sustainability described in this book, peer group projects are used to deepen the knowledge about problem-solving strategies and approaches and to develop solutions for real-world problems. As with almost all design processes, peer group projects are complex, iterative, and ill defined (Maier & Störrle, 2011). As a consequence, the project team will have to cope with a co-evolving understanding of the addressed problem and developed solutions (Maher & Poon, 1994; Dorst & Cross, 2001; Lawson, 1997). Design thinking is used in this context as an approach that enables participants from diverse disciplines with different levels of experience to engage with challenges of sustainability and to develop sustainable solutions.

This new design thinking movement became popular as a business strategy. It should be distinguished from traditional design thinking research, which has a focus on developing theoretical insights into designing as a cognitive act. Although the focus of both areas is somewhat different, both can benefit from taking into account the achievements of the respective other (Badke-Schaub et al., 2010).

Design thinking

Design thinking is an approach that uses the designer's sensibility and methods for problem solving to meet people's needs in a technologically feasible and commercially viable way. In other words, design thinking is human-centred innovation (Brown, 2008).

Design thinking is understood as the interaction of three elements: multidisciplinary teams, working space, and process (Plattner et al., 2009; IDEO, 2011b).

Multidisciplinary teams

The development of technical products requires considering different perspectives. Therefore, it is recommended to select team members with different backgrounds, that is, involving team members from technical disciplines (e.g. mechanical engineering and/or computer science) and members from non-technical disciplines, such as philosophy, economics, etc. In this way, part of the team will have no expertise with existing solutions for the addressed problem. Including economic, humanistic, and technological perspectives by representatives enables a holistic analysis and assessment of the problem to be solved and the solution found. This will help the team to avoid design fixation (Crilly, 2015).

A typical team should encompass three or more people. Although different genders and ages can be useful, all team members should belong to the same hierarchical level, thus having equal rights in the project. Equality is important to avoid inhibition and enables open exchange of thoughts and ideas (Sunstein & Hastie, 2015; Powell et al., 2011).

Workspace

The working environment is considered a crucial factor. The design of workspaces has an important influence on the creativity of the team. The working spaces should ideally have the following features in order to create an informal and joyful atmosphere: bright and colourful rooms, break-out areas for relaxation and communication, issued prototypes, pictures, literature, and other material (Plattner et al., 2009; Brown & Katz, 2009).

Because of the multidisciplinary team composition, team members have no common terminology. Therefore, prototypical implementation and visual communication of ideas is promoted in design thinking projects. Sticky notes are used for quick and easy documentation of ideas. Visual communications, as well as rapid prototyping, ensure that misunderstandings between team members are minimised (McKim, 1972).

Furniture supports the creative work by being easily adaptable to the needs of the team. Whiteboards and tables are preferably mounted on wheels. Materials such as foam, wire mesh, wood, but also Lego bricks and items from the everyday environment, should be available to create prototypes.

Process

The third element of design thinking is the design process. Multiple different process models exist, which vary in the number of phases and the visual representation. They essentially follow the same underlying logic, which will be described hereafter using the process model as proposed by Plattner et al. as one illustrative example (see Figure 8.1). The process describes six phases (Plattner et al., 2009). It starts with a thorough analysis of the problem that motivates the design challenge, the problem's context, and effects on people and the environment along the entire life cycle. Based on an empathic understanding of users and people affected by the existing problem, a variety of solutions are developed. Solutions may be products, services, or a combination of products and services, so-called product–service systems (Müller et al., 2009; McAloone & Andreasen, 2004; Tan, 2010). The process results in the development of prototypes for innovative products, services, product service systems, or business models, enabling testing and collecting feedback.

It is important to note that the design process is not linear or sequential. It is always possible, in every phase and for each activity, to iterate the last activity, the last phase, or even to start the whole process again if it is expected that the iteration will lead to a better understanding of the problem or to a better solution.

During the first three phases (Understand, Observe, and Point of View) the addressed problem will be analysed and reframed. This reframed problem description is called "Point of View". The last three phases (Ideate, Prototype, Test) aim at developing and selecting a suitable, viable, and desirable solution. Starting from the Point of View, solutions are generated, prototypically implemented, and tested. Preferably, intended users and other persons affected are involved in these tests. The six phases of the design thinking process are briefly described in Table 8.1.

Deliverables

The design thinking process typically ends with the creation and test of a prototype representing the developed solution. Depending on the characteristics to be examined during the Test phase, different types of prototypes can be created (see Table 8.2).

The detailing and implementation of a prototype are afterwards done using more conventional design/engineering approaches, usually involving discipline experts. The coupling of the design thinking process with conventional design processes differs greatly depending on the solution's emphasis on using services or different technologies, such as software and mechatronics. For simple solutions that are based purely on services, the design thinking approach may result in a prototype which is near to implementation, whereas for solutions using mechanics, software, or other technologies, the result will be a prototype that allows designers to run a feasibility study preceding a conventional design project (Gericke & Maier, 2011).

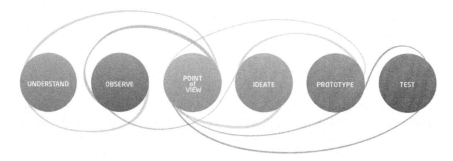

Figure 8.1 Design thinking process
Source: Plattner et al., 2009

Phase	Description
Understand	The starting point of the design thinking process is a brief description of the problem, the 'design challenge'. The design challenge is formulated by the project sponsor, i.e. not by the team itself. The design challenge is phrased as a question, beginning with the phrase 'How might/could we...?'. The Understand phase is intended to create a knowledge base regarding the context of the problem in an open-minded manner. During this phase, the team collects information surrounding the problem. The more diverse the considered information is, the more likely it is to find relevant issues and previously unnoticed possibilities. Identified contradictions serve as a source of inspiration for changing perspectives and in-depth questions. Furthermore, state-of-the-art solutions and trends in the market will be studied.
	One method that enables immersion and gaining empathic understanding of problems is 'self-experience', where team members simulate central problem situations or undergo them in reality, in order to gather first-hand experiences. The gained understanding of the problem is useful for preparing the Observe phase.
Observe	The central activity during the Observe phase is the observation of users and people affected by the addressed problem. Information from the Understand phase is reviewed for its relevance and complemented by new insights gained during observation.
	A variety of methods can be used for observing users, for example, the search for 'workarounds'. Workarounds are improvised solutions developed by users or people affected by the problem.
	Besides observing people, qualitative interviews can be conducted. The conversation has an explorative character without a fixed interview guideline.
Point of View	The Point of View is the central interface between problem definition and solution generation. During the development of the Point of View, all information previously obtained is collected and placed in relation to each other. In doing so, higher-level needs and more abstract statements about the problem can be generated. Reframing the problem (i.e. shifting the boundaries of what is considered as being part of the problem) enables the development of innovative solutions.
	The result of this phase, the Point of View, is a set of criteria for the later evaluation of solution concepts, a description of the target user group, central user needs and requirements.
Ideate	In preparation of ideating possible solutions, based on the reframed problem and the Point of View, a clear problem formulation needs to be created. The problem is formulated as a question beginning with the typical solution-directed question 'How might we...?'
	During the Ideate phase, solutions are generated mainly using the classical brainstorming method (Osborn, 1957). In this phase it is recommended to develop as much ideas as possible.
	After solution generation, the solutions are evaluated comprehensively. The evaluation results in a reduced and prioritised set of ideas. The evaluation of solutions will be based on criteria as formulated in the Point of View. The evaluation should be done from the user's perspective, as the user will later decide whether the solution is satisfying their needs. The teams subjective experiences gained during the Understand and Observe phase support this process. Many of the aspects, based on the empathic understanding of the user, which are relevant for selection, are difficult to communicate, but important for the evaluation. This subjectivity may be interpreted as arbitrariness but it is an important enabler for fast iterations and progress towards a human-centred solution.
Prototype	Prototyping involves building a model for visualising the proposed solution, with simple means such as cardboard, storyboards, or Legos. Prototypes should represent all central characteristics of the solution critical for satisfying the user needs and solving the addressed problem.
Test	A prototype enables testing as a final evaluation. If the prototype fails during tests or if new insights lead to better ideas, the last phases will be iterated. In the Test phase, the prototypes created previously will be tested with the involvement of potential users. The design thinking approach consciously takes into account that users will not immediately accept the first prototype. As it is expected that iterations are required (perhaps including a revision of the Point of View and a repetition of the Ideate phase), not much time should be invested in the creation of prototypes at this stage.

Table 8.2 Different types of prototypes

Phase	Description
Context	A context prototype conveys the interaction between the solution and the surrounding area. This can be, for example, by a film from the perspective of the person to be addressed, illustrating how the new solution is integrated into everyday life, and what value it generates. How the solution works is of secondary importance and is therefore ignored in the implementation of this type of prototypes.
Behaviour	A behaviour prototype conveys the physical, optical, or acoustic behaviour of a solution concept. However, the mechanisms that produce this behaviour can differ from those proposed in the solution concept, as long as this modification will not bias the test results.
Action	Solutions in the form of technical artefacts can be represented as action prototype. These prototypes do not correspond to the final form and do not have all the envisaged functions, but represent the central properties to be realised by the solution concept (i.e. the technical concept, the action, or the mechanism).
Form	Form prototypes offer a means to evaluate aesthetics, shape, or haptic of solution concepts.

Table 8.3 Selected methods applicable during the different phases

Phase	Methods (examples)
Understand	Self-experience, content analysis, mind map, system map
Observe	Photo diaries, interview, DYAD, focus group, card sorting, day in the life, fly on the wall
Point of View	Reframing, personas, empathy map, affinity diagram
Ideate	Brainstorming, design swarm, SCAMPER, gallery method, synectics, analogies
Prototype	Storyboard, scale model, tangible prototype, role playing, dark horse, wireframe, paper prototypes
Test	User tests, interview, user observation, usability test, user diaries

Methods

An overview of methods that are used in the process according to the design thinking approach is described in a variety of publications (IDEO, 2011a; Ulrich & Eppinger, 2007; Gundy, 1988; Jones, 1985; Baxter, 1995; IDEO, 2003; Boeijen et al., 2014; IDEO, 2011b; Osborn, 1957; Roozenburg & Eekels, 1995; Pahl et al., 2007). A small selection of methods, which can be used during the different phases, is provided in Table 8.3.

Design thinking and transformative social learning

This section outlines what the concepts design thinking and transformative social learning have in common and how they can enrich each other for their joint application within the educational framework described in this book. Basically, transformative social learning can guide the application of design thinking, whereas design thinking can provide a learning environment and can contribute with its methods to the learning process.

Social learning can be conceived as a way to organise the learning by individuals, organisations, and communities in social aggregates (Wals, 2007). It is achieved through the interaction of people with diverging interests, beliefs, and worldviews and often described as learning focussed on the process rather than the outcome (Wals, 2007).

Learning is transformative when actors become critically reflective through dialogues on content, process, and premise of perceptions (Taylor, 2009), ultimately transforming their frames of reference, as structures of assumptions and expectations (Mezirow, 2009).

This critical thinking guides the application of design thinking, which in this case also involves questioning the constraints on a solution. Through that, design thinking can contribute to changing paradigms, described by Sterling (2011) as transformative or third order learning.[1]

Interestingly, the processes of social learning (Wals et al., 2009) and design thinking are similarly conceptualised. Both involve sequences of divergent and convergent thinking, and the underlying process structure of analysis, synthesis, and evaluation. Additionally, both processes emphasise the importance of iterations within the process, for allowing reflections and creativity, and for coming up with solutions through new ways of thinking (Wals et al., 2009).

From a theoretical perspective, design thinking can contribute to transformative social learning in many ways. First, as described earlier, design thinking offers a variety of methods in each process phase. Second, it creates a learning environment allowing non-formally trained people to engage with practical problems and opportunities in the real world and with special attention on the workspace. Finally, during the practice of design in general, actors are often faced with "trigger" events, for example, when developed solutions do not achieve the expected results, which challenges their assumptions, potentially leading to perspective transformations (Mezirow, 1990).

However, as Wals (2007) emphasises, the challenge for transformative learning, as the key to a more sustainable world, lies in being "willing and able to question existing routines, norms values and interests". The following sections describe the application of design thinking for transformative social learning in practice and reflect on the process and the results.

Bringing design thinking into action

Whereas understanding the logic, the main concepts, and methods of the design thinking approach is easy, the application and especially facilitation require experience and training. The more experienced the team members are, the more effective the process will become, as the attention will shift from applying a method to creating the desired solutions.

In the context of the program described in this book, the implementation of the approach consists of two parts:

- Workshop format:
  - Introductory presentation of the design thinking approach and process.
  - First application of the approach and selected methods in a one-day workshop.
- Peer group project: Application in a two-semester peer group project.

The two parts of the workshop are directly connected. These two parts are intended to provide a basic understanding of the approach before the project team will apply it on a more complex design challenge in a peer group project guided by an experienced facilitator.

The workshop and the peer group project will be explained in more detail in the following sections and illustrated using examples from past semesters (Boxes 8.1 and 8.2).

Workshop format

The aim of the workshop is to familiarise the participants with the design thinking approach and to introduce some selected methods. The idea is to shift the mind-set of the participants from seeing the challenges of sustainability as large-scale problems, which are almost impossible to solve by a small group of non-experts, to a mind-set where they search for ideas, which may be the basis for large-scale changes. This resonates with the ideas behind social learning for achieving a more sustainable world, which is not pre-determined by experts (Wals et al., 2009).

First experiences in developing solutions for small-scale challenges create the particular attitude, which is required to address bigger challenges in the future. These first experiences motivate and foster the required self-confidence to start bigger projects. The workshop is a means to create these first experiences.

Preconditions

The workshop format described is a full day workshop for groups of up to 30 participants. It is recommended to have one facilitator for each sub-group of five to six participants and a separate working space for each group.

The facilitators should be familiar with the design thinking approach and should have experience using the methods applied during the workshop. The participants should have different backgrounds.

For moderation of the workshop, standard material like whiteboards, paper, and sticky notes are required. In addition, a selection of prototyping material is required, such as paper, cardboard, wire, and foam as well as tools, such as scissors, glue guns, and cutters.

Process and methods

Due to the constrained time frame of having one day (i.e. six to eight hours) for the initial introduction of the design thinking approach and the workshop, the process needs to be adapted and simplified (see Table 8.4).

Table 8.4 Simplified process for the workshop

Phase		Activities	Methods (examples)	Results
Introduction (60 min)		Introduction of design thinking approach		
		Presentation of design challenge		
Workshop	Part 1–60 min **Problem analysis**	Explanation of method(s)	User observation Interviews	Affinity map Reframed problem (PoV)
		Reflection on own experiences Observation on campus	Affinity diagram	
	Part 2–60 min **Ideation**	Explanation of method(s) Idea generation	Brainstorming Gallery method	Set of different of ideas
	Part 3–120 min **Prototyping**	Explanation of prototyping techniques Selection of favoured ideas Prototyping		Prototypes
Presentation of results of all groups				

A possible simplification is to organise the workshop into three blocks, preceded by an introduction of the approach and design challenge and succeeded by a closing session. The first block is focussed on understanding and analysing the underlying problem that motivates the given design challenge. These activities result in the formulation of the Point of View. The second block focuses on the generation of solution ideas. The third block focuses on prototyping favoured ideas. The workshop will close with a presentation of all prototypes and a reflection on the process, the results, and how these results (the prototypes) could be tested.

Although iterations are possible within these blocks, the workshop format restricts iteration across the blocks. The facilitator should discuss during the whole workshop situations where iterations would make sense. It is important that the participants understand that the process will be different for every project and that they should iterate, if they either identify gaps or contradictions in their knowledge base or identify new opportunities for improving the outcome of the activity. At the same time, it is important to respect given time limits.

Each block starts with the introduction of the method(s) used in this phase. The methods should be simple to explain and to facilitate in order to avoid distraction from the design challenge by problems in understanding the method. More advanced methods can be used during the peer group project. Furthermore, it is recommended to use methods which support the documentation of intermediate results, such as affinity diagram and gallery method (Hellfritz, 1978; Pahl et al., 2007); otherwise, the participants invest all their energy in developing an understanding of the problem and addressing hundreds of different aspects without making notes and losing the overview of their previous insights and decisions.

Box 8.1   Example of a workshop: waste segregation on university campuses

Design challenge

Students and staff on university campuses create waste. Although one challenge can be to reduce the amount of waste generated, another typical design challenge can address the recycling of these materials. Efficient recycling starts with the segregation of the waste into different groups of materials, such as paper, plastics, metal, glass, and residual waste. Depending on the country in question, the groups of materials pertaining to segregation systems may vary. Still, they are typically collected in trash bins of different colours to make segregation seamless for users.

University campuses are a melting pot of different nationalities. That means students and staff associate different colours with the same group of material or have different experiences regarding waste segregation (perhaps they have never done this before). A resulting problem is that the waste man agement system does not work properly as groups of materials are mixed, requiring manual segregation of the waste by the facility management, which is unpleasant and expensive.

A typical design challenge for a one-day workshop is formulated as follows: "How might we support that waste is segregated correctly on the campus?"

Process

The participants were found to usually understand the problem immediately and started reflecting about their own experiences and observations with waste segregation systems. As the design challenge addressed a problem form their everyday life, the participants were motivated.

The participants usually identified aspects as those mentioned before quite quickly and identified further causes which may lead to an incorrect waste segregation, such as laziness or lack of awareness, the particular design of the bins, and imprecise explanations of material groups.

If possible, the participants should analyse the waste bins next to the area where the workshop is held. This small shift of perspective from seeing pictures of waste to seeing the actual waste (i.e. taking the perspective of the facility management personal) was useful to make the problem more salient as it enhances the engagement with and empathetic understanding of the issue at hand. The participants understood that not just the people who want to get rid of their waste are affected by any solution developed for that problem. A way to increase this effect was to invite a representative from the facility management team to introduce the design challenge, presenting figures and facts about the volume of waste produced on campus and the related costs.

Results

Depending on the backgrounds, the capabilities of the participants, and the developed Point of View, the solutions varied greatly from changing the colour scheme, re-designing the trash bins, and information campaigns for staff and new students, to social campaigns aiming for solutions which include a reduction of the amount of waste produced.

Peer group project

The peer group projects are an integral part of the program in both semesters to allow transformative social learning by involving collective problem-based inquiry in practice and cycles of action and reflection in diverse learning environments (König, 2015). They provide participants hands-on experience by working in a self-determined way with the possibility to go more into detail on challenges in sustainability and by that to get a deeper understanding of both the topic and also the design thinking approach.

This section discusses what and how design thinking can contribute to peer group projects. Basically, design thinking is one approach among others to conceiving and structuring peer group projects. Due to their strong emphasis on practice, in which basically everything can be investigated, peer group projects especially need guidance on finding a focus. This is where the design thinking process can help, because it requires first analysing the current situation of the design challenge and then finding a new Point of View on the challenge for the whole group by synthesising the findings, before going into ideation for problem solving.

Preconditions

As required by the course program, the peer group project takes two semesters, with nine weeks per semester for the project schedule. The project starts after the workshop. The workshop is planned as an introduction to design thinking and to its process and the application of distinct methods. Compared to the workshop, the peer group project offers the opportunity to address a challenge in more depth by spending more time and performing more activities on the topic.

Each peer group should have its own facilitator, who needs to be knowledgeable about the design thinking approach. The facilitator defines the design challenge, but does not necessarily have to be an expert in the area of the problem. It is based on the facilitator's own observations and is set in accordance with the program coordinator. The challenge should be familiar to many participants and not too restrictive.

No group should have more than seven members. The ideal size is four to six participants. From experience, the composition of the peer group cannot be preserved over the whole year. Therefore, additional resources need to be allocated during planning for the hand-over between groups.

Process and methods

The participants can plan the process by themselves within the framework of the design thinking approach so that they see the project as theirs, are motivated, and take responsibility. The facilitator gives feedback on their planning and provides methodological recommendations.

For the planning of the two semesters, the basic idea is to split the project into two main phases: concept development for the first semester, and refinement and implementation for the second semester. In both semesters, the participants run through the design thinking process. As a result, new group members who join after the first semester are given the chance to influence and refine the pre-developed concept by iterating the design thinking process. The participants are requested to present the results to the other participants of the program at the end of each semester.

Box 8.2   Example of a peer group project: furniture exchange

Design challenge

In the following, one case is reported in which design thinking was applied in an educational and collaborative working environment. The initial challenge of this peer group project was to develop a system for fostering the re-use and recycling of furniture, which students leave behind when moving out of the student dorms run by the university. The operational aim of the solution was to contribute to sustainable development by reducing waste. The participants had to work on a predefined problem with a rough description as a starting point. The problem definition was kept simple by intention in order to leave room for the participants to rephrase their Point of View. The envisioned solution needed to be innovative, desirable, and viable so the participants had an initial direction. These attributes also served as evaluation criteria.

Process

The design thinking process was used as a basic outline of the project. At the kick-off meeting, the group facilitator introduced the topic, set the time frame for the project, and recapitulated the design thinking process. Additionally, the facilitator clarified his or her role as a mentor, supporting methodological and procedural questions, but being neither the leader nor the person in charge of the project. Rather the project responsibility was handed over to the group after the kick-off. To get the project started, the facilitator assigned different roles to volunteers. The roles were coordinator, note taker, moderator of meetings, and IT appointee. In the second semester two new roles were added: one role covered the scientific branch of the project, making connections to the theoretical background of the program. The other role was an observer, who was introduced to boost learning, team cohesion, and self-management by stimulating regular reflections on the project. Moreover, in the second semester no coordinator was defined by the facilitator, because the new group composition seemed equally motivated. Instead more emphasis was put on team building in order to provide a cooperative working environment.

Because design thinking is a human-centred approach, the peer group was advised to conduct a stakeholder analysis. The group also ran two interview studies on future users: first to identify their needs and the requirements of the solution and also to determine the amount of items that can be potentially re-used and recycled, and second to explore their willingness to use the developed solution, described as a preliminary concept. Furthermore, close interaction with other stakeholders, such as the university administration and potential cooperation partners, was sought by the participants to gain new insights during the analysis phase and receive feedback during testing.

As it turned out, the group was "forced" into co-creation by chance, because some of the peer group members lived in student dorms. This provided access to the dorms and to reach more inhabitants easily during field research. The reframing of the challenge by defining the new Point of View after the analysis phases (Understand, Observe, Point of View) proved to be of particular importance, because it allowed the group members to aim for a new direction based on a deeper understanding of the problem and its context. For example, the group found out that not just furniture is left behind. The prototyping of flow charts, videos, and websites enabled the group members to communicate their ideas for collecting feedback and supported the presentation at the very end. Moreover, the participants framed and described the final solution within the business model canvas (Osterwalder & Pigneur, 2013).

Results

Two alternative concepts were developed in parallel in the first semester, because the group was split in two for the ideation phase. During the analysis phases in the second semester, one of these concepts then turned out to be more feasible and was chosen for further exploration. Based on that, the participants developed three variations of that concept. These variations were presented to the university administration and afterwards refined to a single solution.

To utilise the full potential of the design thinking approach, the facilitator reminded the participants regularly to be open for different solutions and avoid jumping on a single solution too early. The results proved that this was critical for making grounded design decisions. Comparing the results of both semesters, the developed solutions became more mature. This supports the principle of design thinking to iterate for refining ideas and concepts. For instance, it was not clear during the first semester if the solution should be non-profit or for-profit. The additional understanding gained during the analysis phase in the second semester highlighted new constraints for the solution, such as legal requirements on redistributing electric devices. Beyond design thinking, participants developed and advanced several skills, such as for communication, project management, ICT, and team work. The fact that some participants even continued working on the project between the semesters showed that design thinking allowed a structured and also engaging approach to solving a problem.

All in all, through the support of the design thinking approach, the final solution was a thorough concept that considered the new Point of View, which was gained through the analysis phases, integrated several existing resources identified by a stakeholder analysis, was tested among different stakeholders and refined several times.

Lessons learnt and recommendations

Design challenge

Although the theme "fail early, fail often" coined by IDEO is valid, it is important that the workshops result in the development of solutions. The sense of achievement that comes with developing one or more solutions for a given design challenge is motivating for the participants and creates the necessary confidence that is required for tackling other, even bigger, design challenges.

Therefore, it is recommended to select design challenges, especially for the workshop, that address problems that were experienced by the participants in the past and thus relate to participants' own lives (opposite to purely academic training tasks) and therefore easily create an intrinsic motivation. Problems that can be observed and experienced on a university campus usually fulfil these requirements. The motivation can be increased when there is a real potential that good solutions will be implemented.

Alternatively, it can be motivating to invite external professionals to present a design challenge. Here the strong relationship to reality and the involvement of external clients is motivating. A challenge is that participants may not possess relevant prior experience pertaining to the problem, which requires close guidance by the external professional.

Independent from its particular source, the challenge addressed during the workshop should not be overly complex. The analysis of the underlying problem and its context should be manageable by the team within the allocated time frame. The peer group projects offer the right format for more complex challenges and are less dependent on prior self-experiences.

Workshops

Prototyping is crucial. Attempts to reduce the duration of the workshop by omitting the Prototyping phase were less beneficial compared to a workshop format that includes prototyping. The results and the experience for the participants are much better when prototyping is included as prototyping allows them to understand the challenges of developing an idea into a solution. The iterations and refinement during the prototyping phase are important and will result in more mature and more convincing solutions.

Peer group project

Design thinking can help in supporting group work by empowering participants through a structured process and methods for collaboration. In an educational setting, design thinking can be considered as learning by doing and offers hands-on experiences in a fault-tolerant environment provided by the university. Because during lengthy projects several actors with different backgrounds, such as students, citizens, and other stakeholders, can work together, this approach accounts for the concept of social learning.

When planning the application of design thinking in a peer group project, it is worthwhile to consider the following aspects: facilitation, coordination, communication, methods, and time. Facilitation is meant to be the role of a guiding person to first kick off the project, but later to take a back seat in order to provide the group room to manoeuvre. Still, in critical times, for instance, if the group is stuck or during conflicts, the facilitator needs to step in. By facilitating reflections on the group, work environment, personal progress, and outcomes throughout design thinking projects, either through reporting or a dedicated role, feedback loops for refinement will come automatically. From experience, the role of the coordinator infused some artificial inequality and was perceived as counterproductive. It is therefore recommended to work in non-hierarchical settings, invest time in team building at the beginning of the project, and foster shared responsibilities through assigning roles and alternating leadership in the various design phases. Clearly, the more heterogeneous the group is, especially with different day-to-day routines, the more important modern communication skills become. These skills include online conferencing, collaborating, and data sharing. Still, face-to-face meetings should not be neglected, especially in the beginning of the project for building trust and for developing a common language among diverse group members. In case there are more than six participants, it may be of value to split the group in two, at least before the Ideation phase, so that everyone can bring in their own ideas and several concepts can evolve in parallel. From a methodological point of view, it is crucial for the group to work over a long period to support everyone in keeping track of the project's progress. For that, it is recommended to use knowledge structuration and visualisation tools such as concept maps. Probably the key to any successful project is to manage time, because design can be performed indefinitely. This includes the reasonable allocation of time up front and its monitoring throughout the project, which is not necessarily the task of the facilitator but should be performed by the team to develop their scheduling skills. By providing a structured process, design thinking can help in managing time.

Results

Although the level of maturity of the solutions was different among the formats, all solutions provided a sense of achievement, motivating participants to search for further challenges as they created solutions without necessarily being experts in the domain of the addressed challenge. Furthermore, design thinking helped in questioning the assumptions of the participants, achieving a mutual understanding, defining a new Point of View, and tailoring the developed solutions.

The developed solutions have a real-life impact, as some of them, such as the "Join the Pipe" project, which aims at replacing plastic water cups on campus with a sustainable solution, were in fact implemented or affected successive decision making about solutions to related problems that will be implemented in the near future (Knopp & König, 2012).

As not all solutions will be developed to a level of maturity that allows implementation, it is recommended to reflect with the participants what would be required to do so, for example, next steps, possible iterations, involvement of experts and external partners, and links to other courses they attend at the university. Furthermore, reflection should include ideas that were not developed further but were radically different and unexpected. Usually, these are the ideas that external partners would be looking for. Here, reflection should concentrate on reasons for excluding them from further development and analysis of their potential for refinement.

Conclusion

Design thinking was applied in a transformative social learning environment, not only to teach students how to design, but primarily for them to learn about challenges of sustainability, as well as how these challenges might be addressed using a human-centred design approach. It served as a means to empower participants during group work to engage closely with problems presented to them while training their abilities to self-manage their agenda in parallel. In this particular endeavour, design thinking has proven to be a good way for the following reasons: first, design thinking strives to involve a diverse team; second, its iterative process allows reflections; and third, the process phases prototyping and testing aim at collecting feedback through tangible results, involving stakeholders in a new way and checking if desired effects can emerge. Additionally, design thinking was found to highly motivate participants to closely engage with the projects at hand and develop responsibility and empathy for both the problem and the developed solution. The experience of being able to come up with creative solutions and seeing how these can have a positive impact in real life was perceived as extremely gratifying and an empowering experience. Participants with no design-related educational or professional background developed confidence in addressing relevant complex challenges posed to organisations, communities, or even society and global economies.

A prevalent organisational issue that proved to be a challenge for the team is for them to self-reflect about the created solution(s) and to assess their multifaceted impacts on the environment, society, etc., particularly when they are in a pre-mature design stage. Although assessing positive impacts already proved to be difficult, due to limited available information, inexperience of participants in the particular domain affected, dynamics of the problem space, and the targeted user group, oftentimes it was even more challenging to apprehend the negative impacts of a newly developed solution. This is particularly due to the fact that although it is usually possible to predict a possible path to success of a thought-up solution, it is considerably more challenging for the participants to apprehend and take into account all possibilities that may lead to failure and long-term negative impacts. This is especially true with regard to the ecological and social dimension. User satisfaction as one major goal of a human-centred design approach does not guarantee that these other dimensions are appropriately addressed, though these might have in fact triggered the search for new solutions to begin with. Re-thinking existing solution concepts, therefore, also requires re-considering and thoroughly investigating relevant effect propagation scenarios.

Design thinking can be a powerful approach (i.e. a tool for peer group work); however, one should not expect that the result will be ready for implementation in every case. For finalisation of newly generated solution concepts as the result of design thinking, it is therefore important to involve experts from relevant domains to shape the championed solution concepts through detailing and refinement.

The main purpose of using design thinking in the context of this educational program is to empower participants to take a different perspective with regard to a given problem, detaching oneself from a purely scientific and technological view and moving towards a human-centred perspective. This is crucial for developing solutions that will be accepted by relevant user groups – despite these being potentially disruptive towards an established lifestyle – and may eventually lead to a lasting positive impact on the way mankind interacts with its environment.

Acknowledgements

The authors would like to thank Pascal Gemmer and Christian Beinke from the Dark Horse Innovation consultancy Berlin for providing insights into design thinking and its application, challenges, and experiences in an industrial context. Further we would like to thank Ulrich Weinberg from d.school Potsdam for enabling insights into teaching design thinking and Jan Glas from GlasJan – Studio for design and design management and for his support in running the workshops.

Questions for comprehension and reflection

1 What needs to be done before starting to develop a solution?
2 What should be considered when choosing a design challenge?
3 What sources can be used to understand the design challenge?
4 What alternative methods exist to capture user needs aside from interviews?
5 What is the benefit of working with a multidisciplinary team?
6 What should be the end point of a design thinking project within the given educational framework?
7 What type of deliverable is created at that point?
8 Which factors are important other than user satisfaction for creating a good solution?
9 Why is it important to question the initial problem description?
10 Why should no more than a few ideas be developed before creating and testing prototypes?
11 Why are prototyping and testing important?

Note

1 Basically, design thinking can also be applied to the two lower orders of confirmative and reformative learning.

References

Badke-Schaub, P., Roozenburg, N. & Cardoso, C. (2010). 'Design thinking: A paradigm on its way from dilution to meaninglessness', in Dorst, K., Stewart, S., Staudinger, I., Paton,

B. & Dong, A. (Eds.) *DTRS8 interpreting design thinking, proceedings of the 8th design thinking research symposium.* Sydney: DAB documents. pp. 39–49.

Baxter, M. R. (1995). *Product design: Texte imprimé.* London and New York: Chapman & Hall.

Beckman, S. L. & Joyce, C. K. (2009). *Reflections on teaching design thinking to MBA students.* Paper presented at Global Forum for Business as an Agent of World Benefit: Manage by Designing in an Era of Massive Innovation, Cleveland, Ohio, USA, Case Western Reserve University.

Boeijen, A. van, Daalhuizen, J., van der Schoor, R. & Zijlstra, J. (2014). *Delft design guide: Design strategies and methods.* Amsterdam: BIS Publishers.

Brown, T. (2008). Design thinking. *Harvard Business Review* 86 (6): 84–92, 141.

Brown, T. & Katz, B. (2009). *Change by design: How design thinking transforms organizations and inspires innovation, 1st edition.* New York, NY: Harper Business.

Charter, M. & Chick, A. (1997). Editorial of the journal of sustainable product design. *The Journal of Sustainable Product Design* 1 (1): 5–6.

Crilly, N. (2015). Fixation and creativity in concept development: The attitudes and practices of expert designers. *Design Studies* 38: 54–91.

Dorst, K. & Cross, N. (2001). Creativity in the design process: Co-evolution of problem-solution. *Design Studies* 22 (5): 425–437.

Gericke, K. & Maier, A. (2011). 'Scenarios for coupling design thinking with systematic engineering design in NPD'. Presented at the *Proceedings of the 1st Cambridge Academic Design Management Conference.* Cambridge: University of Cambridge, Institute for Manufacturing.

Gundy, A. B. Van. (1988). *Techniques of structrued problem solving, 2nd edition.* New York: Van Nostrand Reinhold.

Hellfritz, H. (1978). *Innovationen via galeriemethode.* Königstein: Eigenverlag.

IDEO. (2011a). *Design thinking for educators, 1st edition.* Palo Alto, CA: IDEO. Retrieved from www.designthinkingforeducators.com/

IDEO. (2011b). *Human centered design toolkit: An open-source toolkit to inspire new solutions in the developing world, 2nd edition.* Palo Alto, CA: IDEO.

IDEO. (2003). *IDEO method cards: 51 ways to inspire design.* San Francisco, CA: William Stout.

IDEO. (2015). *The field guide to human-centered design, 1st edition.* San Francisco, CA: IDEO.org.

Jones, J. C. (1985). *Design methods: Seeds of human futures.* New York: Wiley.

König, A. (2015). Towards systemic change: On the co-creation and evaluation of a study programme in transformative sustainability science with stakeholders in Luxembourg. *Current Opinion in Environmental Sustainability* 16: 89–98.

Knopp, A. & König, A. (2012). *Sustainable development 2011–2012 ISCN-GULF Charter Report.* Luxembourg: University of Luxembourg.

Lawson, B. (1997). *How designers think: The design process demystified, 3rd edition.* Amsterdam: Elsevier/Architectural Press.

Maguire, M. (2001). Methods to support human-centred design. *International Journal of Human-Computer Studies* 55 (4): 587–634.

Maher, M. Lou & Poon, J. (1994). Modelling design exploration as co-evolution. *Computer-Aided Civil and Infrastructure Engineering* 11 (3): 195–209.

Maier, A. & Störrle, H. (2011). 'What are characteristics of engineering design processes?', in Culley, S. J., Hicks, B. J., McAloone, T. C., Howard, T. J. & Reich, Y. (Eds.) *International Conference on Engineering Design ICED'11, Impacting society through engineering design, vol. 1: Design processes.* Lyngby and Copenhagen, Denmark: Design Society. pp. 188–198.

Martin, R. L. (2009). *The design of business: Why design thinking is the next competitive advantage.* Boston: Harvard Business Press.

McAloone, T. C. & Andreasen, M. M. (2004). 'Design For utility, sustainability and societal virtues: Developing product service systems', in Marjanovic, D. (Ed.) *8th international design conference – Design 2004*. Glasgow: Design Society. pp. 1545–1552.

McKim, R. H. (1972). *Experiences in visual thinking*. Monterey and California: Brooks and Cole.

Mezirow, J. (1990). 'Transformation theory of adult learning', in Kleiber, P. & Tisdell, L. (Eds.) *31st Annual Adult Education Research Conference (AERC)*. Athens and Georgia: Georgia University, Athens Center for Continuing Education. pp. 141–146.

Mezirow, J. (2009). 'Transformative learning theory', in Mezirow, J., Taylor, E. W. & Associates (Eds.) *Transformative learning in practice: Insights from community, workplace, and higher education*. San Francisco: John Wiley & Sons. pp. 18–31.

Müller, P., Kebir, N., Stark, R. & Blessing, L. (2009). 'PSS layer method – Application to microenergy systems', in Sakao, T. & Lindahl, M. (Eds.) *Introduction to product/service-system design*. London: Springer. pp. 3–30.

Osborn, A. F. (1957). *Applied imagination: Principles and procedures of creative problem-solving*. New York: Scribner.

Osterwalder, A. & Pigneur, Y. (2013). *Business model generation: A handbook for visionaries, game changers, and challengers*. Hoboken, NJ: John Wiley & Sons.

Pahl, G., Beitz, W., Feldhusen, J. & Grote, K.-H. (2007). *Engineering design – A systematic approach, 3rd edition*. Berlin: Springer-Verlag.

Plattner, H., Meinel, C. & Weinberg, U. (2009). *Design thinking: Innovation lernen ; Ideenwelten öffnen*. München: mi-FinanzBuch Verl.

Powell, T. C., Lovallo, D. & Fox, C. R. (2011). Behavioral strategy. *Strategic Management Journal* 32 (13): 1369–1386.

Roozenburg, N.F.M. & Eekels, J. (1995). *Product design: Fundamentals and methods*. Chichester: John Wiley and Sons.

Rumelt, R. (2011). The perils of bad strategy. *McKinsey Quarterly* June: 1–10.

Sterling, S. (2011). Transformative learning and sustainability: Sketching the conceptual ground. *Learning and Teaching in Higher Education* 5: 17–33.

Sunstein, C. R. & Hastie, R. (2015). *Wiser: Getting beyond groupthink to make groups smarter*. Boston, MA: Harvard Business Press.

Tan, A. R. (2010). *Service-oriented product development strategies*. Doctoral Thesis. Kgs. Lyngby, Denmark: Technical University of Denmark.

Taylor, E. W. (2009). 'Fostering transformative learning', in Mezirow, J., Taylor, E. E. & Associates (Eds.) *Transformative learning in practice: Insights from community, workplace, and higher education*. San Francisco: John Wiley & Sons. pp. 3–17.

Ulrich, K. T. & Eppinger, S. D. (2007). *Product design and development*. London: McGraw-Hill Higher Education.

United Nations Department of Economic and Social Affairs. (2013). *World economic and social survey 2013: Sustainable development challenges*. New York: United Nations.

Wals, A.E.J. (2007). Learning in a changing world and changing in a learning world: Reflexively fumbling towards sustainability. *Southern African Journal of Environmental Education* 24: 35–45.

Wals, A.E.J., Van Der Hoeven, N. & Blanken, H. (2009). *The acoustics of social learning: Designing learning processes that contribute to a more sustainable world*. Wageningen: Wageningen Academic Publishers.

Part II

What might transformations look like? Sectoral challenges and interdependence

9 Can ecosystem services help the new agricultural transition?

Nicolas Dendoncker and Emilie Crouzat

Introduction – why our globalized farming system is not sustainable

---

How do the authors engage in/with the science and research stance?

The research question: Can ecosystem services assessments help foster a new agricultural transition?

Main object of research: Transdisciplinary research at the interface between natural and socio-economic science, with a societal aim.

Disciplines: The authors have a background in geography and spatial ecology; hence the focus is territorial, anthropocentric, and systemic. However, elements of other disciplines (e.g. ecological economics, sustainability science, agronomy) are mobilized in the text.

Methods: The chapter is largely qualitative and descriptive, but partly relies on quantitative data.

Main beliefs about the role of science in society: The authors do not believe in an 'objective' science that portrays a simple reality independently of the scientist's personal commitments and values. Rather, awareness of these framing elements, together with their strengths and weaknesses, will help scientists to make a better case for their desired policy. It will also help them to engage in constructive dialogue with scientists and stakeholders holding different or opposing views.

Views on sustainability: The authors believe in a strong sustainability framework which states that economy is a social construct, and that our societies depend on their broader environment. In other words, no economic development is possible without safeguarding the environment.

In this chapter, we argue that today's conventional agriculture is not sustainable not only because it compromises the capacity of future generations to cultivate and produce healthy food in adequate economic, social, and ecological conditions, but also because current agricultural systems do not provide these conditions for today's population (see the unnumbered box above for an overview of how the authors engage with science). For farming systems to be sustainable, they will need to account for the biophysical limits of the planet (Rockström et al., 2009), foster social equity, and fulfil the needs of present and future generations. Ecosystem services (ES) assessments have the potential to be a tool to help the transition towards sustainable agricultural systems.

Symptoms of unsustainability

Across the planet, numerous symptoms of unsustainability can be observed. In Europe, farmers' numbers decrease at worrying rates (Strijker, 2005), and extrapolation of current trends indicate that several European countries may fear a total disappearance of farmers in less than 15 years. The question of what would happen to these presently cultivated areas is worth asking. Meanwhile, cultivated areas increase in many southern countries, often causing loss of natural grassland and forests (Charvet, 2010; Boerema et al., 2016).

The number of European farmers is declining, yet Europe is highly dependent on the outside world to feed itself. Globally, the way in which current agricultural systems are organized leaves around 1 billion people hungry on a daily basis, most of them being farmers (Charvet, 2010). These are located mostly in Asia where the green revolution has not benefited every farmer in a similar fashion. At the same time at least 1.6 billion people are overweight, out of which around 400 million are obese (Charvet, 2010, see also Davila and Dyball, this volume). More worrying is that both the numbers of under-fed and overweight people are increasing. From these numbers only, it can be concluded that the major issue with the current food system is one of distribution and access to food rather than production.

Current conventional agriculture also has major environmental impacts. Land clearing and intensive practices threaten biodiversity. In addition, conventional agriculture requires vast amounts of water and pesticides, and excess fertilizer use decreases water quality. As food is transported over long distances, it relies on cheap abundant fossil fuels. Overall, agriculture is responsible for around 25% of global greenhouse gas (GHG) emissions (Tilman et al., 2011).

A more sustainable agriculture will need to use less energy and will necessarily be more local: this means, for example, that Europe will need all its fertile agricultural land to feed Europeans. Moreover, current trends show that farms tend to enlarge in order to stay competitive in a global market. Consequently, manpower is often replaced by complex machinery and technological inputs, leading to increased GHG emissions and contributing to further degrading the environment (e.g. through soil erosion and habitat simplification). However, studies indicate that small family farms tend to be more productive (e.g. Rosset, 2000). Hence, it seems urgent to stop the decline in farmers' numbers.

In the rest of the chapter, we develop a brief retrospective analysis to better frame the present situation. We then synthesize the impacts of our current agricultural systems and argue for a new transition towards sustainable farming systems. The second part of this chapter explores the potential of the concept of Ecosystem Services to foster a new transition. Starting with a definition and historical perspective over the ES concept, we move on to examining how ES are related to agroecosystems. We then propose an integrated valuation framework to use ES as a transition tool. Finally, we draw some limitations of the concept with regard to the agroecological shift (see later for characteristics of agroecological systems) that should be implemented for increased sustainability.

Retrospective analysis: agricultural revolutions and their broader impacts

Through time, two main ways to increase food production have been and are still implemented to cope with increasing population and demand. The first solution is to increase cultivated areas (i.e. to 'extensify'). This results in deforestation and biodiversity loss and can lead to increased energy and water consumption. Historical evidence shows that this strategy has succeeded in meeting increasing food demands of peasant communities for limited time periods (Mazoyer & Roudart, 2002), but at a given point, when a specific territory occupied by and available to a rural community is fully exploited under prevalent techniques, extensification is no longer an option and another solution needs to be implemented.

The second possible answer is to intensify farming via a change in the dominant agricultural system in place. This happened throughout the history of agriculture through agricultural revolutions (or transitions), which can be defined as drastic and relatively sudden change of the dominant system in place (Mazoyer & Roudart, 2002).

These have occurred in different places at different times, but most started in Europe or the United States and have spread variably throughout the world. As a result, today's agroecosystems represent very varied stages of socio-technical evolution, notably in terms of degrees of intensification and dependence on anthropogenic inputs. Across the earth, farming systems that emerged thousands of years ago but are still practiced today (e.g. slash and burn agriculture) contrast with systems that emerged a few decades ago (e.g. modern moto-mechanised cerealiculture in the United States, Brazil, or Argentina).

Since the Neolithic's revolution, three main agricultural revolutions have occurred: the Middle Ages agricultural revolution, the industrial agricultural revolution, and the scientifico-technical agricultural revolution. Each revolution is characterized by a relatively rapid increase in yields generated by a drastic change in the agricultural system in place, stimulated and preceded by societies in crisis, as they were not producing enough food for a growing population (see Box 9.1).

As Mazoyer and Roudart (2002) argue, it is precisely these states of crisis that led to increased creativity of the societies experiencing these crises and helped them create new niches that could replace the dominant agricultural regime that was in place at the time. As argued earlier and synthesized later, there is no doubt that our global society has entered a new agricultural crisis, and creativity will again need to be mobilized to overcome this crisis. The main difference between the current crisis and those that happened in the past is that whereas the latter occurred at the regional or continental level, the crisis we face now is global.

Box 9.1 Agricultural revolutions

According to Mazoyer and Roudart (2002), three main agricultural revolutions occurred since the tenth century (Figure 9.1):

1   The Middle Ages agricultural revolution is characterized by a set of new investments that led to a complete change of agricultural systems. These relate in particular to animal traction, tools and equipment, and to the

way the soil was worked. For example, the swing plough is abandoned and replaced by proper ploughs, drawn by horses, which improves weeding. The horse's collar is created and the horses draw the harrow, allowing for the creation of smooth seeding beds. The invention of stables improves animals' living conditions and fosters the creation of farmyard manure. The scythe replaces the much smaller sickle, which allows for larger areas of grassland to be cut. All in all, this leads to the creation of open fields, managed with a three-field cropping system and a common herd grazing the fallow land.

2   The industrial agricultural revolution kicks off during the eighteenth and nineteenth centuries and is characterized again by the creation of better tools, including better ploughs. However, the main improvement consists in the replacement of fallow land by 'artificial grassland', made of grasses or legumes, which can be used as fodder crops or green manure. This allows better weed control, fertility renewal, and increased production. Consequently, population and hence manpower increase, leading to the development of the industrial revolution per se. Improved transport systems also allow for greater exchanges between regions, and agricultural areas start to specialize.

3   During the twentieth century, but specially after World War II, the scientific-technical revolution is characterized by increased mechanization and motorization, the synthesis of chemical fertilizer which allows for greater specialization of agricultural activities and the complete separation of crops and livestock systems, and the privately led selection of highly productive yet fragile varieties often requiring higher amounts of pesticides. This last agriculture revolution tremendously increased productivity but also led to negative environmental and social impacts.

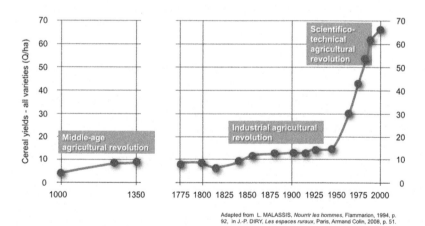

Adapted from L. MALASSIS, *Nourrir les hommes*, Flammarion, 1994, p. 92, in J.-P. DIRY, *Les espaces ruraux*, Paris, Armand Colin, 2008, p. 51.

Figure 9.1 Evolution of cereal yields in France

Synthesis: the need for a new transition

In many territories, the entire farming system is threatened, for many reasons, with perhaps the five main ones being:

1) Rarefication of inputs (fossil energy, phosphorus, water, etc.).
2) Global competition: meaning that any given system has to compete with potentially more efficient systems across the planet.
3) Pressure from other land uses: agriculture is not necessarily the most profitable land use on a given parcel and has to compete with settlements, for example, especially in peri-urban areas.
4) Climate change: global warming (partly caused by agricultural practices), increasing variability in weather events, and increased probability of extreme events make current agricultural practices vulnerable to future changes.
5) Loss of biodiversity and ecosystem services: as we argue later in this chapter, well-functioning agriculture depends on a broad suite of ES. However, as shown by the Millennium Ecosystem Assessment (MEA, 2005), Ecosystem Services are declining in many parts of the world. For example, carbon content levels are critical in many areas, soil is being lost at worrying rates (several tons/ ha/y), and decline of biodiversity is higher in agricultural landscapes.

Many studies show that it is possible to feed a growing population differently and sustainably. Whereas some authors argue that there is still a yield gap between conventional farming and organic farming, for example (Seufert et al., 2012; Ponisio et al., 2014), others find that by ecologically intensifying farming, both food production and Ecosystem Services could be increased (Garbach et al., 2016). Importantly, organic farming does not necessarily mean sustainable farming, as, for example, it does not include a social dimension, and may sometimes lead to increased GHG emissions by replacing herbicides by mechanical weeding, for example. Alternatives to conventional agriculture comprise permaculture, agroforestry, sylvo-pastoralism, or agroecology. Rather than trying to define precisely each of these concepts, which would be out of the scope of this chapter, we focus here on what they have in common – that is, refusing monocultures, respecting nature, and getting inspired by it (biomimicry); striving for autonomy at the farm or landscape scale; and aiming for greater resource efficiency in terms of water and energy for greater economic efficiency. Among these concepts, agroecology arguably has the broadest meaning, as it refers to a science, a set of practices (see Box 9.2.), but also a social movement (Wezel et al., 2009).

Box 9.2   Main characteristics of agroecological farming systems

1   Maintenance of a vegetation cover as an effective soil and water conserving measure, met through the use of no-tillage practices, mulch, and cover crops;
2   Provision of a regular supply of fresh organic matter through the addition of manure, compost, crop residues, and the promotion of soil biotic activity;

3   Enhancement of nutrient recycling mechanisms through the use of live-stock systems based on nitrogen fixing legumes;
4   Promotion of pest regulation through biological control agents achieved by introducing and/or conserving natural enemies and antagonists and by developing an ecological infrastructure.

In sum, these systems are based on ecosystem services, provided by agroecosystems.

In sum, the key words of the transition to agroecological systems are resilience (as the capacity of a system to overcome a shock), diversity, and autonomy. Our current agricultural systems are not resilient. They work as a chain of interdependent links. If one of the links breaks, the whole food system fails (Servigne, 2014). If oil supply stops, our conventional farming systems stop as well. In order to be resilient, agricultural systems need to diversify and strive for autonomy. Autonomy aims at closing cycles, transforming, and perhaps selling agricultural products locally or producing organic farmyard manure. For a producer, being autonomous also means choosing whom to sell the products to. A resilient system is also a diverse system. In an agroforestry farm, cereals can be grown under trees and protected by hedges. Each of these system elements plays a different role in the agricultural landscape, accomplishing different functions (e.g. the hedges protect against wind, regulate the microclimate, serve as habitat for crop auxiliaries; the trees also provide shelter but may also provide wood, etc.). As developed later, these functions provide Ecosystem Services to the farming systems themselves and to society.

As discussed by Davila and Dyball (see this volume), myriad conflicting priorities held by parties with different power and economic agency has led to massive inequities (e.g. in access to food) and created a global 'wicked problem'. To engage with this, there is a need to better understand the interactions between the social, technological, economic, and environmental dimensions of the conventional 'productionist' agricultural system as well as the interdependencies and feedbacks between these dimensions, developing a system's perspective.

In order to do so, new conceptual and yet operational tools are needed to direct attention at the interface of society, well-being, economy, and the environment. In the next sections of this chapter, we discuss the potential of the Ecosystem Services concept to act as an operational tool to foster the needed agricultural transition.

Ecosystem services to foster a new agricultural transition: strengths, opportunities, and limits

What are ecosystem services?

At the interface between social and ecological systems, Ecosystem Services (hereafter ES) have been proposed to make explicit "the benefits people obtain from ecosystems" (MEA, 2005). They are defined as "the direct and indirect contributions of nature to human wellbeing" (TEEB, 2010) and stress human dependency on natural processes (Diaz et al., 2006, Diaz et al., 2015). ES are generally classified into three main categories: provisioning (e.g. food, water, etc.), regulating (e.g. pollination, climate regulation, etc.), and cultural (e.g. landscape aesthetics, spirituality, etc.) ES (MEA, 2005).

Broad historical analyses of the concept can be found in Crouzat (2015) or Gomez-Bagghethun et al. (2010). Early mentions of the concept date back to the 1970s, under the terminology 'nature's service' (Westman, 1977). The concept of ecosystem services was introduced in the late 1970s and early 1980s by authors such as Westman (1977) and Ehrlich and Ehrlich (1981), building on earlier literature highlighting the societal value of nature's functions. The term was first used as a means to raise awareness of the global biodiversity loss and ecosystem degradation (Lamarque et al., 2011). A growing body of literature has since then made use of the concept. Its influence has spread from the academic sphere into the policy and economic fields with as major milestones two worldwide initiatives to assess and value the contributions of ecosystems to human wellbeing: the Millennium Ecosystem Assessment (MEA) in 2005 and The Economics of Ecosystems and Biodiversity (TEEB) in 2010. Thus, in some 30 years, ES turned from a metaphoric to a heuristic concept (Abson et al., 2014) and further to a "concrete, tangible and measurable" object (Barnaud & Antona, 2014). Iconic of this reification into an explicit decision and policy tool (de Groot et al., 2010) is the initiation of the Intergovernmental Platform on Biodiversity and Ecosystem Services (IPBES http://ipbes.net/work-programme.html) in the early 2010s. Likewise, the EU biodiversity strategy to 2020 encourages member states to develop indicators of Ecosystem Services to be included in national accounts (see Chapter 15). In sum, Ecosystem Services increasingly act as a boundary (or bridging) object between science, society, and policy (Arnaud de Sartre et al. 2014; Abson et al., 2014).

Importantly, as Jacobs et al. (2013) state, the research field and concept of ecosystem services (ES) is rooted in strong sustainability thinking. The explicit link between sustainability and ES assessments stresses the importance of accounting for the three main value domains of ES: ecological sustainability, social fairness, and economic efficiency. Conclusively, the final goal of ES valuation is to achieve a more sustainable resource use, contributing to the well-being of every individual now and in the future by providing an equitable, adequate, and reliable flow of essential ecosystem services to meet the needs of a burgeoning world population (Jacobs et al., 2013).

Ecosystem services in agroecosystems

Many ES are required to obtain well-functioning agroecosystems that are in turn able to provide other ES to their beneficiaries For example, agroecosystems (and a fortiori farmers) will benefit from a living soil rich in organic matter, which will help increase production and benefit society. The presence of pollinators and crop auxiliaries can also increase agricultural productivity.

However, as Peeters et al. (2013) mention, since the middle of the nineteenth century, a large part of the services provided by ecosystems before the Industrial Revolution has been replaced by techniques relying on a massive use of fossil fuel. For instance, the artificial synthesis of nitrogen has replaced symbiotic nitrogen fixation by legumes, crop protection by pesticides has replaced the effect of pest and disease regulation by complex living communities, and mechanization has replaced manpower and draught animals.

The use of these artificial inputs and techniques has not only replaced ecosystem services, but they have also increased production and induced negative impacts on the environment, and, arguably, on society. They provoked pollution and bio-diversity losses that in turn decreased the supply of ecosystem services essential to farming itself.

The issue is that in a free market economy, farmers will perceive the benefits of high yields generated by chemical fertilizers, but generally do not pay the environ-mental costs generated by the loss of nitrogen in water tables or in the atmosphere (the so-called externalities); in other words, there is a non-respect of the polluter pays principle, one of the key principles of sustainability.

Conversely, externalities can also be positive. For example, well-maintained grass-lands store vast amount of carbon hence contributing to mitigate climate change, which benefits to the broader society. As this service is not recognised and currently not paid (it escapes the market), it is produced in a sub-optimal quantity by farmers.

The current economic logic leads 'rational' farmers to maximise provisioning services (for which there is a market) at the expense of other categories of ES (for which there is no market). Note that this focus on food as prime and sole objec-tive was the aim of the green revolution and continues to influence farming today.

In spite of this, internalizing externalities and/or creating a market for non-provisioning ES, a process referred to as the commodification of nature, will likely not be sufficient to ensure sustainable farming and may even reinforce current unsustainability issues such as access to resources and power asymmetries (Kallis et al., 2013).

The question then becomes: Can the ES concept be used as an operational tool to foster a transition to sustainable agroecosystems? If so, under which conditions?

ES valuation as a transition tool

Valuation of nature and its services has become central to an increasing amount of academic literature (Fisher et al., 2009; Seppelt et al., 2011). This proliferation has been stimulated by policy initiatives such as the European Biodiversity Strategy to 2020, the Aichi targets, the Sustainable Development Goals, and the Intergovern-mental Platform on Biodiversity and Ecosystem Services (IPBES). Under these umbrellas, national and local ecosystem service assessments and valuations are thriv-ing (e.g. UK NEA, 2011; Santos-Martín et al., 2014; Jacobs et al., 2015).

At the local level, numerous attempts to internalize environmental externalities are already occurring across the planet under Payment for Environmental Services (PES) schemes, which can be considered the main attempt to operationalize the ecosystem services concept. It is important to note that we do not see environ-mental services as equivalent to ecosystem services. Although they are often used as synonyms, we see the main difference related to the former referring to the action of a human agent benefiting another agent, whereas the latter refers to the con-tribution of ecosystems to human well-being. In this respect, PES can be defined as the "payment of an economic agent for a service given to other agents, through an intentional action aiming at preserving, restoring, or increasing a given envi-ronmental service" (Karsenty, 2011). Agrienvironmental schemes (AEM) are one example of PES in the European Union.

Although such instruments can play a role in improving environmental gov-ernance, they face a series of limitations. Muradian et al. (2013) argue that such instruments are not independent from the institutional setting in which they take place. For example, pressure groups might have a large influence on the design of payment schemes, shaping their effectiveness and distributional outcomes. PES can sometimes act as incentive for perverse strategic behaviour (Banerjee et al., 2013). In addition, some authors are concerned by the shift PES induce from a polluter pays principle to a beneficiary pays principle (Pirard et al., 2010). Most importantly, Muradian et al. (2013) argue to shift emphasis from designing PES to tackling the ultimate causes of environmental degradation, deeply rooted in structural power inequalities.

In general, monetary valuation of ES face a series of limitations, from methodological pitfalls to failing to represent the broad realm of values associated with the natural world. Indeed, Ecosystem Services deliver a broad range of benefits and hence affect many actors of agricultural landscapes: from co-producers and managers of ES (e.g. farmers, foresters) to ES beneficiaries (e.g. local inhabitants, tourists). In order to design policies that encourage sustainable landscape management, an integrated valuation framework including a broad set of values and stakeholders seems particularly relevant.

Assessing and valuing ES implies accounting for normative (what should be) and cognitive (what is) complexities and uncertainties. An integrated valuation framework is needed to reveal the diversity of values that can be attributed to ES. It is integrated if it offers a way to articulate between the different value domains (e.g. biophysical, social, monetary) and inclusive if it does so by involving the broad set of stakeholders concerned with the valuation case (Dendoncker et al., 2013).

Over the last years, many place-based case studies (either conducted by researchers or institutions of the civil society) have tried to assess locally produced ES. Many invoke improved decision-making as a vindication for their research. However, it is unclear whether these have actually led to improved landscape management (Laurans et al., 2013).

We do believe, however, that if they meet a series of conditions, integrated and inclusive ES valuation initiatives may lead to increasingly sustainable (agricultural) landscapes. This means that they could improve the environment, reduce inequalities, and account for and maintain value plurality. In Box 9.3, we provide an example of how social and biophysical ES valuations have been integrated to optimize ES delivery through a participatory land consolidation exercise.

Box 9.3   Using the ecosystem services concept for priority setting in a land consolidation scheme

ES mapping and valuation at the local scale can be used to set priorities and guide decision making to optimize ES provision. This example describes how ecosystem service maps were combined with biophysical models and social valuation data to serve a participatory land-consolidation plan of three municipalities in Wallonia, Belgium. It is co-constructed by the administration, scientists – led by the French consultancy Biotope – and local stakeholders. The project's objective is to design a replicable methodology based on hands-on experience in a first case study. Figure 9.2 describes the methodological framework further.

After predefining a list of locally relevant ES and a typology of ecosystems, biophysical assessment and social valuation are carried out. The biophysical assessment includes mapping and quantification of selected ecosystem services based on indicators obtained from a hydrological model and scenario development of potential ecosystem service supply. Social analysis comprises stakeholder analysis, societal valuation according to these stakeholders, participatory validation of the biophysically mapped ecosystem services, and participatory mapping of ecosystem service demand. These supply and demand maps are then used to guide participatory comparison of land-consolidation actions. For instance, maps of biophysical indicators were compared with demand maps to highlight locations for which there is potential improvement of supply.

Accounting for local stakeholders' demands and suggestions, technical experts of land consolidation then suggest potential measures (e.g. installation of new hedgerows, creation of new water retention basins, new flower strips along a walking path, etc.) to be implemented in the final land consolidation plan. This example clearly demonstrates that ES integrated valuation is used as a central means in combination with various other data, methods, and actions to achieve a broader objective shared by various stakeholders and lead to improved local environmental conditions for the increased well-being of local communities.

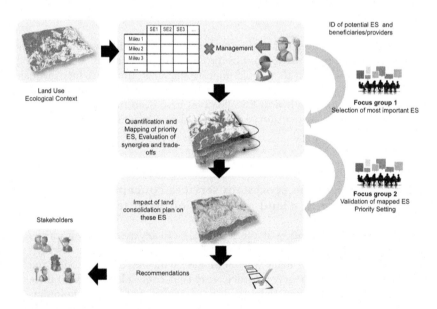

Figure 9.2 Methodological framework for integrated valuation of ecosystem services to set priorities for land consolidation in Wallonia

Source: After Jacobs et al., 2016a

ES and agroecology: limitations of the ES assessment approach

The example in Box 9.3 seems to depict an ideal situation where the ES tool helped shaped a collaborative, future-oriented process: using an integrated ES valuation scheme leads to an improved ES delivery while increasing credibility, salience, and legitimacy of the ES approach, as well as social learning or even arguably empowerment of local stakeholders. In the depicted study area, it will likely boost the economy by improving landscape quality and accessibility, hence enhancing tourism; it will limit water damages by preventing floods; and land consolidation itself should lead to reduced energy consumption by bringing cultivated fields closer to the farms. However, it should be noted that impacts on agricultural practices remain limited, as these are beyond the objectives and scope of a land consolidation exercise. Similarly, although agroenvironmental measures may improve ES delivery, most of them contribute little to changing the broad set of unsustainable agricultural practices (e.g. a farmer may apply a buffer strip to limit erosion and flood risk yet plant an annual monoculture of wheat next to it). The way ES assessments are designed and the specific issues they address are thus critical in engaging in a collective transformation of (agro-)ecosystems.

In general, integrated valuation of ES faces a series of challenges, including fragmented policy and governance fields to target, fragmented science fields to combine for comprehensive assessments, and difficulty to account for equity issues in the context of power imbalances (see Jacobs et al., 2016b for a broader discussion).

It can also be argued that the ES assessment framework, by focusing on instrumental values of nature, may be too restricted as it fails to incorporate other types of values (e.g. inherent value of nature, patrimonial values, etc.), though these are sometimes considered related to 'cultural ecosystem services'.

Even if all values are accounted for, the issue of how to make the final decision remains. Valuation exercises always take place in a given institutional setting (Vatn, 2005; Dendoncker et al., 2013). Because environmental resources are often common and complex goods, this institutional setting should ideally favour social rationality and communicative action, requiring a societal perspective is taken and that the procedure must be able to treat weakly comparable or incommensurable value dimensions (Vatn, 2005; Martinez-Alier, 1998). At the global level, some authors argue that new institutions and more resources devoted to environmental governance are needed (Norgaard, 2010).

At the local level however, the increase in place-based actions and public support for change raises hope. Arguably, place-based territorial applications of transformative research could provoke local regime shifts in agriculture. Co-constructed actions between science, society, and policy may lead to greater changes. The operational potential of integrative and inclusive ES assessments to foster the transition to agroecology remains to be strengthened, however.

Conclusion

In this chapter, we highlighted why our current conventional agricultural systems are in crisis. They are unsustainable as they rely on fossil energy and limited resources (e.g. water, phosphorus, etc.), are not resilient against climate change, and generate biodiversity and ecosystem services losses. Sustainable agricultural systems will need to be designed for autonomy, resilience, and diversity. Because it may bring together a broad range of values, and hence local actors, integrated valuation of ecosystem services has the potential to act as a tool to foster a transition to sustainable agriculture. Although there are local cases where ES assessments have led to increased ES delivery and social learning, it has not been demonstrated that ES assessments could lead to more systemic changes in agroecosystems by increasing economic efficiency, improving the environment, and also increasing equity by accounting for and dealing with power asymmetries. Moreover, at the global level, it is likely that for agroecological systems to replace the current dominant regime, wider institutional changes at larger scales should be implemented.

Questions for comprehension and reflection

1   Think of several symptoms of unsustainability of our current agricultural systems. Can these be linked to Ecosystem Services?
2   What is the concept of Ecosystem Services used for, and what are the limitations of the Ecosystem Services approach?
3   Under which conditions could Ecosystem Services assessments contribute to a transition to sustainable agriculture?

References

Abson, D. J., von Wehrden, H., Baumgärtner, S., Fischer, J., Hanspach, J., Härdtle, W., . . . Walmsley, D. (2014). Ecosystem services as a boundary object for sustainability. *Ecological Economics* 103: 29–37.

Arnaud de Sartre, X., Castro, M., Dufour, S. & Oswald, J. (2014). *Political ecology des services écosystémiques*. Brussels: Peter Lang.

Banerjee, S., Secchi, S., Fargione, J., Polasky, S. & Kraft, S. (2013). How to sell ecosystem services: A guide for designing new markets. *Frontiers in Ecology and the Environment* 11: 297–304.

Baptist, F., Degré, A., Grizard, S., Maebe, L., Pipart, N., Renglet, J., . . . Dendoncker, N. (2016). *Elaboration d'une méthodologie d'évaluation des incidences sur l'environnement de l'aménagement foncier s'appuyant sur la notion des services écosystémiques*. Rapport général. Direction Générale Opérationnelle de l'Agriculture, des Ressources Naturelles et de l'Environnement.

Barnaud, C. & Antona, M. (2014). Deconstructing ecosystem services: Uncertainties and controversies around a socially constructed concept. *Geoforum* 56: 113–123.

Boerema, A., Peeters, A., Swolfs, S., Vandevenne, F., Jacobs S., Staes, S. & Meire, P. (2016). Soybean trade: Balancing environmental and socio-economic impacts of an intercontinental market. *PLOS-one* 11 (5): e0155222. doi:10.1371/journal.pone.0155222

Charvet, J. P. (2010). *Atlas de l'agriculture*. Paris: Autrement.

Crouzat, E. (2015). *Etudes des compromis et synergies entre services écosystémiques et biodiversité: Une approche multidimensionnelle de leurs interactions dans le socio-écosystéme des Alpes françaises*. Thèse de doctorat, Université de Grenoble.

De Groot, R. S., Alkemade, R., Braat, L., Hein, L. & Willemen, L. (2010). Challenges in integrating the concept of ecosystem services and values in landscape planning, management and decision making. *Ecological Complexity* 7: 260–272. doi:10.1016/j.ecocom.2009.10.006

Dendoncker, N., Keune, H., Jacobs, S. & Gomez-Baggethun, E. (2013). 'Inclusive ecosystem service valuation', in Jacobs, S., Dendoncker, N. & Keune, H. (Eds.) *Ecosystem services: Global issues, local practices*. New York: Elsevier. pp. 19–28.

Díaz, S., Demissew, S., Joly, C., Lonsdale, W. M. & Larigauderie, A. (2015). A Rosetta Stone for nature's benefits to people. *PLoS Biology* 13 (1): e1002040. doi:10.1371/journal. pbio.1002040

Díaz, S., Fargione, J., Chapin III, F. S. & Tilman, D. (2006). Biodiversity loss threatens human well-being. *PLoS Biology* 4: 1300–1305.

Ehrlich, P. R. & Ehrlich, A. H. (1981). *Extinction: The causes and consequences of the disappearance of species*. New York: Random House.

Fisher, B., Turner, R. K. & Morling, P. (2009). Defining and classifying ecosystem services for decision making. *Ecological Economics* 68: 643–653.

Garbach, K., Milder, J., DeClerck, F., Montenegro de Wit, M., Driscoll, L. & Gemmill-Herren, B. (2016). Examining multi-functionality for crop yield and ecosystem services in five systems of agroecological intensification. *International Journal of Agricultural Sustainability* 15 (1): 11–28. doi:10.1080/14735903.2016.1174810

Gomez-Bagghethun, E., de Groot, R., Lomas, P. L. & Montes, C. (2010). The history of ecosystem services in economic theory and practice: From early notions to market and payment schemes. *Ecological Economics* 69 (6): 1209–1218. doi:10.1016/j.ecolecon.2009.11.007

Jacobs, S., Burkhard, B., Van Daele, T., Staes, J. & Schneiders, A. (2015). 'The Matrix Reloaded': A review of expert knowledge use for mapping ecosystem services. *Ecological Modelling* 295: 21–30.

Jacobs, S., Dendoncker, N. & Keune, H. (2013). 'No root, no fruit – Sustainability and ecosystem services', in Jacobs, S., Dendoncker, N. & Keune, H. (Eds.) *Ecosystem services: Global issues, local practices*. New York: Elsevier.

Jacobs, S., Dendoncker, N. & Verheyden, W. (2016a). 'Why to map? Guidelines for critical and effective ecosystem service mapping', in Burkhard, B. & Maes, J. (Eds.) *Mapping ecosystem services*. Sofia: Pensoft.

Jacobs, S., Dendoncker, N., Martin-Lopez, B., Barton, D. N., Gomez-Baggethun, E., Boeraeve, F., . . . Washbourne, C.-L. (2016b). The new valuation school: Integration of diverse values of nature in land use decisions. *Ecosystem Services* 22 (B): 213–220.

Kallis, G., Gomez-Baggethun, E. & Zografos, C. (2013). To value or not to value? That is not the question. *Ecological Economics* 94 (C): 97–105.

Karsenty, A. (2011). Coupler incitation à la conservation et investissement. *Perspective* 7: 1–4.

Lamarque, P., Quétier, F. & Lavorel, S. (2011). The diversity of the ecosystem services concept and its implications for their assessment and management. *Comptes Rendus Biologies* 334 (5–6): 441–449.

Laurans, Y., Rankovic, A., Billé, R., Pirard, R. & Mermet, L. (2013). Use of ecosystem services economic valuation for decision making: Questioning a literature blindspot. *Journal of Environmental Management* 119: 208–219. doi:10.1016/j.jenvman.2013.01.008

Martinez-Alier, J., Munda, G. & O'Neill, J. (1998). Weak comparability of values as a foundation for ecological economics. *Ecological Economics* 26 (3): 277–286.

Mazoyer, M. & Roudart, L. (2002). *Histoire des Agricultures du Monde: Du Néolithique à la Crise Contemporaine*. Paris: Points Éditions.

MEA. (2005). *Ecosystems and human well-being: Current states and trends*. Washington, DC: Island Press.

Muradian, R., Arsel, M., Pellegrini, L., Adaman, F., Aguilar, B., Agarwal, B., . . . Urama, K. (2013). Payments for ecosystem services and the fatal attraction of win-win solutions. *Conservation Letters* 6 (4): 274–279. doi:10.1111/j.1755-263X.2012. 00309.x

Norgaard, R. B. (2010). Ecosystem services: From eye-opening metaphor to complexity blinder. *Ecological Economics* 69 (6): 1219–1227. doi:10.1016/j.ecolecon.2009.11.009

Peeters, A., Dendoncker, N. & Jacobs, S. (2013). 'Enhancing ecosystem services in Belgian agriculture through agro-ecology: A vision for a farming with a future', in Jacobs, S., Dendoncker, N. & Keune, H. (Eds.) *Ecosystem services: Global issues, local practices*. New York: Elsevier.

Pirard, R., Billé, R. & Sembrés, T. (2010). Upscaling payments for environmental services (PES): Critical issues. *Tropical Conservation Science* 3: 249–261.

Ponisio, L., M'Gonigle, L. K., Mace, K. C., Palomino, J., de Valpine, P. & Kremen, C. (2014). Diversification practices reduce organic to conventional yield gap. *Proceeding of the Royal Society B* 282 (1799): 20141396. doi:10.1098/rspb.2014.1396

Rockström, J., Steffen, W., Noone, K., Persson, Å., Chapin, F. S. III, Lambin, E., . . . Foley, J. (2009). Planetary boundaries: Exploring the safe operating space for humanity. *Ecology and Society* 14 (2): 32. Retrieved from www.ecologyandsociety.org/vol14/iss2/art32/

Rosset, P. (2000). The multiple functions and benefits of small farm agriculture in the context of global trade negotiation. *Development* 43 (2): 77–82.

Santos-Martín, F., Montes, C., Martín-López, B., et al. (2014). *Ecosystems and biodiversity for human wellbeing – Spanish National Ecosystem Assessment.* Madrid: Ministerio de Agricultura, Alimentación y medio ambiente.

Seppelt, R., Dormann, C. F., Eppink, F.V., Lautenbach, S. & Schmidt, S. (2011). A quantitative review of ecosystem service studies: Approaches, shortcomings and the road ahead. *Journal of Applied Ecology* 48: 630–636. doi:10.1111/j.1365–2664.2010.01952.x

Servigne, P. (2014). Nourrir l'Europe en temps de crise: vers des systèmes alimentaires résilients. *Editions Nature & Progrès*. pp. 190

Seufert,V., Ramankutty, N. & Foley, J. A. (2012). Comparing the yields of organic and conventional agriculture. *Nature* 485: 229–232.

Strijker, D. (2005). Marginal lands in Europe – Cause of decline. *Basic and Applied Ecology* 6 (2): 99–106.

TEEB. (2010). *The economics of ecosystems and biodiversity: Mainstreaming the economics of nature: A synthesis of the approach, conclusions and recommendations of TEEB.*

Tilman, D., Balzer, C., Hill, J. & Befort, B. L. (2011). Global food demand and the sustainable intensification of agriculture. *PNAS* 108 (50): 20260–20264.

UK NEA. (2011). *The UK national ecosystem assessment: Understanding nature's value to society: Synthesis of the key findings.* Cambridge: UNEP-WCMC.

Vatn, A. (2005). Rationality, institutions and environmental policy. *Ecological Economics* 55 (2): 203–217.

Westman, W. E. (1977) How much are nature's services worth?, *Science* 197, 960–964.

Wezel, A., Bellon, S., Doré, T., Francis, C., Vallod, D. & David, C. (2009). Agroecology as a science, a movement and a practice. A review. *Agronomy for Sustainable Development* 29 (4): 503–515.

10 Food systems and human ecology

An overview

Federico Davila and Robert Dyball

Introduction

A major sustainability challenge is to feed the world's population whilst reducing environmental impacts, narrowing inequities in food access, and meeting global nutritional needs (Lawrence et al., 2010; Ingram, 2011). If this challenge is to be met, processes operating between key interacting factors must be successfully managed. These factors include the social and economic (Carolan, 2016; Dethier & Effenberger, 2012), environmental (Vermeulen et al., 2012), and health and wellbeing (Friel & Ford, 2015) and occur across the food chain from production, manufacture and processing, to distribution, retail, and end consumption. Because these factors dynamically interact to drive changes in each other, it is preferable to think of food systems rather than chains (Ericksen, 2008; Ingram, 2011). The goal of a food system is, or should be, to regularly and reliably make appropriate food available at a specific scale, be it a household, town, or nation. We add the words 'should be' to flag that the purpose or goal of food systems is actually contested, as discussed later.

Current food systems are failing many people and communities around the globe. A billion people cannot regularly and reliably obtain minimally adequate calorific food intake, and 2 billion more do not achieve adequate nutrient consumption needed for good health. Another 2 billion suffer from overconsumption of inappropriate foodstuffs and suffer from a range of health issues associated with being overweight or obese, often also in conjunction with inadequate nutrient intake (Ingram et al., 2016; Westengen & Banik, 2016; Friel & Ford, 2015). Typically, the solution to poor health and wellbeing outcomes from food system failure has been sought through applied agronomics, aimed at increasing volumes of food produced, in combination with agricultural and trade policies designed to facilitate free markets and trade (Carolan, 2016; Lee, 2012). This approach has been called the 'productionist paradigm'(Lang, 2010; Lang & Heasman, 2004). Not only has the productionist approach failed to meet the challenge of feeding the world's population, but it has also produced social and environmental ills of its own. These negative outcomes range from the poor incomes and low social status of many of the world's food producers and rural communities to the fact that agriculture globally is one of the major drivers of biodiversity loss, nutrient loss and land and water degradation, and climate change (Carolan, 2016; Deutsch et al., 2013; Ericksen, 2008; Ingram, 2011). Policy and decision making at all scales from local, regional, national, and international levels is urgently needed to address these critical yet persistent health and wellbeing effects of inadequate food system outcomes and to halt and reverse associated environmental damage.

Understanding the behaviour of food systems is further confounded by the systemic uncertainty brought by the knowledge, beliefs, and judgements of people. At the level of individual households and consumers, people's values and belief systems influence how they produce, purchase, and consume food, including what they see as a 'normal' expectation of comfort, choice, and cost of foodstuffs year round. This expectation of entitlement, at least in affluent nations, demands that food systems make a wide range of foodstuffs available irrespective of the realities of local or regional seasonal agricultural production (Clapp, 2015; Porter et al., 2013). Satisfying this demand requires constant stocking of foodstuffs sourced from highly flexible globally distributed inventories and is inherently energy intensive, wasteful, and uncaring of justice and sustainability issues (Christensen, 2015).

At the level of governments and private corporations, dominant beliefs in, for example, what is seen as the role of business and industry, technology, free markets, and trade influence how food systems are conceived as optimally operating (Westengen & Banik, 2016; Barling & Duncan, 2015). Decision makers, and those with the power to influence them, strive to create legal, institutional, and market mechanisms that reflect and give effect to these priorities. Furthermore, many global agribusinesses activities sit beyond the reach of any sovereign jurisdiction and are subject to little accountability, other than the corporation's own sense of responsibility (Christensen, 2015; Kalfagianni, 2015). These myriad and often-conflicting sets of priorities, beliefs, and value-judgements held by various parties with differing power and economic agency interact across and between scales (Berbés-Blázquez et al., 2016). The ultimate result is massive inequities in the availability of food to feed the world's population, and myriad environmental and social justice problems, generating a global 'wicked problem' (Brown et al., 2010).

Wicked problems have endemic features that limit the capacity of conventional science, operating as 'problem solving', to contribute towards their resolution. Wicked problems typically are not so much 'solved' as rendered 'manageable' or 'acceptable' to those engaging with them. To address wicked problems we need novel ways of framing and designing interventions that deal with social and environmental domains and identify the root causes of sustainability problems (Abson et al., 2017; Lövbrand et al., 2015). To address them, we need a form of science that is capable of handling both quantitative and qualitative variables in the same frame, as people's beliefs or opinions about the problem are as important as its factual elements.

We need a science that does not eschew normative judgements about how just or acceptable the situation is: that is comfortable with defending what should be. And we need a science that embraces the knowledge and values of the broader community that are affected by problems and whose opinion about proposed interventions must be genuinely taken into account. Such a science would be fundamentally democratic and synthetic. Funtowicz and Ravetz (1991) coined the term 'post-normal science' for the form of science need to help inform decision making in these circumstances where 'facts are uncertain, values in dispute, stakes are high, and decisions are urgent' (Funtowicz & Ravetz, 1993, p. 744). They also coined the term 'extended peer community' for the class of affected stakeholders who must be politically engaged in the co-production of the knowledge needed to inform mutually acceptable and prudential policy directions. With high and contested decision stakes, inherent uncertainty, and significant ethical dimensions and political power imbalances, the challenge of justly and sustainably feeding the world's population sits squarely in the domain of interdisciplinary post-normal science. What is needed to operationalize rigorous post-normal science for understanding global food systems is a conceptual framework that enables comprehensive understanding of the nature of the problem as a whole. Such framework needs to be logical to all relevant actors, preserve policy relevance, advocate for stakeholder inclusion, and inform decision making at the level of the specific local contexts where the problems manifest (Ericksen, 2008; Ravetz, 2006; Foran et al., 2014; Wittman et al., 2016; IPES Food, 2015).

Here we present the systems thinking framework developed in *Understanding Human Ecology* (Dyball & Newell, 2015) as a framework for grounding a post-normal scientific understanding of sustainable food systems. This framework allows us to analyse the influence that underlying discourses have on a system's behaviour and outcomes and to compare two or more systems in terms of their common structure. Applying human ecology to food systems allows us to capture how different dominant discourses and degrees of social power influence system outcomes and affect social arrangements, human well-being, and ecosystem health. This in turn draws attention to the need to challenge and change these belief systems if we are to generate new food systems with different structures and so with more just and sustainable outcomes.

Within this context, we have two aims for this chapter:

1   To demonstrate how human ecology helps in identifying the influence that different discourses have on perspectives and solutions for food system challenges.
2   From this identification, to propose a future food systems research agenda that acknowledges and integrates governance and politics, including issues of power.

We first present the human ecology framework and apply it to a food system problem space. We then provide an overview of two competing sets of food discourses and compare how they create differing meaning, judgement, and behaviour to influence food systems. The overwhelmingly dominant discourse we address is that of 'food security', with its conventional definition of being the situation 'when all people at all times have access to sufficient, safe, nutritious food to maintain a healthy and active life' (FAO, 2015). With the application of the human ecology framework, we will show how this discourse privileges food systems that can be structured to be neither just nor sustainable and liable to not deliver the expected health outcomes to consumers. We contrast this with a discourse of 'food sovereignty', which is focused on national and community-level rights and inclusion in food decision-making processes (Wittman et al., 2010a; Wittman et al., 2011). Food sovereignty remains more marginalized than food security as a discourse, yet we argue that it offers a way of re-conceptualizing the goal of food systems, specifically to give voice to the dimensions of justice and sustainability that the food security discourse disempowers. We then organize themes from the literature that studies food issues and solutions in relation to the ecosystems, human wellbeing, and institution variables in the framework. We conclude the chapter by arguing that food systems can benefit from social science work that looks at food governance and politics, as they are driven by the discourses analysed throughout the chapter.

The theory: human ecology framework

We employ a dynamic systems framework drawn from human ecology as it allows us to holistically and comprehensively understand the behaviour of complex human–environment systems. Such an approach draws specific attention to the influence of the dominant discourses driving change in such systems. Ostrom (2010) highlights the crucial role that frameworks play in generating meaning and understanding of complex situations. Frameworks 'organize diagnostic and prescriptive inquiry and provide a general list of variables that should be used in analysis. The elements contained in the framework help the analyst identify the central questions that need to be addressed' (Ostrom, 2010, p. 5). The human ecology framework helps reveal what community and stakeholders seek to include and what to exclude in the problem situation under analysis. It can then promote debate about what the problem is but also what should be done about it through interpreting the role of different discourses.

The framework deliberately constrains itself to the consideration of a limited number of key interacting variables. This is crucial to enable the participating peer community to not get lost in the complex detail of the problem situation, but to illustrate the 'non-linear effects caused by feedback and accumulation, and focus on the endogenous dynamics generated within well-defined boundaries' (Newell & Siri, 2016, p. 93). The resulting diagrams are heuristic devices to simply and clearly reveal different understandings of how the situation is understood and to 'see where each other is coming from': a crucial first step to collaboration and co-production of knowledge and policy. The models facilitate discussion of 'dynamic hypothesis', defined as 'a causal structure that is proposed to explain the behaviour of a system in terms of endogenously generated feedback effects' (Dyball & Newell, 2015, p. 66). Consequently they guide democratic and collective debate of the 'what about/what if' questions asked by post-normal science (Ravetz, 1997).

'Variables' are the objects (understood to include non-physical objects such as a 'discourse') that the particular problem situation is composed of, with the definitional meaning that those objects can be present in greater or lesser amounts. In food systems it is, for example, the amount of food that a person has access to that might be one variable of concern. The amount of a variable changes dynamically over time, so we might be concerned not just that a person has sufficient food on a given day, but whether the pattern of change in the amount they have over time is regularly and reliably sufficient or periodically insufficient. Our focus then would turn to the processes that are changing those volumes over time to see what intervention might remove the problem of periodic insufficient supply. The diagrams illustrated here are then 'snapshots in time' capturing the state of the system, as evidenced by the amount or extent of its variables, at a given moment. However, the system is dynamic and interacts and adapts across time, and so it is the patterns or trajectory of change that is important. Too much focus on the state of the system, such as the number of fish at a given time, can be deceptive. If the rate of fishing exceeds the rate of replenishment through breeding, then the fishery is in an unsustainable downward trajectory towards collapse even if its population appears inexhaustibly large.

Figure 10.1 presents a human ecology framework based on Dyball and Newell (2015). The four variables shown are the fewest number of the most abstract and

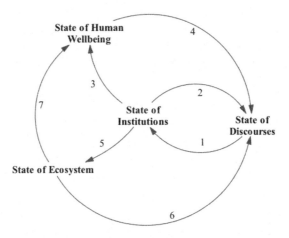

Figure 10.1 The four generic labels for variables in human–environment systems and feedback

generic categories of variables (including immaterial things like discourses) that require consideration in any human–environment situation.

These variables are labelled as follows:

- State of Ecosystem: the quantity or extent of a variable that indicates the state of the environment at any point in time. At certain levels some people may experience the amount of this variable as a problem. For example, the amount of some particulate may accumulate in the atmosphere to a level where it is considered a pollutant. Equally, the problem could emerge if the level of the variable deteriorated below a certain threshold, such as the decline in terrestrial stocks of available phosphorus. A provisional list of the key variables of concern, and associated ranges of levels of anthropogenic impacts which are deemed acceptable, are proposed in an influential publication on the concept of 'planetary boundaries' (Rockstrom et al., 2009).
- State of Human Wellbeing: the quantity or extent of a variable that indicates the state of an individual or community's physical and psycho-social wellbeing. The level of these indicators would allow judgements as to whether the individual or group's standard of living was sufficiently above some threshold that they could be said to be living well. A provisional list of universal human health needs is offered by Boyden (2016). This list acknowledges that there are both bio-physical subsistence thresholds common to all humans, such as minimal nutrient intake for good health, and culturally relative standards of adequacy unique to a particularly society. For example, there is no biological requirement to wash or have a number of different clothes suitable for different occasions, but in many societies it would be socially unacceptable not to. Furthermore, although these variables may have minimal thresholds, they do not necessarily have obvious maximum limits, such as the point at which the

community has too much love. Boyden reminds us that living well is driven both by the absence of stressors, such as the absence of sources of fear in our community, and the presence of mitigating factors, such as the presence of convivial social networks (Boyden, 2004, pp. 67–68).

- State of Institutions: the quantity or extent of a variable that indicates the state and effectiveness of both formal and informal rules that are structuring the interactions and the collective behaviour of the community in question. An institution is

> a persistent, reasonably predictable arrangement, law, process, custom or organization structuring aspects of the political, social, cultural or economic transactions and relationships in a society. Institutions allow organized and collective efforts toward common concerns and the achievement of social goals. Although by definition persistent, institutions constantly evolve.
>
> (Dovers, 2001, p. 5)

We note that institutions at different levels can enable collective endeavour towards desirable social goals, but also blind or obstruct reform (Fischer et al., 2012).

- State of Discourse: a discourse is a set of ideas that stimulates human activity and collective action (Dryzek, 1997). State of discourse is represented by indicators that capture how the situation is being framed and what collective meaning is interpreted from a given set of signals about the state of the situation. Discourses influence the behaviour and goals of systems, largely because they establish institutions that are intended to give effect to whatever collective response they promote as prudent ('wise') in the circumstances. Not all individuals in a society share the same discourse, and any two people may have a greater or lesser degree of a 'shared conceptual repertoire', which can be a major obstacle to collaboration (Dyball & Newell, 2015, p. 53). This lack of a shared conceptual repertoire is almost inevitable when an extended peer community that includes scientists, policy makers, extension officers, and farmers undertake to collaborate on reducing food insecurity. It is another reason the framework uses only a few, simply labelled, variables and processes, because these generic and basic concepts are more likely to have shared understanding and meaning. However, the ability to frame the discourse at a national or international level is a crucial dimension of social power (Lakoff, 2004). For many complex problems the enduring solution lies in recognizing and subverting the power of the dominant discourse so as to reframe the discourse and restructure the system (Meadows & Wright, 2009).

These variables interact with each other through processes that feedback to constrain each other's behaviour, represented in the framework by arrows. Links one, three, and five represent individual and collective activity that function to change the quantity or extent of the variables to which they point. Links two, four, and six are observation processes whereby the individual or community receives signals informing them about the change in the quantity or extent of affected variables. This may cause learning and adaptive change in the dominant discourse, which then would feed back to manifest as new collective action and drivers on the affected variables. Entrenched power and policy resistance may mean the signals are too weak to change the discourse. Link seven is the only process link that is not mediated through social institutions. It represents the direct effect of changes to environmental variables on human health and wellbeing. It can be thought of as the 'co-benefits' (or burdens) that action to change the state of the environment has on changing the state of human health and wellbeing. An example would be the co-benefit of protecting riverine habitat for endangered fish on the quality of water drawn from that river for human consumption.

The framework promotes understanding of human–environment interactions as primarily feedback systems in which the overall behaviour of the whole emerges from the interactions between its parts. This is important for decision making, as one cannot understand the behaviour of such systems by studying the behaviour of the parts taken in isolation. It follows that any policy intervention design to affect a part of such a system in isolation of the whole is liable to fail. We have to study the system as a whole. However, we need to do so in such a way as to not be overwhelmed by its complexity and retreat to ineffective partial approaches. The framework provides a means of understanding human–environment systems comprehensively. By promoting 'feedback guided analysis' we can reveal the way that the systemic structure of problematic systems is acting to constrain how the parts of the system drive change over time. This then allows us to consider what the case is and what should be the case for any human–environment situation and to suggest points of successful and lasting intervention by changing the structure of the situation and consequently its behaviour.

In this view the meaning of the word sustainability becomes a description of the characteristic rate of change in the value of key indicator variables over time, such that the variables are not accumulating (or declining) towards their relevant safe thresholds. The three principles of sustainability set out by Dyball and Newell (2015) are:

1   A process that consumes a non-renewable resource is sustainable as long as the rate at which it uses that resource does not exceed the rate at which a renewable resource (used sustainably) is substituted;
2   A process that consumes a renewable resource is sustainable as long as the rate at which it uses that resource does not exceed the rate of regeneration of the resource;
3   A process is sustainable as long as the rate at which it generates a pollutant does not exceed the rate at which that pollutant can be recycled, absorbed, or rendered harmless in the environment (Dyball & Newell, 2015, p. 94).

The principles of justice enshrined in human ecology (Christensen, 2015) demand that these principles are met in such a way that all members of the community achieve a level of consumption that enables at least a minimally dignified level of health and wellbeing.

Summary of the value of a human ecology framework

Some values of the human ecology framework are:

1   It provides an operational definition of sustainability that managers can apply to any context, and it conjoins biophysical sustainability with standards of justice and fairness.

2   It surfaces the pernicious role of often unseen internal feedbacks operating between sectors that are often seen as separate (e.g. urban planning, freeway construction, and health) and encourages managers of those sectors to collaborate towards common goals.

3   It focuses attention on a problem indicator (e.g. declining food security) as a symptom emerging from the system structure and ensures policy interventions are drawn towards changing the structure that is causing the symptom.

4   It distinguishes between the state of the system at a point in time and its change process. That means it allows for the range of states over time to be explored.

5   It operates with a few accessible concepts that enable shared understanding within the community or policy-making group as a foundation to the co-production of knowledge and decision making.

6   It can reveal how different actors in different contexts relate through shared common feedback structures. Thus, actors can learn from each other's successful interventions, even if the specific elements are different.

7   By focusing on changing state change behaviour by changing system structure, it helps avoid ultimately futile policy interventions that attempt to change the state of a single variable in isolation from the broader system of which it is a part.

8   It draws attention away from simple cause-and-effect relationships that are the proximate explanations of change to the ultimate drivers of change.

9   In most human-ecological systems, this quest for finding solutions that address the ultimate drivers of change involve identifying the 'goal' or 'purpose' of the system and the power of the discourse that legitimizes that goal.

In the next section we apply this human ecology framework to the kinds of problems endemic to food systems, drawing attention to the dominant discourse that is ultimately responsible for how food system problems are being framed.

The human ecology of food systems

In this section we focus on food security and food sovereignty as two major food discourses that influence the food systems' behaviour. We note that, conventionally understood, the former fails to adequately address the broader social, justice, and environmental aspects of food systems (Lee, 2012; Wittman et al., 2010b). These aspects are the central concern of the latter discourse, and at a scale where governance of food systems becomes possible. However, the food sovereignty discourse is not without its shortcomings, notably in contexts where food shortages are endemic, its stance on global trade, and idealistic visions of smallholder farmers feeding the world (Bernstein, 2014; Jansen, 2014; Aerni, 2011; Burnett & Murphy, 2013). Both social systems and ecosystems are affected by food activities driven by the respective discourses, and both co-exist (Jarosz, 2014). Consequently, we discuss how the discourses propose solutions in the form of sustainable agriculture, sustainable intensification, and agroecology and the research opportunities available to explore how both discourses co-exist in food governance systems.

State of food discourses

Here we introduce food security and food sovereignty as two discourses that currently exist in food systems research and policy debates. Food security discourse is associated with technical, positivist approaches to tackling hunger through a mix of technological advancements and providing economic access to food (Maye & Kirwan, 2013; Jarosz, 2014). These ideas are globally prevalent in agricultural policies, research programs, and social activities throughout the world (Lee, 2012). Food security literature is focused on increasing food production to meet projected increases in population by 2050 (Maye & Kirwan, 2013), and thus promotes and legitimizes supporting trade policies, corporate investments, and policies into specific sectors geared primarily to the economically efficient increases in volumes and distribution of food. Some of the propositions within the food security discourse include:

- Re-evaluating trade practices to ensure food availability throughout the world,
- Expanding deliberation across stakeholders,
- Increasing private labelling and governance systems,
- Developing biotechnology and sustainable intensification practice,
- Achieving nutritional opportunities, and
- Having greater dialogue between actors across scales (Candel, 2014; Maye & Kirwan, 2013).

An alternative perspective is provided by food sovereignty. Defined as the right of nations and peoples to control their own food systems, including markets, production models, food cultures, and environments (Wittman et al., 2010b), food sovereignty came from civil society organizations, notably the peasant farmer group La Via Campesina. It promotes concern over the food security discourse being used to support large-scale agricultural development policies and technical solutions to world hunger (Desmarais, 2007), and its language has been used to mobilize alternative food production systems and civil society networks throughout the world (Wittman et al., 2010a). At its core, the food sovereignty discourse sees food as a fundamental human right rather than solely a market commodity.

The main propositions of the food sovereignty discourse include:

1   Treating food as a human right
2   Promoting equitable agrarian reform
3   Protecting natural resources
4   Reorganizing food trade
5   Ending hunger
6   Social peace
7   Democratic control over food policies

Food security and food sovereignty have been presented as contradictory and opposing discourses. The fundamental difference is that food security aligns with the interest of economic growth and global agricultural markets (Lee, 2012; Westengen & Banik, 2016). Food sovereignty is concerned with decision-making processes, cultural diversity, and environmental wellbeing. Presenting the discourses as opposing, however, is unhelpful in pursuing meaningful interventions in food system activities that lead to human and environmental wellbeing (Clapp, 2014). Instead of an 'either/or' argument, what is really needed is critical integration and empirical analysis of how both discourses can co-exist across scales (Jarosz, 2014; Clapp, 2014; Edelman, 2014; Leventon & Laudan, 2017). The debates on transitioning to sustainable food systems that include human rights and sustainable production concerns provide a platform for analysing how both discourses influence human wellbeing, institutions, and ecosystems.

State of human wellbeing in food systems

The health and wellbeing of consumers are affected by lack of balance in food choices and consumption habits, poor dietary intake, and obesity problems which stem from the commodification of food (Lawrence et al., 2010). Human wellbeing in food systems also includes the socio-economic states of food producers (Carolan, 2016). The framing of food security as a commodity production problem to be primarily addressed through economic efficiency measures has prevented these broader human wellbeing issues from being addressed (Lang & Heasman, 2004; Westengen & Banik, 2016; Wittman et al., 2010b).

Health outcomes in food systems also relate to how nutrient and calorific deficiencies are being met, as they are critical to human development (IFPRI, 2015). Some of the major challenges include:

- Undernutrition, which poses threats to cognitive functioning, immunity, growth, and reproductive outcomes. People in low-income countries are most at threat from undernutrition, with child undernutrition creating long-term human development challenges (IFPRI, 2015).
- Excessive consumption can lead to over-intake of calories, yet still be nutrient deficient. This phenomenon has become increasingly common in industrialized countries over the last few decades, and rapidly growing middle-income economies are seeing upward trends in the percentage of their citizens who suffer poor health from excessive and unbalanced food intake. Even low-income countries can have significant sub-populations of over-consumers, resulting in the so-called 'double burden' on their health care systems of having to cope with both over- and under-consumption (The World Health Organization, 2016).
- A mix of factors, ranging from easy access to energy-intensive foods and poor levels of physical activity, has contributed towards the negative health consequences of excessive consumption (Carolan, 2016).

People acting on the market-focused ideas from the food security discourse have largely prioritized the production of staple commodities to meet market demand (Lee, 2012). Total food output has outpaced human population growth, largely through the increased use of technology and policies supporting food trade. Agricultural technologies include mechanization and industrialization to increase production efficiency, fertilizers to increase soil and crop productivity, pesticides and herbicides to reduce losses, animal veterinary and feed improvements to boost growth rates and muscle mass, and genetic modification of crops (Ehrlich & Harte, 2015). These technological advancements have allowed for increased staple commodity production, such as corn, sugar, and soy. These bulk commodities feed into agri-businesses and are processed to make a range of products, which enter the markets like any other retail commodity. This distribution of food produced has been facilitated by an expansion in global trade, which has served to make a wide range of foods available relatively cheaply, although this cheap food has not realized the promise of adequate access to the world's poor, and by negatively affecting local producers, arguably reduced local food security (Wittman et al., 2010a).

The consequence of being able to mass-produce food items from staple crops has been to make a wide range of food products available that prompt comfort, convenience, and cheapness. As the food security discourse treats all products of a particular kind as 'like' all others of that kind of product, it cannot discern between product types on the grounds of their healthiness and offers little insight into the promotion and overconsumption of highly processed, nutritionally poor, and energy-dense food products. Within this discourse, the only way consumers interact with food systems is as economic agents, making food choices on the grounds of perceived value and preference. The health outcome of these choices is seen as the individual's responsibility, or the concern of the health industry, not a food security issue.

Alternatively, the food sovereignty discourse values the production of culturally appropriate localized food, providing consumers with a food options that reflect the constraints of regional conditions. Here, the focus is on producing a range of products for consumption through domestic markets, rather than staple commodities for global trade. Community networks, including non-commercial produce from sites such as urban gardening, encourage the distribution and consumption of local and culturally appropriate seasonal foods. These informal networks can also enable feedback between producers and consumers, creating adaptive behaviour to balance ecosystems' health and human wellbeing outcomes (Davila & Dyball, 2015). However, the focus on local foods and seasonality can be problematic for many of the world's poor and consumers who live in countries that do not have sufficient productive agricultural land available. For example, smallholder farmers who are dependent on seasonal commodities for their income face 'hungry seasons' when they are unable to buy food from markets and their production is insufficient to meet their household's demands (Bacon, 2015). Many dense urban populations would simply be unable to meet their consumption needs from their regional hinterlands, or would have to accept a highly monotonous, potentially nutrient deficient, diet were they to try (Porter et al., 2014). It is then neither possible nor desirable that the food sovereignty discourse, as currently conceptualized as local production servicing local demand, be globally extended.

State of food-producing ecosystems

Food-related activities across the entire food system are estimated to contribute between 12% and 19% of global anthropogenic greenhouse gas emissions (Vermeulen et al., 2012). From production, to processing and manufacture, to distribution and retail and consumption, food system activities both affect and are in turn affected by water resources, biodiversity decline, land use changes, nutrient cycling, and ecosystem services (Vermeulen et al., 2012; Ingram et al., 2016; Deutsch et al., 2013). For example, land used for agricultural production is expected to expand by 110 million hectares in the coming decades in emerging economies (Alexandratos & Bruinsma, 2012). This is likely to result in food production being pushed into land areas with high conservation value (Montesino Pouzols et al., 2014) and further intensification of existing food production landscapes. Intensification has serious detrimental effects; for example, the production of staple crops such as sugarcane, palm oil, rubber, and coffee have high impacts on biodiversity loss as high and increasing global demand drives expansion of plantations and mono-cropping (Chaudhary & Kastner, 2016).

The food security discourse productivity focus has created a range of perceived solutions to environmental problems that fit within its over-arching narrative. For example, the term 'sustainable agriculture' has become dominant in governments, civil society, and private groups involved with food. A challenge lies in the fact that there has been a focus on attempting to identify what sustainable agriculture looks like whilst failing to fully analyse the social relations that impact the development, uptake, and potential success of change in practices (Velten et al., 2015). Another solution to environmental impacts of food production is 'sustainable intensification'. This solution focuses on producing more food within existing landscapes. There is a risk of focusing only on one variable within the food system (maximizing production) whilst ignoring other domains of sustainability (such as human wellbeing), thus not being a truly sustainable solution (Loos et al., 2014). These technical examples indicate that there is scope to integrate different cultural and social contexts into solutions and focus on more than solely improving food production.

The food sovereignty discourse differs in the solutions it proposes for problems in agricultural production and to improve ecosystem wellbeing. Food sovereignty literature and activists often use the language of agroecology. Agroecology is a knowledge-intensive way of producing food that maximizes on farm natural resources, closes nutrient cycles, reduces or avoids waste and losses, and reduces dependence on external inputs (Tomich et al., 2011; Altieri, 1987; Gliessman, 1990). The mix of nutrient cycling, provision of ecosystem services, seasonality of production, and on-farm social relations created by agroecological farming present a way in which addressing a problem in one part of the system (production) can have wide implications in other parts of the system (such as food diversity in markets) and contribute to environmental wellbeing. Agroecology has expanded to more than an approach towards food production, and has created a social movement centred on restructuring how people relate to their food systems (Wezel et al., 2009). Inherently holistic, agroecology extends to include both producers and consumptions as agents within the food system, and thus has the potential to break down the typical conflict between the rural–urban divide and replace it with a common alliance of mutual recognition of their co-dependence (Stuart, 2014). The political potential of more deeply connecting urban consumers with rural producers is high and can be facilitated by making both consumers and producers active agents in their food system (Davila & Dyball, 2015)

State of food institutions

Global food institutions have historically focused on agricultural supply, neglecting environmental and social concerns within food systems (Lee, 2012; Barling & Duncan, 2015). The World Trade Organization (WTO) has played a major role in creating rules for food flows across nations, yet the legally binding agreements have been criticized as harming poorer countries. The long history of agricultural protection policies has disadvantaged producers from the developing world. Agricultural development programs from global institutions like the World Bank and aid agencies have focused on prioritizing access to food and developing coping mechanisms for poor people. A dominant belief has been the idea that increasing agricultural productivity is the best tool for smallholder farmers to escape poverty, yet this has not occurred evenly throughout the world (Dethier & Effenberger, 2012). The Food and Agriculture Organization (FAO) has been responsible for creating global guidelines and metrics on food security and nutritional outcomes; however, they do not provide legally binding agreements for domestic institutions.

Domestic agricultural institutions include national governments that provide support and regulations to domestic food systems. They make strategic decisions based on their national interest, and are free to ignore many of the global, non-legally binding food guidelines global institutions put forward. Consumers in food importing nations, such as Japan, for example, have their food production standards determined by jurisdictions over which they have no influence other than by refusing to purchase from certain sources (Dyball, 2015).

Another major institutional player includes the private corporations involved in agricultural inputs and retail (Fuglie, 2016). Significant power has accrued to retailers who are able to source their inventories from almost anywhere in the world, effectively putting producers around the globe in competition with other, with orders placed with the producer willing to supply at the lowest price, with associated negative impacts on social and environmental standards. Corporations have increased the control they have over the flow of food throughout the world through owning seeds, agrochemicals, food distribution outlets, and a large number of processed products. This has had implications on the types of nutritious foods available to consumers and the way farmers produce their food, as well as the food choices available to urban consumers (Carolan, 2016; Clapp, 2015).

Alternatively, informal institutions prevalent in food sovereignty discourses provide opportunities for learning, agroecological knowledge extension, and diversification of food diets (Wittman et al., 2011). For example, farmer-to-farmer knowledge exchange networks and field schools can act as a platform to learn agroecological production practices and reduce dependence on farm inputs and single-commodity incomes (Altieri & Toledo, 2011).

Food sovereignty is frequently attributed to local and national scales, civil society, or alternative movements that seek to break from the norm in their immediate food systems (Wittman et al., 2010a). However, food sovereignty interests have made it into formal institutions such as the Committee on Food Security within the FAO and a number of national governments (Brem-Wilson, 2015; Hospes & Brons, 2016). Despite lack of concrete evidence of the impact of this inclusion into formal institutions, the importance here is the fact that a civil society–driven discourse has permeated global governance systems, showing the opportunities of diversifying discourses in food systems.

Framework application

Having presented the state of human wellbeing, ecosystem, and institutions under the two food discourses, we can now contrast the behaviour and goals of the two food systems. From this, we will suggest two future areas of research based on the themes identified in this chapter.

These diagrams shown in Figures 10.2 and 10.3 indicate how different discourses influence the different variables in the human ecology framework. They show what the proponents of each discourse believe the goal of a food system ought to be and the power that they have to try to structure the food system to meet that goal. The diagrams also reveal each discourse's 'blind spots' – those variables that they do not see as 'part of' their system, but which in reality are affected by actions taken by the discourse's followers. This is in keeping with the basic system principle that you cannot change just one thing in a complex system. Cross-sector feedback practically guarantees that a cascade of consequences will follow from your intervention. The problem that bedevils policy makers who are not alert to this fact is that the

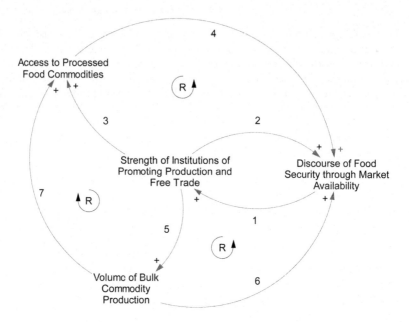

Figure 10.2 The food security discourse. All main feedback loops are positive, showing how advocates in this discourse's narrative believe more and more food becomes available, hence achieving universal security. The processes are discussed in Table 10.1.

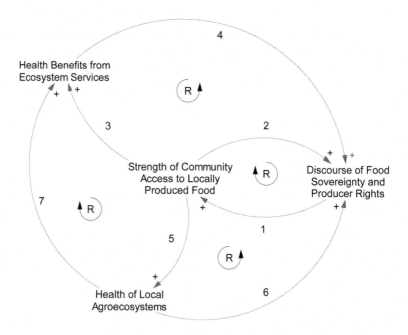

Figure 10.3 The food sovereignty discourse. Like the food security discourse, feedback loops in the food system are all positive, as the discourse itself believes them to be. Table 10.2 describes the feedback processes represented by the links and processes that the food sovereignty discourse observes and values.

Table 10.1 Processes that the food security discourse observes and values

Link number	Process represented by the link
1	This link is positive. The food security discourse holds that food security is best achieved by increased production volumes. Consequently, it acts to create institutions empowered to promote production and open markets to free trade.
2	This link is positive. Observations of the strength and effectiveness of free trade institutions further reinforce the discourse of food security through increased production. Any observations of production inefficiencies or barriers to trade are corrected.
3	This link is positive. Consumer access to processed food commodities moves in the same direction as the strength of institutional support for food commodity systems.
4	This link is positive. The food security discourse is reinforced by observing levels of consumer access to food commodities. The discourse corrects for shortfalls between level of existing levels of access and complete access. It is neutral or antagonistic to food secured by means other than commodity system, because these are 'inefficient' in its narrative. The strength of the food security narrative is unaffected by other measures of health and wellbeing, such as obesity, as it does not monitor those variables. Other discourses, such as the population health discourse, would monitor variables such as obesity, and for that discourse this link would be negative, indicating the decrease in health from overconsumption of processed food commodities.
5	This link is positive. The presence of strong markets institutions for bulk commodities drives agriculture systems to increase their production of those commodities. Farms combine into larger systems, industrialize, mechanize, and become input intensive under a narrative of increased efficiency. Smaller farms that cannot compete go out of business. This may well include national producers as global free markets are uncaring where on the planet the food is produced.
6	This link is positive. The food security discourse is reinforced by large volumes of economically efficient produce feeding into the production of cheap, convenient food commodities. There may be some concerns that food security has been harmed where primary production has moved offshore and the nation become import dependent for key staples, but free trade agreements are seen as the solution to this. Other side effects of industrial food production, such as biodiversity loss, over-fertilization, animal welfare, or producer's pay and conditions are not monitored.
7	This link is positive. High yield produce from intensive farming systems directly correlates to the volumes of food commodities produced. Many other environment-to-human health links are negative, such as the degrading of other ecosystem services under intensive agriculture. However, these negative feedbacks are not considered aspects of food security and are not monitored.

Table 10.2 Processes that the food sovereignty discourse observes and values

Link number	Process represented by the link
1	This link is positive. The food sovereignty discourse holds that food is the sovereign property of the producer and the region of production. Food is not a commodity – it is a right. Producers and consumers are to participate in food systems as mutually interdependent active agents. The discourse promotes local community markets and other spaces where this exchange can take place.
2	This link is positive. Observations of the vibrant community markets where consumers and producers meet in convivial relationships with mutual solidarity reinforce the discourse of food sovereignty. Little attention is placed on either consumers or producers who are, for whatever reason, unable to participate in this market – for example, for reasons of distance. If the region or nation is food import dependent, then, of necessity, some proportion of food consumption will not be produced locally.
3	This link is positive. Consumer access to healthy, locally produced, and culturally appropriate food choices moves in the same direction as the strength of institutional support for food local markets.
4	This link is positive. The food sovereignty discourse is reinforced by observing levels of consumer access to healthy food choices. Furthermore, the community members whose wellbeing is being monitored include the producers themselves. Consequently, the sovereignty discourse is equally reinforced by the economic and social wellbeing of producers and takes steps to correct the system if their rights are being violated. If the community is dependent for a proportion of its food consumption by producers external to the system, then this political concern for the producers' wellbeing is extended to them, even if it cannot be directly enforced.
5	This link is positive. The food sovereignty discourse places value on all ecosystem services, not merely its productive capacity. The health of agricultural landscapes is enhanced through such approaches as low-input, free-range, soil-first farming techniques, which are heavily reliant on the local knowledge and aptitude of the farmer.
6	This link is positive. The food sovereignty discourse directly monitors the health of the agricultural lands under its jurisdiction. Where it is forced by reasons of production shortfall to import food, it attempts to also monitor the health of the landscapes that produce that imported food through, for example, trustworthy labelling and traceable provenance.
7	This link is positive. The community experiences the direct health benefit from healthily produced uncontaminated and lightly processed food. It also directly benefits from a range of other ecosystem services that healthy farming landscapes provide, such as biodiversity refuges, water filtration, and carbon sequestration, as well as the cultural value of being surrounded by vibrant regional farming communities.

desired and intended consequences of your action typically appear immediately, whereas the unintended and undesired outcomes typically emerge after a delay.

We have provided an example in Box 10.1 of how detailed understanding of a food system, in this case from the Philippines, can be used to populate the framework variables. We have used evidence from our applied qualitative research experience and understanding of the literature to label the variables with Philippines food system information.

Box 10.1   The human ecology of Philippine food systems

The Philippines: an overview

The Philippines comprise 7,000 islands occupying 300,000 square kilometres. Approximately 100 million people inhabit the Philippines, half of which remain in rural areas.

State of human wellbeing

- Agriculture is a major employment sector and land use activity in the rural Philippines. Poverty is three times higher in Filipino agricultural households than in non-agricultural households, and two-thirds of the poorest Filipinos depend on agriculture as their main income (UNDP, 2013).
- The main Filipino diet consists of rice, fish, and vegetables. As access to foreign imports has grown, the consumption of starchy roots has declined and the consumption of fats and oils has increased.
- The Philippines has severe levels of stunting (33.6%) and underweight (20.2%), and medium severity in wasting (7.3%) amongst children aged 0 to 5 years. The nutrition Millennium Development Goal was not met, with the final report stating that malnutrition prevalence for children under 5 was 20.2%, failing to meet the target of 13.6%.
- One out of every ten Filipinos still relies on household food production and cannot purchase additional foodstuff to meet other nutritional needs.

State of ecosystems

- Food in the Philippines is produced in mega-biodiverse landscapes.
- The intensification of agriculture has had impacts on the country's biodiversity. Of the 167 different mammal species, over 60% are endemic, and 65% of the over 10,000 plant species are also endemic (Goldman, 2010).

- Since Spanish colonization in 1565, forest cover has decreased from 90% to 18% (Wagner et al., 2015).

State of institutions

- A range of socio-economic challenges, including weak governance, corruption, lobbying, and increased human population, inhibits biodiversity conservation action in the Philippines. Furthermore, there are little policy synergies between pursuing agricultural expansion and self-sufficiency and stemming biodiversity losses.
- National policies promote agricultural expansion and intensification, leading to severe environmental degradation (Coxhead et al., 2001).
- Policy support for staple commodities such as rice, sugar, and maize have narrowed the focus on rural development and failed to create a diversity of livelihood opportunities (UNDP, 2013).
- Land reform is a major issue, with smallholder farmers being tenants and having little influence in agricultural policies.

State of discourses

- A focus on staple commodities has driven agricultural policies and extension programs.
- Food has been framed as a commodity, and the food system in the Philippines shows the focus on sugar, coconut, and bananas as the main interest.
- A single-commodity focus has deprived rural areas from diversifying production and generating diverse economic opportunities from agricultural landscapes.
- Farmers have little capacity to influence political and policy processes that affect their food system.

The process represented by the arrows is discussed in Table 10.3.

The analysis of the Filipino smallholder food system portrayed in Figure 10.4 and discussed in the accompanying Table 10.3 suggests key points of intervention. One is to strengthen feedback process L4 so that farmers have more influence on government policy, for example, by actively lobbying for support for agroecological training. At the same time, switching away from policy focus on commodification to one on income diversification, innovation, and local value-adding would make process L3 a positive link. As a consequence, feedback loops L1, L3, and L4 would become reinforcing (as all the links would have positive polarities) and the current balancing loop which is trapping farmers in poverty would be replaced by one in which their income grew to at least above a minimally acceptable dignity threshold.

Another intervention point is to strengthen process L6, for example, by revealing cross-sector feedback from agricultural policy to tourism. If this weak link were

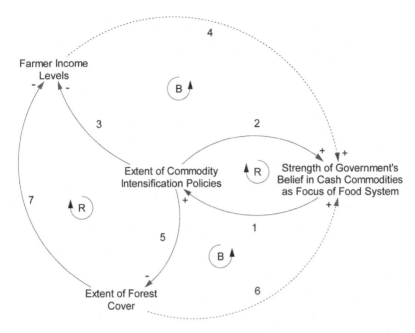

Figure 10.4 The Philippine food system for smallholder cash commodities. Farmer incomes are trapped at low levels by undesirable balancing (B) feedback loops. Forest cover declines as they are driven to pursue land clearing to try and break out of their commodity traps. Signals to the government that could moderate this systems behaviour are weak (indicated by the dashed lines). The central reinforcing (R) feedback of the belief in the policy of cash commodities drives the system into decline.

strengthened, this would stop processes L1 and L5 from functioning as an influence chain, driving forest cover down, and become a balancing feedback loop. The extent of forest cover would then be balanced against other demands but maintained around a stable state. If the positive feedback loop described earlier were to be created to lift farmers out of poverty, this balancing loop would place a brake on how affluent they could then become, or at least how much additional affluence could be achieved without degrading forest cover.

The ultimate goal would be to change process L5 so that it was not a negative. This would create a biosensitive economy in which increased human wellbeing did not come at the expense of a trade-off against environmental health and associated ecosystem services. If this were possible, the co-effects loop linking processes L1, L5, L7, and L4 would all be positive, and environmental health would increase as human wellbeing increased. This could happen under a bio-mimicking circular economy in which living energy pathways drove the increased rate of nutrients through closed cycles, allowing the overall carrying capacity to increase. Such an economy would only be possible with a significant shift in the dominant discourse, from one that believed that a growth economy was the only path to community wellbeing to a biosensitive discourse that believed in living in harmony with a

Table 10.3 Processes driving the Filipino food system

Link number	Process represented by the link
1	This link is positive. Two decades of policy orientation to increased productivity demonstrate the strength of the government's belief in commodities (Cororaton & Corong, 2009).
2	This link is positive. Observations of declining agricultural productivity during the last two decades are responded to with increasing effort to strengthen institutions charged with enacting intensification policies.
3	This link is negative. As intensification efforts go up, farmers' incomes go down. Rural incomes are especially vulnerable, with over 30% spending over half their income on food. Figures for poor nutrition, especially in children under the age of 5, are representative of the negative effects on the state of health and wellbeing.
4	This link is positive, but weak. *If* farmer income went up, it would reinforce the government policy. *In fact* income is going down and that *should* drive policy in the same direction. That is, falling farmer income ought to cause the government to change its stance, but does not. This represents the weak influence rural smallholders have on policy and government decision making.
5	This link is negative. As the policy of commodity intensification goes *up*, the behavioural response is the activity of land clearing, and so the forested land area goes *down*. The state of this variable is currently 18% of land cover and falling.
6	This link is positive, but weak. *If* forest cover were to go *up* under commodity intensification programs, that ought to drive policy in the *same direction* (i.e. strengthen it). *In fact* forest cover is going down, so that ought to weaken the policy. Lost biodiversity ought to concern policy makers because, for example, this could negatively affect tourism and associated income. However, as agriculture and tourism are seen as different policy sectors, the signal to the agricultural policy makers is weak.
7	This link is negative. As the farmers have few other options to try to escape their commodity trap other than to expand areas of production to increase total volumes, any efforts to increase forest cover would negatively affect their income. The consequence is ongoing farmer efforts to clear forest cover as one of the few strategies left to them to increase their income. With delay there is highly likely to be harmful consequences for the farmers of this strategy as a range of ecosystem services are lost. Farmers may be aware of this, but the short-term demands of their immediate perilous state of wellbeing do not give them the luxury of taking this longer view into account.

human-modified natural process. This would achieve what Ravetz (2006, p. 281) calls a 'revolution in consciousness' because it would reject the process of increasing linear resource appropriation as a requisite for the increasing the material basis of what constitutes 'living well'. Such a revolution would overcome the 'contradiction' of trying to solve the problem of poverty and natural resource degradation by the further application of the process of expropriation that created the problem in the first place.

It is one thing to describe future food systems that have the structure discussed previously, and quite another thing to plausibly imagine how the politically revolutionary changes to create it might come about. The reason for adopting an approach compatible with post-normal science was due to instrumental science's ability to say what is, yet its impotence in saying what ought to be, let alone how society might get there. Aligning with post-normal science, human ecology, as discussed earlier, is overtly normative in that it seeks to critically reflect on problem situations, precisely to change them into some better situation, both socially and environmentally. However, human ecology has not typically engaged with analysis of the role of political power in structuring systems as they are and the capacity of those wielding that power to resist change. We cover this issue in the next section, as political and power relations in a food system play a critical role in the system's behaviour.

Future research

In this chapter, we have framed food systems as a wicked problem and human ecology as a framework for pursuing interdisciplinary research and practice. The use of this framework can focus on mediating between the uncertainty of facts, competing values, and urgent actions needed for sustainable futures. To show the value of human ecology, we concentrated on two dominant discourses that influence the state and outcome of a food system. What is evident from the vast literature on food security and food sovereignty is that one discourse (food security) is dominant amongst public institutions, corporations, and food producers. The dominant focus on commodity production has made other discourses focused on ecological and human wellbeing subordinate, preventing them from meaningfully influencing the outcome of a food system. In this concluding section, we propose that in order to expand research into food discourses in food systems, critical attention needs to be given to issues of power and social relations. Doing so will require a range of methods capable of capturing human activities and behaviours and the broader connections between discourses and power in specific community groups. Such methods and analysis will need to be specific to the context in which a selected food system operates, yet findings and analysis will need to be relevant to the broader global debates presented in this chapter.

Here, we will argue that the following areas of study are needed in future food systems research:

- Food governance as a process that influences ecosystems, human wellbeing, and institutional behaviour. A focus on process can help explore how food sovereignty can expand from a subordinate to a dominant discourse.
- Taking an explicit political angle to food systems, and more broadly, sustainability science. Politics is the process whereby discourses are formed and changed,

and as a change in the dominant discourse is a necessary component of moving towards just and sustainable futures, politics is a major driver of sustainability.

Both of these areas of study are critical for sustainability because they extend beyond the interest of academics and researchers – they encompass the interests of the diverse group of stakeholders concerned with the outcome of the food system. This is critical for future pathways in transdisciplinary research, which require the identification of problems and solutions by a range of stakeholders and the formation of democratic extended peer communities, discussed in the earlier sections of this chapter.

Food governance

Governance deals with the processes and structures that influence individual or collective action that lead to the realization of a collective goal (Young, 2002). With major environmental change occurring at unprecedented rates, governance needs to become more adaptive and reflexive to deal with uncertainty and unexpected system behaviour (Hospes & Brons, 2016).

Food governance is carried out by the different institutions and actors presented in this chapter and is influenced by the competing discourses. The interactions between these groups offer an opportunity to explore the competing beliefs and discourses held by different agents (Candel, 2014). Governance is inherently a social process, and as such it would be adequate to study it within a human ecological framework. Future research can focus on discourse co-existence and how this transfers to individual and institutional actions. Case studies can be used to understand how different discourses co-exist in specific food systems. More broadly, the study of the social dimensions of sustainability is critical, given historical traditions of focusing on biophysical changes whilst ignoring socio-cultural values (Lövbrand et al., 2015).

The use of frameworks concerned with human and ecological interactions can contribute to the broader evidence base of how social systems can adapt to and manage global environmental change (Lövbrand et al., 2015). Empirical qualitative data will need to be gathered through inter- and transdisciplinary methods to capture the diverse ways of framing food problems and solutions in specific contexts. Researchers will need to work with extended peer communities to negotiate agreed normative standards for improving both human and environmental health, seen as co-extensive aspects of what it means to 'live well' (Dyball, 2010; Dyball, 2012). To achieve this and have tangible social applicability, research needs to embed contextual and political realities into its analysis. For this, explicitly politicizing food systems research can help inform future food system governance and policies that lead to environmental and social equity outcomes.

Politicizing food systems research

As food is often produced on biodiverse landscapes by lower socio-economic–resourced groups, politics is a major issue. The literature on the politics of natural resources extensively documents how peoples' interactions with one another influence environmental outcomes (Robbins, 2012; Zimmerman & Bassett, 2003). The political nature of food systems is at the core of the food sovereignty literature

(McMichael, 2009); however, there is much potential to bring this into the realms of human ecology and broader sustainability science. Doing so will broaden the disciplinary perspectives required to tackle sustainability problems.

Whereas the food security literature is often apolitical and reports on environmental and social outcomes, the food sovereignty literature has taken a much more explicit focus on the politics of food. The issue of decision-making control and power is a major one throughout different scales in food systems (Hospes & Brons, 2016). The power of key food corporations, trade systems, and retailers to dominate the flow of staple commodities, and associated prices, increasingly disadvantages all producers, and small-scale producers especially. As mentioned, much of this private commodity control is simply unregulatable sitting beyond any sovereign jurisdiction (Christensen, 2015). Without critical analysis of the social and power relations in a system, we are likely to leave unchanged the traditional ways of understanding and addressing problems, and hence covertly support the power relations of the status quo.

Exploring political issues in sustainability requires bridging the technical understandings from the sciences with the critical social analysis of the social sciences. Studying and critiquing how social behaviours and assumptions inform how we understand, act, and adapt to environmental change is a major research opportunity (Lövbrand et al., 2015). Bringing these social issues to transdisciplinary research forums can have ramifications for stakeholders, dependent on their social status within that system. Problem mapping frameworks, such as human ecology, help highlight the tensions caused by power in a system and bring shared understandings on possible future interventions. The critical social sciences provide ideas from Marxism, constructivism, critical theory, Foucauldian concepts of power and knowledge, and feminist perspectives to document how historical contexts have crafted social structures influence behaviour (Stevenson, 2015). Developing integrative research that uses this disciplinary and theoretical diversity can further inform the study of the social drivers within food systems.

Conclusion

In this chapter, we have:

1   Demonstrated how human ecology helps identify the influence that different discourses have on perspectives and solutions for food system challenges,
2   Proposed a future food systems research agenda that acknowledges and integrates governance and politics, including issues of power.

We achieved our first objective through highlighting how food discourses align with the concept of applied post-normal science and with different literatures concerned with human wellbeing, ecosystems, and institutions. Global food systems have created a wide availability of relatively cheap, convenient, and satisfying food products, but this has created unhealthy eating habits, and food has failed to reach the most vulnerable. Tensions are prevalent between pursuing agroecological production that can expand food outputs and achieve biodiversity outcomes versus continuing highly technical production that meets global market demand and increasingly reflects the power of global retailers. Institutions modelled on the principles of dominant discourses play a critical role in facilitating food activities and have the potential of influencing all producers, processors, distributors, and consumers. To have this influence altered so it restructures food systems to promote more just and sustainable outcomes, enhanced understandings of food governance is needed, and particularly the role of dominant discourses.

We achieved our second objective by arguing that governance is a process that brings together different perspectives on an issue and can be enhanced to maximize the value of different proposed solutions. Wider discussions into governance for sustainability and the value of critical social research in the context of environmental change make food governance research a critical research avenue to pursue.

Food systems scholarship is rapidly growing and can be enhanced through more deeply integrating critical social science analysis within its concerns for sustainability. The human ecological framework used in this chapter provides an analytical tool for drawing links between literatures and specific elements of a system of interest. The food literature indicated that explicitly embedding power and politics into analysis is crucial, as it is a major driver of how people engage with food activities. The use and expansions of these critical human ecological and social science methods will be of value to the future of the food systems research agenda's ability to meaningfully discuss and inform food governance and policy debates and action.

The complexity of food systems requires the integration of disciplines to analyse and intervene in problems (Foran et al., 2014). This integration of disciplinary lines of enquiry can aid the development of inter- and transdisciplinary research processes needed to address food problems, and more broadly, sustainability challenges. Human ecological analysis offers a platform for identifying links between ecosystems and human wellbeing and can be enhanced by those disciplines that can contribute towards bringing political and power analysis into central consideration.

Questions for comprehension and reflection

1   Use the human ecology framework presented in Figure 10.1 and elaborated on in this chapter to show the influence of different food discourses in your country, continent, or community on food system behaviour. Can you see a difference in the food system behaviour under the influence of the different discourses?
2   Can you think of other discourses that exist in other sustainability issues? For example, are there competing discourses related to addressing climate change or biodiversity loss?

References

Core references are marked with two asterisks (★★)

Abson, D. J., J. Fischer, J. Leventon, J. Newig, T. Schomerus, U. Vilsmaier, H. von Wehrden et al. 2017. Leverage points for sustainability transformation. *Ambio* 46 (1):30–39.

Aerni, P. (2011). Food sovereignty and its discontents. *African Technology Development Forum* 8 (1): 23–40.

Alexandratos, N. & Bruinsma, J. (2012). *World agriculture towards 2030/2050: The 2012 revision.* ESA Working paper No. 12–03. Food and Agriculture Organization of the United Nations. Retrieved from www.fao.org/docrep/016/ap106e/ap106e.pdf

Altieri, M. (1987). *Agroecology: The scientific basis of alternative agriculture.* Boulder: Westview Press.

Altieri, M. & Toledo, V. M. (2011). The agroecological revolution in Latin America: Rescuing nature, ensuring food sovereignty and empowering peasants. *Journal of Peasant Studies* 38 (8): 587–612.

Bacon, C. M. (2015). Food sovereignty, food security and fair trade: The case of an influential Nicaraguan smallholder cooperative. *Third World Quarterly* 36 (3): 469–488.

Barling, D. & Duncan, J. (2015). The dynamics of the contemporary governance of the world's food supply and the challenges of policy redirection. *Food Security* 7 (2): 415–424.

Berbés-Blázquez, M., González, J. A. & Pascual, U. (2016). Towards an ecosystem services approach that addresses social power relations. *Current Opinion in Environmental Sustainability* 19: 134–143.

Bernstein, H. (2014). Food sovereignty via the 'peasant way': A sceptical view. *The Journal of Peasant Studies* 41 (6): 1–33.

Boyden, S. (2004). *The biology of civilization.* Sydney: University NSW Press.

Boyden, S. (2016). The biohistorical paradigm: The early days of human ecology at The Australian National University. *Human Ecology Review* 22 (2): 25–46.

Brem-Wilson, J. (2015). Towards food sovereignty: interrogating peasant voice in the United Nations Committee on World Food Security. *The Journal of Peasant Studies* 42 (1): 73–95.

Brown, V. A., Harris, J. A. & Russell, J. Y. (2010). *Tackling wicked problems through the transdisciplinary imagination.* London and Washington, DC: Earthscan.

Burnett, K. & Murphy, S. (2013). What place for international trade in food sovereignty? *Food Sovereignty: A Critical Dialogue.* Conference paper No. 2. International conference Yale University, September 14–15.

Candel, J. L. (2014). Food security governance: a systematic literature review. *Food Security* 6 (4): 585–601.

★★Carolan, M. (2016). *The sociology of food and agriculture.* New York: Routledge.

Chaudhary, A. & Kastner, T. (2016). Land use biodiversity impacts embodied in international food trade. *Global Environmental Change* 38: 195–204.

Christensen, C. B. (2015). Two kinds of economy, two kinds of self – Toward more managable, hence more sustainable and just, supply chains. *Human Ecology Review* 21 (2): 3–22.

Clapp, J. (2014). Food security and food sovereignty: Getting past the binary. *Dialogues in Human Geography* 4 (2): 206–211.

Clapp, J. (2015). Distant agricultural landscapes. *Sustainability Science* 10 (2): 305–316.

Cororaton, C. B. & Corong, E. L. (2009). *Philippine agricultural and food policies: Implications for poverty and income distribution.* Research Report No. 161. International Food Policy Research Institute.

Coxhead, I., Shively, G. & Shuai, X. (2001). 'Agricultural development policies and land expansion in a southern Philippine watershed', in Angelsen, A. & Kaimowitz, D. (Eds.) *Agricultural technologies and tropical deforestation.* Wallingford: CABI. pp. 347–365.

Davila, F. & Dyball, R. (2015). Transforming food systems through food sovereignty: An Australian urban context. *Australian Journal of Environmental Education* 31(Special Issue 01): 34–45.

Desmarais, A. A. (2007). *La via campesina: Globalization and the power of peasants.* Winnipeg: Fernwood Publishing.

Dethier, J. J. & Effenberger, A. (2012). Agriculture and development: A brief review of the literature. *Economic Systems* 36 (2): 175–205.

Deutsch, L., Dyball, R. & Steffen, W. (2013). 'Feeding cities: food security and ecosystem support in an urbanizing world', in Elmqvist, T., Fragkias, M., Goodness, J., Güneralp, B., Marcotullio, P. J., McDonald, R. I., Parnell, S., Schewenius, M., Sendstad, M., Seto, K. C. & Wilkinson, C. (Eds.) *Urbanization, biodiversity and ecosystem services: Challenges and opportunities.* Netherlands: Springer.

Dovers, S. (2001). *Institutions for sustainability*. Working/technical paper. ANU Research Publications.

Dryzek, J. (1997). *The politics of the Earth: Environmental discourses*. Oxford: Oxford University Press.

Dyball, R. (2010). 'Human ecology and open transdisciplinary inquiry', in Brown, V., Harris, J. A. & Russell, J. Y. (Eds.) *Tackling wicked problems: Through the transdisciplinary imagination*. London: Earthscan. pp. 273–284.

Dyball, R. (2012). 'Human ecology', in Kundis Craig, R., Pardy, B., Copeland Nagle, J. & Schmitz, O. (Eds.) *The Encyclopedia of Sustainability: Vol. 5: Ecosystem management and sustainability*. Great Barrington: Berkshire Publishing. pp. 13–20.

Dyball, R. (2015). From industrial production to biosensitivity: The need for a food system paradigm shift. *Journal of Environmental Studies and Sciences* 5 (4): 560–572.

Dyball, R. & Newell, B. (2015). *Understanding human ecology: A systems approach to sustainability*. London: Routledge.

Edelman, M. (2014). The next stage of the food sovereignty debate. *Dialogues in Human Geography* 4 (2): 182–184.

Ehrlich, P. R. & Harte, J. (2015). Food security requires a new revolution. *International Journal of Environmental Studies* 72 (6): 908–920.

Ericksen, P. (2008). Conceptualizing food systems for global environmental change research. *Global Environmental Change* 18 (1): 234–245.

FAO. (2015). *The state of food insecurity in the World 2015, Rome, FAO*. Retrieved from www.fao.org/3/a-i4646e.pdf

Fischer, J., Dyball, R., Fazey, I., Gross, C., Dovers, S., Ehrlich, P. R., . . . Borden, R. J. (2012). Human behavior and sustainability. *Frontiers in Ecology and the Environment* 10 (3): 153–160.

Foran, T., Butler, J.R.A., Williams, L. J., Wanjura, W. J., Hall, A., Carter, L. & Carberry, P. S. (2014). Taking complexity in food systems seriously: An interdisciplinary analysis. *World Development* 61: 85–101.

Friel, S. & Ford, L. (2015). Systems, food security and human health. *Food Security* 7 (2): 437–451.

Fuglie, K. (2016). The growing role of the private sector in agricultural research and development world-wide. *Global Food Security* 10: 29–38.

Funtowicz, S. O. & Ravetz, J. R. (1991). 'A new scientific methodology for global environmental issues', in Costanza, R. (Ed.) *Ecological economics: The science and management of sustainability*. New York: Columbia University: pp. 137–152.

Funtowicz, S. O. & Ravetz, J. R. (1993). Science for the post-normal age. *Futures* 25 (7): 739–755.

Gliessman, S. R. (1990). *Agroecology: Researching the ecological basis for sustainable agriculture*. New York: Springer-Verlag.

Goldman, L. (2010). *A biodiversity hotspot in the Philippines, WWF*. Retrieved from www.world wildlife.org/blogs/good-nature-travel/posts/a-biodiversity-hotspot-in-the-philippines

Hospes, O. & Brons, A. (2016). 'Food system governance: A systematic literature review', in Kennedy, A. & Liljeblad, J. (Eds.) *Food systems governance: Challenges for justice, equality and human rights*. London: Routledge: pp. 13–42.

IFPRI. (2015). *Global food policy report 2014–2015*. International Food Policy Research Institute, Washington, DC.

Ingram, J. (2011). A food systems approach to researching food security and its interactions with global environmental change. *Food Security* 3 (4): 417–431.

Ingram, J., Dyball, R., Howden, M., Vermeulen, S., Garnett, T., Redlingshöfer, B., . . . Porter, J. R. (2016). Food security, food systems, and environmental change. *Solutions* 7 (3): 63–73.

**Ingram, J., Ericksen, P. & Liverman, D. (Eds.) (2010). *Food security and global environmental change*. London: Earthscan.

IPES Food. (2015). *The new science of sustainable food systems: Overcoming barriers to food systems reform*. International panel of experts on sustainable food systems.

Jansen, K. (2014). The debate on food sovereignty theory: Agrarian capitalism, dispossession and agroecology. *The Journal of Peasant Studies* 42 (1): 213–232.

**Jarosz, L. (2014). Comparing food security and food sovereignty discourses. *Dialogues in Human Geography* 4 (2): 168–181.

Kalfagianni, A. (2015). 'Just food': The normative obligations of private agrifood governance. *Global Environmental Change* 122 (2): 174–186.

Lakoff, G. (2004). *Don't think of an elephant: Know your values and frame the debate*. Vermont: Chelsea Green Publishing.

Lang, T. (2010). Crisis? What crisis? The normality of the current food crisis. *Journal of Agrarian Change* 10 (1): 87–97.

Lang, T. & Heasman, M. (2004). *Food wars: The global battle for mouths, minds and markets*. London: Earthscan.

Lawrence, G., Lyons, K. & Walkington, T. (Eds.) (2010). *Food security, nutrition and sustainability*. Oxon: Earthscan.

Lee, R. P. (2013). The politics of international agri-food policy: discourses of trade-oriented food security and food sovereignty. *Environmental Politics* 22 (2):216–234.

Leventon, J. & Laudan, J. (2017). Local food sovereignty for global food security? Highlighting interplay challenges. *Geoforum* 85: 23–26.

Loos, J., Abson, D. J., Chappell, M. J., Hanspach, J., Mikulcak, F., Tichit, M. & Fischer, J. (2014). Putting meaning back into 'sustainable intensification'. *Frontiers in Ecology and the Environment* 12 (6): 356–361.

Lövbrand, E., Beck, S., Chilvers, J., Forsyth, T., Hedrén, J., Hulme, M., Lidskog, R. & Vasileiadou, E. (2015). Who speaks for the future of Earth? How critical social science can extend the conversation on the Anthropocene. *Global Environmental Change* 32: 211–218.

Maye, D. & Kirwan, J. (2013). Food security: A fractured consensus. *Journal of Rural Studies* 29 (0): 1–6.

McMichael, P. (2009). A food regime genealogy. *The Journal of Peasant Studies* 36 (1): 139–169.

Meadows, D. H. & Wright, D. (2009). *Thinking in systems: A primer*. London: Earthscan.

Montesino P., Federico, Toivonen, T., Di Minin, E., Kukkala, A. S., Kullberg, P., Kuustera, J., Lehtomaki, J., Tenkanen, H., Verburg, P. H. & Moilanen, A. (2014). Global protected area expansion is compromised by projected land-use and parochialism. *Nature* 516 (7531): 383–386.

Newell, B. & Siri, J. (2016). A role for low-order system dynamics models in urban health policy making. *Environment International* 95: 93–97.

Ostrom, E. (2010). 'Institutional analysis and development: Elements of the framework in historical perspective', in Crothers, C. (Ed.) *Historical developments and theoretical approaches in sociology/social theory*. UNESCO: EOLSS Publications: pp. 261–288.

Porter, J. R., Dyball, R., Dumaresq, D., Deutsch, L. & Matsuda, H. (2013). Feeding capitals: Urban food security and self-provisioning in Canberra, Copenhagen and Tokyo. *Global Food Security*, 3 (1): 1–7.

Ravetz, J. (1997). The science of 'what-if?' *Futures* 29 (6): 533–539.

Ravetz, J. (2006). Post-normal science and the complexity of transitions towards sustainability. *Ecological Complexity* 3 (4): 275–284.

Robbins, P. (2012). *Political ecology*. Oxford: Blackwell Publishing.

Rockstrom, J., Steffen, W., Noone, K., Persson, A., Chapin, F. S., Lambin, E. F., . . . Foley, J. A. (2009). A safe operating space for humanity. *Nature* 461 (7263): 472–475.

Stevenson, H. (2015).'Alternative theories: Constructivism, Marxism and critical approaches', in Harris, P. (Ed.) *Routledge handbook of global environmental politics*. Oxon: Routledge: pp. 42–55.

Stuart, D. (2014). *Barnyards and Birkenstocks: Why farmers and environmentalists need each other*. Washington, DC: Washington University Press.

Tomich, T. P., Brodt, S., Ferris, H., Galt, R., Horwath, W. R., Kebreab, E., Leveau, J. H., Liptzin, D., Lubell, M. & Merel, P. (2011). Agroecology: A review from a global-change perspective. *Annual Review of Environment and Resources* 36 (1): 193–222.

UNDP. (2013). *2012/2013 Philippine Human Development Report*. Accessed from: www.ph.undp.org/content/philippines/en/home/library/human_development/2012-2013_PHDR.html

Velten, S., Leventon, S., Jager, N. & Newig, J. (2015). What is sustainable agriculture? A systematic review. *Sustainability* 7: 7833–7865.

Vermeulen, S. J., Campbell, B. & Ingram, J. (2012). Climate change and food systems. *Annual Review of Environment and Resources* 37 (1): 195–222.

Wagner, A., Yap, D.L.T. & Yap, H.T. (2015). Drivers and consequences of land use patterns in a developing country rural community. *Agriculture, Ecosystems & Environment* 214: 78–85.

Westengen, O.T. & Banik, D. (2016). The state of food security: From availability, access and rights to food systems approaches. *Forum for Development Studies* 43 (1): 113–134.

Wezel, A., Bellon, S., Doré, T., Francis, C., Vallod, D. & David, C. (2009). Agroecology as a science, a movement and a practice. *A review, Agronomy for Sustainable Development* 29 (4): 503–515.

Wittman, H., Chappell, M. J., Abson, D. J., Kerr, R. B., Blesh, J., Hanspach, J., Perfecto, I. & Fischer, J. (2016). A social–ecological perspective on harmonizing food security and biodiversity conservation. *Regional Environmental Change* 17(5): 1–11.

Wittman, H., Desmarais, A. A. & Wiebe, N. (Eds.) (2010a). *Food sovereignty: Reconnecting food, nature and community*. Winnipeg: Fernwood Publishing.

Wittman, H., Desmarais, A.A. & Wiebe, N. (2010b). 'The origins and potential of food sovereignty', in Wittman, H., Annette Aurélie, D. & Wiebe, N. (Eds.) *Food sovereignty: Reconnecting food, nature and community*. Winnipeg: Fernwood Publishing: pp. 1–12.

Wittman, H., Desmarais, A. A. & Wiebe, N. (Eds.) (2011). *Food sovereignty in Canada: Creating just and sustainable food systems*. Halifax: Fernwood Publishing.

The World Health Organization. (2016). *The double burden of malnutrition, WHO*. Retrieved from www.who.int/nutrition/double-burden-malnutrition/en/

Young, O. (2002). *The institutional dimensions of environmental change: Fit, interplay and scale*. Cambridge: The MIT Press.

Zimmerman, K. & Bassett, T. (2003). *Political ecology: An integrative approach to geography and environment – Development studies*. New York: The Guilford Press.

11 Energy

Physical and technical basics

Susanne Siebentritt

Introduction

This chapter gives a brief description of the physical laws underlying any generation and usage of energy. It outlines the forms of energy that we use and the technological options that are available to generate useful energy for human use. It finishes with a brief outlook into a potential sustainable energy system of the future.

What is energy?

When we talk about energy, we think about electricity to run our appliances and the machines in industry, about heat to keep our rooms warm, to cook our food and to enable industrial processes, as well as about fuel for transport or electricity generation. Thus energy is the life blood that not only keeps our lifestyle and our societies running, but is also necessary to cover our very basic human needs. On the other hand, this also means that what we really need is not energy per se, but the services that are enabled by energy and are not possible without energy.

The laws of thermodynamics

Energy is a physical quantity. The science of energy is thermodynamics. A generally understandable description of its basics can be found in Boyle et al. (2004) and MacKay (2009). There are two very fundamental laws in thermodynamics: the law of energy conservation and the law of energy conversion. These laws are not like man-made laws; they are concise and powerful ways of describing fundamental properties of physical systems. Some are approximate (Hooke's for elasticity, Boyle's for gases), and some have been modified through advances in science (Lavoisier's law of conservation of mass, now mass-energy after Einstein). The two laws of thermodynamics have been much debated and reinterpreted, but they still stand.

The law of energy conservation states that energy is conserved; we can neither create nor destroy energy – we can only convert it from one form of energy into another. If you are using a light to read this text (and you are not using a candle), then the energy that's now in the form of light was in the form of electricity before. This electricity was generated in a generator, which transforms mechanical energy (i.e. rotation) into electricity. And so on. . . (see "Forms of energy and conversions" later). The important conclusion for this discussion is this: we cannot make energy; we can only transform it from one form into another. Nevertheless, we speak about energy generation when we transform energy into useful and versatile forms. For this "generation" we always need some resource, some other form of energy.

The law that governs energy conversion has been given several different statements. In its fundamental form it states that it is not possible to transform heat completely into mechanical energy or useful work. In a more general way it states that no energy transformation is 100% efficient, that we always lose something in the form of heat. Thus from the first law we know that we cannot create energy and the energy we use comes from another form of energy. But the second law tells us that during the transformation we lose some of the energy, heating the air around us. We need to understand that energy has different degrees of usefulness. The useful part of energy is called 'free energy' or exergy (Kjelstrup et al., 2015; Brockway et al., 2016); the non-useful part is described by entropy. The usefulness of energy can be described by the following example. The energy contained in the movement of air molecules in a 20m² office at 20°C is more than the energy stored in three standard 12V car batteries. Whereas the energy in the air only keeps us warm, we can use the energy in the batteries to start our car, cook our lunch or run our computer. The reason is that even though their quantities are the same, the quality – or usefulness – of the energy in the air and in the battery is different. With every energy conversion we lose some of the useful energy and transform it into heat, hence the efficiency of every conversion is lower than 100%.

We notice that when energy is converted for the services we require, this happens at different rates. Our car may travel, or our kettle may boil, slowly or quickly. We describe this rate of energy conversion as 'power', and we measure it as 'kilowatt' or 'horsepower' (!). Then the measure of energy (which is what we pay for) is that of power delivered through a period of time, or 'kilowatt-hour'. In electricity systems, the immediate demand is for power, and the technology is oriented around predicting and meeting that demand as it varies through the day. Centralised generating stations, with large equipment, will have a high 'power density', making them more economical but also more vulnerable in the case of accidents.

Forms of energy and conversions

As we have just seen electricity is the most versatile form of energy. Therefore it is worth looking at the conversion efficiencies of various means to generate electricity. Currently most of the world's electricity comes from coal- or gas-fired power stations, and their efficiency is typically around 30%, that is, 30% of the chemical energy contained in the coal or gas ends up in the electricity grid; 70% of the energy is lost into the atmosphere as heat.[1] Nuclear power stations transform about 30% of the fissionable energy contained in the uranium into electricity. Renewable energy sources do not use any kind of fuel to generate electricity, but they convert the energy contained in the movement of air molecules in the wind or the energy contained in sunlight. The conversion efficiency of big wind turbines is about 35%; that of solar cells is between 15 and 20%.[2] To take a look at other forms of energy: when using a conventional electric hob about 30% of the electrical energy is actually used to heat the food. Or as an example for transport: a car transforms about 20% of the chemical energy contained in the gasoline into movement of the car; the rest is lost as heat into the cooling water and the exhaust.

To provide its electricity, heat and transport needs humankind uses ever more primary energy. Primary energy includes the sources of energy that are provided by nature. Currently about one-third of the world's energy consumption is covered by oil and coal, each, about a quarter by natural gas, 7% by hydroelectric power, 4% by nuclear power and 3% by renewable sources (IEA, 2015; BP, 2016). Over the last decades the total energy consumption has increased by about 50% every 20 years. In industrialised countries about one-third of the energy is lost in conversion and transport of energy. Of the two-thirds that are finally used about one half is used for heating (rooms, food, industrial processes), about one-third for transport and the rest for light, appliances and machines (Boyle et al., 2004).

Figure 11.1 summarises the use and the sources of energy. The central role of electricity is based on its versatility. Generally electricity is generated by a genera- tor which converts mechanical rotational energy into electricity, by rotating a wire coil in a magnetic field. The generator can be rotated by a turbine, which in turn is driven by steam, which is generated by heating water by either a coal, oil or gas fire, by nuclear fission or by sun light. For a more detailed description see e.g. (Boyle et al., 2004; Wikiversity, 2016). The generator can also be rotated by the movement of water in hydropower station or by the movement of the wind in a wind turbine. A completely different form the electricity generation is photovoltaics, where sun- light is directly converted into electricity by the use of semiconductor devices (i.e. solar cells) without any moving parts (Laboratory for Photovoltaics, 2009).

Although currently the main use of electricity is to run appliances, to provide light and for transport, electricity can also be used to provide heat. In most cases heat is, however, provided by burning fuel (coal, gas, oil). Alternatively the heat of the sun can be used directly in the form of solar thermal collectors, solar cookers and passive solar buildings. Transport relies mainly on fuel (gasoline, diesel, kero- sene, bunker fuel); only a few percentage points of the energy used for transport are provided by electricity (electric trains, trams and cars). Alternatively, we can use biofuels made from plants. Their growth is, however, in competition with food production, and the energy yield per area is much lower than, for example, with

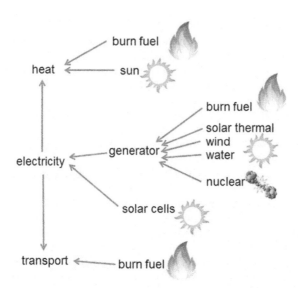

Figure 11.1 Energy use and sources

photovoltaics (see e.g. Geyer et al., 2013). Therefore, alternative solutions based on the use of renewable electricity for fuel or gas production have been proposed: power-to-gas or power-to-fuel.

Sustainability considerations

To discuss the sustainability of the various energy sources, we have to answer the following questions:

- Is it available for the foreseeable future? Let's say not only for our grandchildren, but for their grandchildren as well, that is, is the source available for the next 150 to 200 years?
- What is the impact on the environment in terms of emissions and land use?
- What is the risk in case of accidents or misuse?

Other questions to consider are more concerned with the convenience and economy of the use of energy sources:

- Is it controllable, that is, can we turn it on and off whenever we need to?
- Can it be stored?
- Can it be transported?

From the previous section it can be seen that the vast majority of our energy needs is covered by burning fossil fuels. Their availability is limited. The availability can be measured in the production-over-resources ratio. This is not a prediction of how long a resource will last, but only how long the known resources will last at the current rate of consumption. This is about 50 years for oil and gas, 110 years for coal (BP, 2016) and 85 years for uranium (World Nuclear Association, 2016). These estimates are highly uncertain, as new technologies of discovery, extraction and processing can alter the 'known resources'. In the case of nuclear, alternatives to uranium have been proposed. Renewable energies, wind, water and sun, on the other hand, are available without limitations for the foreseeable future.

Emissions are chemicals or radiation that is released into the atmosphere either during the mining or during the operation of power stations. Probably the most dramatic effects are those caused by the emission of carbon dioxide and other greenhouse gases. They lead to the heating of the atmosphere and the oceans and to climate change (IPCC, 2013). Carbon dioxide is released by the burning of fossil fuels: oil, gas and coal, where coal has the highest rate of emissions relative to the energy contained (US Energy Information Administration, 2016). A dramatic reduction of carbon dioxide emissions is necessary to stabilise the climate, indicating the necessity to avoid using the remaining fossil fuel resources (IPCC, 2014).

The effect of radioactive emissions from nuclear power stations is heavily debated. One undeniable emission from nuclear power stations is radioactive nuclear waste, which can be controlled in the short term, but no country in the world has a final solution for treating or storing this waste.

For the question of land use one has to take the whole production chain, from the mining to the power station, into account. Although conventional power stations themselves have only limited impact on the landscape, their fuel supply changes landscapes significantly by the operations of mining or oil and gas drilling. Wind, water, sun and biofuel, on the other hand, have to use rather large areas for electricity generation. This effect is particularly severe for large hydropower installations which can encompass huge areas, and for biofuel, if it is not grown in a sustainable way. Even windfarms are opposed by some, with claims that that they damage the scenery, emit annoying vibrations and kill birds and bats. In retrospect we can say that the industrialised countries had great good luck in the supply of extremely cheap fossil oil, whose only immediate harm was to the societies that supplied it. These complications reflect in a way the first law of thermodynamics: we cannot make energy, we can only transform it. And we also use something in the transformation: fuel or land or the atmospheric balance. These changes of the landscape require intense societal discussions, participation of all stakeholders and inevitable compromises. As Barry Commoner said in his fourth law of ecology: there's no such thing as a free lunch. The amount of risk due to accidents or misuse is clearly related to the power density during the energy conversion. In power stations with a high power density like large nuclear, fossil or hydro-power stations, an accident or attack has a much higher impact than in power stations with a low power density, like photovoltaic or wind generators. In nuclear power stations there is the additional impact of radioactive contamination with dramatic short- and long-term effects. Although the long-term effects are difficult to quantify, their amount can be estimated from the fact that decades after the Chernobyl disaster Belarus and Ukraine have been spending between 5% and 20% or their national budgets on disaster recovery (IAEA, 2005).

Summarizing this short assessment: the only sustainable solutions are renewable energy sources – mostly wind and sun, water and biofuel with certain reservations. And they are not without impact in the environment. Thus it is mandatory for a sustainable energy system to significantly reduce the amount of energy that is used.

Sustainable energy systems of the future

Various models have been put forward to evaluate the possibility of nearly 100% supply of the world with renewable energies (Ecofys, 2011; Umweltbundesamt, 2013; Greenpeace, 2015). These studies come to the conclusion that a renewable energy supply by 2050 is possible, given the necessary investments into the energy system.

A central element of renewable energy system is electrification, because electricity can be produced in a sustainable way by wind and photovoltaics much more easily than fuel. Thus a much larger portion of transport and heating than today will rely on electricity. And electricity will have to be the major source of energy. Because wind and sun are, by nature, decentralised, these future systems will require a different electricity grid than today. Today's electricity grid is based on a few large producers and distributes the electricity in an increasingly fine one-way web to the consumers. In the future more and more electricity will be produced by medium-sized and small power generators, often close to or at the point of consumption. This development is described by the notion of 'prosumers': entities that both produce and consume electricity. The future grid will consist of interconnected, more or less equivalent clusters of prosumers with much less weight of central power stations. Because electricity as such cannot be stored, the future grid will also have to supply storage possibilities in the form of batteries, of power-to-gas installations with gas storage and of water pump reservoirs. It might also be necessary to adapt temporal electricity consumption patterns to the supply, for example, run energy-intensive machinery during the day when photovoltaic power is available. The control of consumption, storage and production will require a smart grid, which not only transports and distributes electricity but also measures and regulates production and consumption. Because the smart grid deals with enormous amounts of sensitive data, it has to be protected against cyber-attacks. Issues of the protection of sensitive personal data, require intense societal discussions and participation.

A second central element of all renewable scenarios is a dramatic reduction of the use of primary energy. This requires substantial investments into energy savings and energy efficiency.

As discussed previously, in the end what we need is energy services, not energy per se. Thus increased effort will be necessary to increase the efficiency of energy conversion and of the processes that provide the services we need.

Conclusion

Energy is not free. Making energy available for human use always has impacts on resources, landscape and the atmosphere or the oceans, as well as on the societies, local and regional, that provide the resources for conversion. This is based on physical laws, and no future innovation will solve this problem. Our societies and our lifestyle depend critically on the supply of energy. The only sustainable option for this energy supply is renewable energy. Several scenarios demonstrate that a 100% renewable energy system is possible. However, it requires enormous investments. Societal discussions and participation are necessary to decide on the future energy systems.

If I may end with a personal conclusion: the technical solutions for a sustainable energy system are available. Their implementation is absolutely necessary and should be put in place as quickly as possible. However, to reduce the pressure on the ecosystem, further changes in terms of a much more resource-preserving lifestyle, particularly in the industrial countries, will be needed. This is often described as increasing sufficiency, not only efficiency (Samadi et al., 2016).

Questions for comprehension and reflection

1   Why do we need energy? Or could we live without energy?
2   What is the impact of various energy sources, and how sustainable are they?
3   What can be done to increase efficiency?
4   What can you personally do to increase sufficiency?

Notes

1 Modern gas-fired power plants can have up to 45% efficiency.
2 With focussed sunlight, efficiencies of more than 40% are obtained.

References

Boyle, G., Everett, B. & Ramage, J. (Eds). (2004). *Energy systems and sustainability*. Oxford: Oxford University Press.

BP British Petroleum. (2016). *BP statistical review of world energy 2016*. Retrieved from www. bp.com/statisticalreview

Brockway P. E., Dewulf J., Kjelstrup, S., Siebentritt, S., Valero, A. & Whelan, C. (2016). *In a resource-constrained world: Think exergy, not energy*. Brussels: Science Europe. Retrieved from http://scieur.org/brochure-exergy

Ecofys. (2011). *The energy report – 100% renewable energy by 2050*. Retrieved from www.ecofys.com/files/files/ecofys-wwf-2011-the-energy-report.pdf

Geyer, R., Stoms, D. & Kallaos, J. (2013). Spatially-explicit life cycle assessment of sun-to-wheels transportation pathways in the U.S. *Environmental Science & Technology* 47 (2): 1170–1176. doi:10.1021/es302959h

Greenpeace. (2015). *Energy Revolution 2015*. Retrieved from www.greenpeace.org/international/Global/international/publications/climate/2015/Energy-Revolution-2015-Full.pdf

IAEA International Atomic Energy Agency. (2005). *Chernobyl's legacy: Health, environmental and socio-economic impacts*. Retrieved from www.iaea.org/sites/default/files/chernobyl.pdf

IEA International Energy Agency. (2015). *Key world energy statistics 2015*. Retrieved from www.iea.org/publications/freepublications/publication/key-world-energy-statistics-2015.html

IPCC Intergovernmental Panel on Climate Change. (2013). *Climate change – The physical science basis*. Retrieved from http://ipcc.ch/pdf/assessment-report/ar5/wg1/WG1AR5_SPM_FINAL.pdf

IPCC Intergovernmental Panel on Climate Change. (2014). *Climate change 2014: Mitigation of climate change*. Retrieved from www.ipcc.ch/pdf/assessment-report/ar5/wg3/ipcc_wg3_ar5_summary-for-policymakers.pdf

Kjelstrup, S., Dewulf, J. & Nordén, B. (2015). A thermodynamic metric for assessing sustainable use of natural resources. *International Journal of Thermodynamics* 18 (1): 66–72.

Laboratory for Photovoltaics. (2009). *The principle of a solar cell*. Retrieved from http://wwwen.uni.lu/research/fstc/physics_and_materials_science_research_unit/research_areas/photovoltaics/research/solar_cell_principle

MacKay, D.J.C. (2009). *Sustainable energy – Without the hot air*. Cambridge: UIT.

Samadi, S., Gröne, M.-C., Schneidewind, U., Luhmann, H.-J., Venjakob, J. & Best, B. (2016). Sufficiency in energy scenario studies: Taking the potential benefits of lifestyle changes into account. *Technological Forecasting and Social Change* – available online 3 October 2016. doi:10.1016/j.techfore.2016.09.013

Umweltbundesamt. (2013). *Germany in 2050 – A greenhouse gas-neutral country*. Retrieved from www.umweltbundesamt.de/sites/default/files/medien/376/publikationen/germany_2050_a_greenhouse_gas_neutral_country_langfassung.pdf

US Energy Information Administration. (2016). *How much carbon dioxide is produced per kilowatt hour when generating electricity with fossil fuels?* Retrieved from www.eia.gov/tools/faqs/faq.cfm?id=74&t=11

Wikiversity. (2016). *Power Generation/Steam Power*. Retrieved from https://en.wikiversity.org/wiki/Power_Generation-Steam_Power

World Nuclear Association. (2016). *World Nuclear Power Reactors & Uranium Requirements and Supply of Uranium*. Retrieved from www.world-nuclear.org

12 Urban energy transitions through innovations in green building

Julia Affolderbach, Bérénice Preller and Christian Schulz

Introduction

Recent debates on climate change have increasingly focused on cities as a strategic spatial scale to implementss climate change mitigation and adaptation strategies. Within this context, green building and the way the built environment interfaces with urban structures and services have become significant levers of action for cities to reduce greenhouse gas emissions and become climate change leaders (Bulkeley et al., 2011). Approximately 30% to 40% of final energy consumption is linked to buildings and, as a consequence, the building sector has been identified as one of the most relevant sectors to reduce CO_2 emissions (UNEP, 2011). Although green building is largely associated with technological innovations, building design and the way elements are embedded within the overall urban fabric, a shift towards green building in cities largely depends on modes of sustainable governance. Relevant dimensions include support of and for green policies and incentives, institutional support through resource centres, think tanks, certification bodies, and training, aspects of inclusivity both in the planning process as well as the later use of (and access to) buildings and to a considerable extent on lived sustainability (i.e. the ways individuals interact with and use buildings). This latter dimension of possibly changing user behaviour and consumer lifestyles seems to be absent from most of the energy scenario studies, as Samadi et al. (2016) revealed in their assessment of a series of internationally influential studies and policy programmes. Like other scholars (e.g. Sachs, 1999; Princen, 2003; Schneidewind & Zahrnt, 2014), they plead for a stronger conceptualization of sufficiency oriented policy approaches and differentiate persuasive instruments (e.g. through education and communication) from incentive based (price/tax policies) and more coercive approaches (limits, bans).

A proper understanding of green building, then, requires consideration of a whole range of aspects, including technological, institutional, procedural and sociocultural innovations. Against this background, this chapter investigates conditions and drivers behind innovations resulting in green building in selected city regions. The term 'green building' is here used as an umbrella term for all activities related to sustainable construction (i.e. the green building sector as a whole). It is thus not limited to the physical building (i.e. a single residential or commercial project/neighbourhood), but applies a more comprehensive understanding of building activities, including the political and regulatory context and all relevant actors and stakeholders involved. This chapter embeds green building transitions within the recent literature on sustainability transitions that pays attention to the spatial dimensions of green transitions, including regional variations and multi-scalar linkages within and between cities and regions (Truffer & Coenen, 2012). Research on sustainability transitions rooted in Transition Studies focuses on technological innovations and modernization processes to understand drivers and processes towards low-carbon economies and societies more generally. The chapter uses the Multi-Level Perspective (MLP) heuristic (see Box 12.1) to understand regional trajectories of innovations in green building based on a more neutral and open understanding of sustainability transitions that seeks to grasp all sorts of more or less beneficial transitions, including parallel and uneven processes, exclusions and less successful innovations. It does so using a broadened transition studies perspective that breaks free from its technocentric focus. It incorporates green building innovations that go beyond the technical or procedural realm to encompass organizational, social and cultural changes that can play a role in transition processes. The chapter places particular emphasis on the role of both individual and institutional actors as agents of change. As such, knowledge generation, transfers and learning processes amongst practitioners, experts and decision makers, both in the building sector and at the urban policy level, are considered to be central to understanding green building innovations and developments in city regions.

Box 12.1   Sustainability transitions and Geels's (2002) multi-level perspective

The problem: human-induced climate change requires significant rethinking and economic restructuring to achieve reductions in CO_2 emissions. Sustainability transition research analyses how societies can achieve a transition towards a more sustainable future. The core assumption of the approach is that such restructuring requires technological innovations and that these innovations result from an interplay between social and technological processes. Frank Geels developed his multi-level perspective (MLP) as an analytical framework to understand and explain socio-technical transitions integrating an institutional perspective (including actor groups and framework conditions) with temporal dimensions.

The approach: Geels distinguishes between three mutually dependent levels: niche, regime and landscape. Niches lie at the micro-level and act as test beds for innovations and new socio-technical constellations. They usually consist of spaces that are protected from rules and structures at the higher scales of the regime and landscape (e.g. exemptions from certain regulations or free market forces). The regime level acts as the meso-level of socio-technical systems and describes predominant organizational standards and norms, whereas the landscape describes the macro level, consisting of the most persistent structures such as cultural norms and values. Successful niche innovations can evoke changes at the regime and landscape level, but change can also be triggered by changes at the landscape level (e.g. environmental disasters can lead to an increased environmental awareness of the general public).

Criticism: although the MLP provides a strong heuristic model, a number of limitations have been criticized, including its technocratic focus (i.e. a narrow definition of innovations bound to technology and linked to this

> limited acknowledgement of socio-political dimensions of (sustainability) transitions). Geographers in particular have criticized the lack of spatial considerations and a tendency to conflate the three levels with spatial scales (e.g. Coenen & Truffer, 2012).

The conceptual framework presented in this chapter requires an appropriate methodological design which allows us to reconstruct origins and trajectories of green building innovations. Here, the implementation of the research framework is itself guided by different knowledge exchanges and learning processes. The next section introduces and discusses two specific tools, the World Café and the Delphi approach, that seem particularly well suited for knowledge generation and data collection of complex and multi-actor processes common in (urban) sustainability transitions. Based on practical experiences with the two tools, the benefits and challenges of the methods are critically discussed as well as their potential contribution to critical reflection. In respect to green building, including policies and societal and technological innovations, the two approaches are well suited to help explore new actor constellations and policy arrangements as well as their potential for initiating mutual learning processes between researchers and practitioners but also amongst practitioners themselves. The latter aspect will be illustrated through case study examples from four city regions (Freiburg in Germany and Vancouver in Canada as leaders in green building and Brisbane in Australia and Luxembourg City in Luxembourg as more recent adopters of green building) and discussed against the backdrop of current debates about the use of participatory action research (PAR).

Research perspectives and instruments

Transition studies and green building

Over the last two decades, the growing field of Social Studies of Technology (or 'Transition Studies') has resonated increasingly in sustainability related research, notably in the energy and mobility sectors. Initially developed in the field of technological innovation, empirical case studies of sustainability related issues, and particularly those using Geels's multi-level perspective (MLP, see Box 12.1. and Chapter 6), have proliferated to an extent that Transition Studies are now frequently adopted and used as a normative perspective (e.g. as Strategic Niche Management). In the literature, there is a tendency to assume a directed process (transition) towards a pre-determined finality (sustainability) that runs the risk of ignoring multi-directional and diffuse aspects of transitions (e.g. in the case of the agrofuel boom in North America and Europe which turned out to be questionable in terms of environmental and social sustainability). Further, and with the adoption of the MLP in human geography, scholars have started to criticize the rather 'a-spatial' approach dominating transition studies research (for overviews see Coenen & Truffer, 2012; Rohracher & Späth, 2013; Hansen & Coenen, 2015; Murphy, 2015). Whereas classic case studies in transition studies have focused on national or regional innovation systems, critics have argued in favour of non-essentialist, relational (networked) rather than territorial (nation-states, regions) understandings of space that account for the mobility and 'travel' of ideas and innovations (sometimes over long distances) not only in the form of transferable best practices and cookie cutter models, but also in various forms of knowledge, practices and experiences through very personalized channels (Affolderbach & Schulz, 2016). Whereas processes at the landscape level are often associated with higher spatial scales (e.g. national politics and regulations), there is no spatial hierarchy between the different levels in the MLP. For example, the focus on city regions as discussed in this chapter understands cities as defined by processes and actors within and outside of the city itself which may all co-constitute the landscape, regime and niche level. As such, a city may act as local niche and/or may encompass a number of localized niches that may depend on non-local factors.

Figure 12.1 shows a possible adaptation of the MLP concept to energy-related green building transitions. If one considers the 'orthodox' building sector as the well-established regime and green building initiatives rather as burgeoning niche phenomena, both levels are exposed to overarching changes in the energy sector and in climate change policies (Schulz & Preller, 2016).

Recent changes in climate mitigation policies at the European level, for example, have led to substantial changes in the EU energy policy 'landscape' which have an immediate impact on the various socio-technical regimes, in particular, the building sector. The 2010 amendment of the Energy Performance of Buildings Directive (EU, 2010) and the 2012 Energy Efficiency Directive (EU, 2012) have set ambitious goals to be transposed via the member states' legislations. Amongst others, they foresee that all new buildings must be nearly zero-energy buildings by 31 December 2020 (public buildings by 31 December 2018). This new regulation not only puts national governments under pressure to implement these standards within the given time period; it also causes adaptation processes in the building sector as it challenges many of the current routines that define the mainstream building

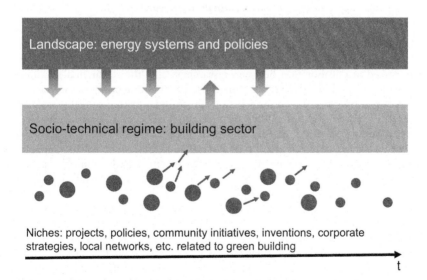

Figure 12.1 The Multi-level perspective (MLP) adapted to the building sector

Source: Schulz & Preller, 2016, p. 274, based on Geels, 2002, p. 1263

sector. These standards, however, do not fall from (Brussels') sky but are the outcomes of long negotiation processes, which are partly built on pioneering standards developed in local contexts. The City of Freiburg's stringent low-energy standards, for example, established in the early 1990s and further developed since have been incrementally incorporated into Germany's federal building standards and as such indirectly serve as a reference for the current EU scheme (Fastenrath & Braun, 2016). This evolution goes beyond the mere techno-administrative notion of policy upload (usually from member state to EU-level) as it comprises a particular niche context where path-breaking urban policies could emerge and consolidate before they became a role model.

Although the regime–landscape interface is less of an interest here, the focus is on both the emergence of niche initiatives and their articulation with the incumbent regime. This means the research design needs to take into account the related actor constellations and institutional settings at all three levels, including the respective power topographies, barriers and driving forces. The wider understanding of the term innovation broadens the focus to look beyond radical, disruptive changes in the production process, as it might be typical for technology oriented transition studies. Organizational, social and cultural innovations might occur rather incrementally, but not without leading to fundamental changes in the longer run.

As arenas for possible innovations of this kind, it seems suitable to focus on micro-case studies (such as the Freiburg standard) and to reveal their individual trajectories through in-depth investigation. In order to carefully select the most relevant cases for empirical purposes, a narrow interaction with experts and stakeholders involved should be conceived as a participatory research approach. The following section will introduce the basic notions of participatory approaches and will present two promising instruments.

Participatory approaches in urban sustainability research

Recent trends towards more participatory approaches in both policy making and research, which have been coined as "participatory turn" (Aldred, 2011), offer valuable tools to sustainability research. Here, the notion of 'knowledge co-production', understood as collaboration between researchers and 'the researched' at different stages of the research process, has gained particular momentum in the social sciences. It is substantiated by arguments on the complex nature of reality compared to scientific theory (Callon, 1999), practical application or 'utilisation' of research (Hessels & van Lente, 2008, p. 741, Martin, 2010, pp. 211–212) and the socially transformative stance adopted by action research (Pain, 2004). Participatory approaches are based on the key premise that knowledge is embedded within the practices and everyday experience of all those directly involved and/or affected, including practitioners and civil society (Bergold & Thomas, 2012; Borg et al., 2012). As such, it challenges traditional concepts of expertise and knowledge generation, predominantly understood as a single-sided knowledge generation in academia and research centres, with practitioners being considered as mere recipients of scientific knowledge produced outside their everyday realm and then 'transferred' from the scientific world for application at a later stage. In contrast, participatory approaches offer promising opportunities for both the researchers and the research participants in terms of knowledge generation in general and scientific advances in particular, specifically when it comes to deliberate co-production schemes. This is especially relevant to environmental policy and sustainability issues, which require "a scientific practice which can cope with uncertainty, with value plurality and with the decision-stakes of the various stakeholders of the problem at hand" (Hessels & van Lente, 2008, p. 744), due to their complex and dynamic interactions with broader social, economic and physical processes (Blackstock et al., 2007). It also allows insights into motivations behind and actions taken by those involved in sustainability transitions that are highly dependent on the context and situated knowledge, but also on individual trajectories, personal networks and values. In order to better understand green innovations and sustainable transitions, it thus seems necessary to complement 'traditional' qualitative research methods with participative elements.

Participatory Action Research

Research approaches labelled as Participatory Action Research (PAR) are usually driven by two core motivations: 1) a progressive understanding of the roles of both the researcher and the researched, and 2) the ambition to generate results that have an impact on 'the real world'.

As to the first, PAR deliberately tries to overcome the hierarchical distinction between the researcher's position and the researched community as a study object. The object of such research is no longer seen as a mere source of information to be explored with an appropriate methodology. Rather, it considers the 'researched' as a partner in a collaborative research endeavour (the "P" in PAR) where specific knowledge carried by the researched is recombined with knowledge acquired by the researcher. Both sides are thus engaged in the co-construction of new knowledge (Hessels & van Lente, 2008; Kindon, 2010).

Second, the wish for relevance in societal debates often comes with a normative stance taken by the researcher, thus engaging with a particular agenda to "make the world a better place" (the "A" in PAR). Not surprisingly, PAR is often practiced in highly politicized fields of research such as development, feminist and environmental justice studies. Frequently, PAR researchers see themselves as parts of a movement (e.g. a dedicated NGO) to which they deliberately contribute their research as a means of empowerment (Kindon et al., 2007; Mason et al., 2013).

This chapter is based on research that has been strongly inspired by the collaborative aspects of PAR (the "P"), whereas the latter, more political dimension (the "A") has been more marginal in framing research on green building transitions. The participatory research approach presented here could be described as 'Interactive Transition Research' (ITR) that acknowledges the key role of the constituencies (here the stakeholder communities within the green building sector) in the co-production of new knowledge which then might have a direct impact on the respective field (for more details on PAR and ITR see Preller et al., 2017).

World Café approach

One of the methods used to co-produce knowledge with the researched communities was to host World Café events with a range of local sustainable building practitioners (Box 12.2).

Box 12.2   World Café

The World Café smethod was developed in the mid–1990s by Juanita Brown and David Isaacs and consists of a group intervention that encourages an open dialogue between participants by relying on unconstrained and interactive conversations. It is operationalized by splitting participants across tables of four to five where they are invited to tackle a specific question. Participants then progress through several conversation rounds with additional questions as they are asked to circulate and mix across the different tables (TheWorldCafé, 2008). The content of each conversation round is further retained and passed on to the next group by a fixed table host and eventually complemented by a final plenary discussion to ensure sharing and connecting of the generated information amongst the totality of participants. Through this "recombination" of knowledge (Brown, 2001, p. 3), reflexive processes amongst participants can be initiated and may lead to a collective understanding of an issue. This includes shared tacit knowledge which may contribute to creating joint "ownership" of the sessions' outcome (Brown, 2001; Fouché & Light, 2010; Prewitt, 2011).

In contrast to other group interventions, the method attempts to create a rather informal setting by conveying the atmosphere of a café through the use of symbolic items like tablecloths, the availability of drinks and food or even more playful tools as the possibility to write or visualize ideas directly on paper tablecloths. This framework is supposed to encourage participants to act as they would during an informal and relaxed meeting at a café (Jorgenson & Steier, 2013). It aims at fostering the dialogic exchange between participants who should feel less in the role to 'make their point' but rather to listen openly and to accept other standpoints in order to engage in a constructive discussion on the given topic.

World Cafés are used by various types of public, private and non-governmental organizations, in rather different contexts for very diverse purposes, including learning (Anderson, 2011). Their objectives vary and include the following aims:

- To empower communities through joint learning and the creation of shared knowledge (Fouché & Light, 2010; Sheridan et al., 2010, for a critical discussion see also Aldred, 2011),
- To facilitate collaboration and communication within an organization (Tan & Brown, 2005; Prewitt, 2011),
- To stimulate innovation through networking and relationship building (Fouché & Light, 2010),
- To improve sales of a product (Brown & Isaacs, 2005, p. 31, quoted in Aldred, 2011, p. 68).

Also, different labels are in use to designate similar techniques (e.g. Knowledge Café, Conversation Café or Innovation Café); some organizers even invent individual labels for particular purposes (Prewitt, 2011). Nevertheless, the potential to encourage active participation of a wide range of participants and to help them to overcome their traditional understanding of meeting formats is common to all the different types of 'Café-style' methods. All usages allow the temporary suspension of "ordinary interactional routines" (Jorgenson & Steier, 2013, p. 390). This particularly includes hierarchical relationships within an organization, as, for instance, the application of World Cafés within the Singapore Police Force shows (Tan & Brown, 2005).

The method thus helps to reveal more diversified, inclusive and changing understandings of a specific topic. It explicitly seeks for the diversity of perspectives held by the participants involved "[rather than] over-stating consensuality" (Aldred, 2011, pp. 62–63). Following the four main objectives of World Cafés – constructive dialogue, relationship building, collective discoveries and collaborative learning – the method also allows us to produce highly practical and contextually adaptable outcomes for researcher and researched alike.

Delphi techniques

The Delphi approach shows a series of similarities to the World Café workshops (e.g. interactive approach, composition of expert panels) (Box 12.3). Technically speaking, its main difference compared to the World Café can be found in its incremental, usually two-stage, approach aimed at validating findings from previous rounds of data collection. Usually both rounds are run anonymously, but openings towards more interactive formats are becoming more frequent.

The use of Delphi techniques in the socio-environmental sciences has so far been relatively limited. Among the exceptions are the so-called "spatial Delphis" that use mental maps and interactive Geographic Information System techniques to collaboratively gather expert knowledge about spatial phenomena, environmental impacts, territorial trends and related development strategies (Balram et al., 2003; Vargas-Moreno, 2008; Evrard et al., 2014). Orthodox Delphi techniques, which over the last years have been applied in multiple fields and in a very flexible manner, can be similarly applied to sustainability transition research. They also allow the combination with other methods such as focus groups, interviews or document analysis. For example, Landeta et al. (2011) propose a "Hybrid Delphi" when

---

Box 12.3   Delphi techniques

In methodological terms, the Delphi approach was initially motivated by the search for reliable forecasting techniques in areas of limited knowledge (e.g. technological risks, marketing studies), as a decision-making tool ("policy Delphi") or as a consensus-making procedure among stakeholders (Evrard et al., 2014). Given the variety of uses, Rowe and Wright (2011) prefer talking about "Delphi techniques" instead of a single "Delphi method". The common idea of the various applications is "to obtain a reliable group opinion from a set of experts" (Landeta et al., 2011), be it for scenario building (forecast) or be it for the validation of research results. In both cases, the researchers filter and categorize information obtained to give expert panels the opportunity to comment on preliminary results and to discuss the most intriguing aspects in more depth.

combining face-to-face exploration via focus groups with a more formalized two-stages Delphi based on questionnaires (non–face-to-face).

Case study experience

The green building sector is an emerging, rapidly growing and promising transition field (IPCC 2014) with new actor constellations and institutional arrangements, pioneering initiatives and complex articulations between the corporate, public and civil society realms (Schulz & Preller, 2016). In order to retrace how climate change–led innovations in the building sector occur and become mainstreamed, context-specific learning paths and development trajectories are especially relevant, as are the key factors and actors that have been instrumental to these changes. As indicated earlier, analyses of innovation processes are not limited to technological change and specific building projects, but should consist of a co-evolutionary perspective, taking into account interrelated organizational, procedural, legislative and other innovations bringing together a variety of views and interpretations to analyse the phenomena under study. This will be illustrated through empirical experiences of transition processes towards low-carbon economies in the building sector in four city regions: Vancouver, Freiburg, Brisbane and Luxembourg.

Research design

The research project on green building involved the two described methods (World Café and Delphi) that were combined in an incremental and cross-fertilizing manner, complementing other methods such as expert interviews and document analysis. To initiate contact as well as to involve as many expert voices as possible in each case study region, field research was kicked off by four successive workshops – one in each case study region – inspired by the World Café and Delphi techniques that consisted of experts representing different aspects and institutions concerned with green building (Table 12.1). For each workshop, three discussion rounds were set

Table 12.1 Field of expertise and affiliation of World Café participants

Workshop	Number of local participants (+ researchers)	Sectors represented and participant affiliations
Vancouver	14 (+5)	Architects, engineer and design firms, developers, think tanks, research institutes, NGOs, municipality, energy provider
Luxembourg	27 (+7)	Architects, engineer and design firms, private and public developers, interest and professional associations, research institutes, NGOs, ministries (sustainability, economy, housing), national energy consultancy
Freiburg	10 (+7)	Architects, engineer and design firms, public developers, research institutes, municipality, energy provider
Brisbane	10 (+5)	Architects, engineering and design firms, research institutes, NGOs, municipality, state ministry, regional administration

up. Each round focused on a specific dimension of the sustainable building sector following the project's co-evolutionary approach: actors and organizations, building projects and framework conditions (encompassing institutional aspects like legislation, socio-economic aspects, etc.). Following the first workshop experience, a fourth discussion table was added to address challenges and barriers to the development of sustainable building practices, as it had been an important and recurring topic of exchange amongst participants.

Follow-up of the workshops and further communication involved the dissemination of a report summarising the main outcomes to the participants in form of a Delphi-inspired questionnaire, where participants were asked to critically reassess and validate the transition factors that emerged from the workshops. The information from the questionnaires were used as guidance to determine a number of key aspects for in-depth qualitative micro-case studies in each of the four city regions, covering selected green building policies and programmes, influential organizations and actors, as well as built environment projects. This step was backed up through document analysis and semi-directive interviews. The workshops further helped to open doors to relevant interview partners and generated necessary background knowledge and references. The research design thus combined and complemented elements of two participatory methods with more 'orthodox' qualitative research methods. This incremental procedure was designed to assure a high level of reflexivity both of the researchers as well as the researched group. To this end, final workshops were held in Freiburg and Luxembourg allowing the participants of the former rounds to critically reflect on shared knowledge and to validate final interpretations of the data collected. Participants were also encouraged to disseminate and ensure transmission of the results to eventually allow further utilization within the researched community.

Obviously, such an approach requires a high commitment of the participants and their availability over the project's life span. In order to facilitate buy-in to the research endeavour, participants of the first Delphi round, as well as later interviewees, were kept informed about the project's progress via newsletters and (where possible) personal communication.

Outcome and discussion

The research design described allowed the involvement of a diverse range of stakeholders in the case study regions at a very early stage of the project. It not only helped to identify the most relevant micro-case studies in each region, but also to develop a first understanding of the individual trajectories and main issues at stake. Without being able to go into further detail, the respective narratives can be summarized as follows:

Vancouver. Early greening initiatives in Vancouver dating back to the 1970s highlight the narratives of strong links between nature and residents. Early environmentalism (e.g. the creation of Greenpeace), geographic distance to political centres of power and the liberal political position of the Canadian west coast led to a progressive/ alternative attitude shared by many inhabitants. Most changes have been driven by legislation involving the local up to the provincial scale (e.g. Vancouver's building codes and British Columbia's carbon tax). More recent trends show that Vancouver's current policies in terms of local sustainability and green building are increasingly marked by an explicit aspiration for green leadership at a global scale, actively promoted by the local government and public administration (as, amongst others, illustrated by the municipality's Greenest City Action Plan), but relying largely on public participation and changes at the individual level (Affolderbach & Schulz, 2017).

Freiburg: Similar to Vancouver, Freiburg's green building policies are rooted in a particular 'myth', here around the 1970s anti-nuclear power movement which is seen as the nucleus of the subsequent development to becoming Germany's 'green capital'. Drivers and motivations that initially evolved from a focus on energy and housing shortage and included clear social concerns of affordability and liveability have shifted towards a 'green economy' discourse with a strong extrospective and competitive dimension that is now disconnected (Freytag et al., 2014) from its original community roots.

Luxembourg: Although Vancouver's and Freiburg's historical legacies are missing in the Grand-Duchy, the more recent efforts to catch up with international trends in terms of green building have been predominantly triggered by economic motivations. The focus in Luxembourg is largely technology oriented which frames green building as a possible way to further diversify Luxembourg's mono-structured economy. It hence suggests an underlying ecological modernization and green technology logic. The focus lies on single lighthouse projects, whereas ecological, urbanistic and social aspects are mostly absent.

Brisbane: The transition pathways in Brisbane's building sector are characterized by ambivalence. Whereas a significant shift towards 'greener' office buildings occured in the central business district (CBD), there has been comparatively little momentum in the relatively conservative residential sector. With the exception of a small number of cooperatives or NGO-based initiatives, no particularly progressive developments took place in the residential sector, partly due to policy discontinuities. In the office building market, however, building rating tools have proven to be important drivers of sustainability transitions. Rating tools such as 'Green Star', developed by the Green Building Council of Australia, have played an important role as 'green' office buildings are becoming increasingly mainstream in Australia's CBDs. Developers, institutional investors (e.g. pension funds) and public authorities take Green Star certifications as a guarantee for (economically) sustainable long-term investments.

As to the actual application of the World Café format, challenges related to the generation of an interactive dialogue and the emergence of a shared understanding at some of the tables where participants' contributions remained quite detached from each other. Some participants expressed the feeling of having repeated themselves between the successive rounds (tables). This might be related to the choice of topics and questions implying quite descriptive and informative responses. Several authors insist therefore on the importance of carefully crafting the Café questions (Brown, 2001; Prewitt, 2011), as well as of facilitation skills of the Café host(s), in order to deal with group dynamic (Prewitt, 2011). The maturity of the community dealing with the subject at stake during the Café might also be given explanatory power, as stronger dynamics were at work within the two case studies with a longer record of climate change mitigation within the building sector (Vancouver, Freiburg). Overall, participants provided extremely positive feedback as they perceived the methods as being both inspiring and efficient. Most notably, they appreciated the 'side effect' of informally and openly engaging in dialogues with other actors outside their usual settings and agendas.

Another secondary impact, which occurred quite frequently, involved facilitation with contact requests and matching of potential cooperation partners between the case study regions. At least one senior expert from Luxembourg and one from

Freiburg organized fact-finding and networking trips to Vancouver; one senior civil servant from Freiburg asked for advice prior to a consultancy trip to Brisbane. One expert from Vancouver was invited to speak at a high-profile business conference in Luxembourg as a result of his involvement in the research project. These unplanned outcomes contributed to direct knowledge exchange between the case study regions, which could also be framed as "policy mobility" (Peck & Theodore, 2010; McCann, 2011). The idea of policy mobility was developed by urban studies scholars to better operationalize the relational dimension of city regions and their interconnectedness with global networks of ideas and experts. The concept's potential to strengthen the spatial concepts inherent to current transition studies research in general and the MLP in particular are obvious, but cannot further be outlined here (for an overview see Affolderbach and Schulz (2016).

The closing workshops held in Freiburg and Luxembourg confirmed the participants' strong interest in the findings from the other regions. Moreover, they gave reassuring feedback as to preliminary interpretations of the respective case study findings and indicated current and possible future trends.

Conclusion

This chapter had two objectives: (1) to argue in favour of a broadened understanding of the transition studies framework and MLP to analyse sustainability transitions and (2) to present useful instruments to implement the approach in research practice. Both points were illustrated using the example of urban green building transitions. The multi-level perspective provides a useful analytical tool for researchers to address sustainability transitions, but more attention needs to be placed on individual actors and spatial dimensions. The brief sketches of green building transitions in Vancouver, Freiburg, Luxembourg and Brisbane reveal different transition pathways in the four case study regions that reflect the interplay between different levels. For example, the landscape level in Vancouver and Freiburg was characterized by relatively strong environmental concern of residents, which provided a favourable climate for innovations in green building. Since the 1990s, this has been systematically taken up by local politicians and helped sharpen Freiburg's pioneering role. Here, the actor-centred perspective also helped reveal motivations, objectives and positions on green building, including actor networks and knowledge exchanges between different places. It also identified imbalances and biases (e.g. towards energy efficiency while social implications seem neglected). The application of MLP to this sector showed the spatial complexity of interpersonal networks and biographical trajectories, which indicate the potential for more relational perspectives, as it could be provided by the policy mobility approach. As outlined in more detail in Affolderbach and Schulz (2016), recent research on policy mobility (PM) has identified a number of dimensions of knowledge diffusion, learning and innovation that may not only complement the MLP perspective, but also help to overcome some of the latter's conceptual and empirical limitations.

Apart from the "Where?", that is the aforementioned spatial complexity challenging the mere territorial understanding of MLP and PM's intrinsic relational conception of space, further complementary dimensions include (see Table 12.2): (1) The object of study ("What?"), the creation of new knowledge within specific socio-technical processes (MLP) against PM's interest in the mobility and adoption of already existing knowledge; (2) PM's explicit focus on individual actors and their biographies and trajectories, compared with MLP's stronger interest in the role of

Table 12.2 Synopsis of main conceptual dimensions of policy mobility and transition studies

	Policy mobility	Transition studies
What?	Mobility/transfer of knowledge	Knowledge creation
	Socio-spatial(-political) processes	Socio-technical processes
Who?	Individuals & actor groups	Actor networks and institutional structures
How?	Learning, adaptation and mutation	Radical niche innovation
Where?	Relational	Localized

Source: Affolderbach & Schulz, 2016, p. 1950

structures and formal institutions ("Who?"); and (3) The "How?", that is the aim to understand the respective patterns of radical niche innovations (MLP) and of diffusion and adaptation mechanisms (PM).

In terms of the presented research approach, the interactive formats, notably the World Café or workshop sessions, proved to be valuable in at least three ways:

A) As a highly efficient means to gather information from key actors which was immediately commented, complemented and thus shared by other participants;
B) As a way to establish robust networks which were helpful in other stages of the project (expert interviews in micro-case studies, validation of findings);
C) As a platform that generated dialogue between stakeholders in a neutral, "non-threatening environment" (Fouché & Light, 2010) where the usual institutional standpoints did not have to be defended as it is the case in public forums.

As such, the World Café and Delphi methods can offer effective tools to not only gather a large amount of information, but also as accompanying and strengthening framework for subsequent micro case studies and their in-depth analysis. Further, the techniques can be used to gather, filter and analyse findings, which – together with the outcome of the micro-case studies – are then resubmitted to participating experts and become subject to critical discussion. These interactive formats go beyond what interviews or group discussions could have revealed. They create their own internal dynamics and help to bring particular facets of a problem to the forefront that are easily overlooked in more conventional settings. These dynamics tend to persist after the actual event and can lead to ongoing engagement between the researchers and the 'researched' and thus may nurture co-production of new knowledge over longer periods (and can be the starting point for new joint endeavours, e.g. in applied projects).

As with all research methods, a high degree of self-reflexivity and critical evaluation is required to make best use of the methods' potential and to obtain reliable and unbiased findings. For the methods presented, a particular focus has to be placed on adequate moderation approaches and communication skills. In particular, World Café sessions risk failure or yielding only little novel information if conceived and executed inappropriately. Upstream methodological training and accompanying reflection leading to continuous improvement and adjustment are therefore highly desirable, if not a prerequisite, when engaging with these techniques.

Acknowledgements

This chapter is based on experiences and insights collected as part of the Green-Regio research project jointly funded by the National Research Fund Luxembourg and the German Research Foundation (INTER_DFG/12–01/GreenRegio, 07/2013–06/2016). We would like to thank the two institutions for their financial support as well as all research participants in Brisbane, Freiburg, Luxembourg and Vancouver for sharing their valuable time and insights.

Questions for comprehension and reflection

1   What are the main drivers of green building transitions?
2   How can the MLP approach be applied to sustainability transitions in sectors such as transportation, waste and food?
3   What are some of the reasons behind different forms of stakeholder engagement in urban climate change initiatives around the world?
4   What are the main characteristics of participatory research approaches?
5   What particular role can green building certificates play?

References

Affolderbach, J. & Schulz, C. (2016). Mobile transitions: Exploring synergies for urban sustainability research. *Urban Studies* 53 (9): 1942–1957.

Affolderbach, J. & Schulz, C. (2017). Positioning urban green building initiatives: Vancouver's Greenest City 2020 Action Plan. *Journal of Cleaner Production* 164 (15): 676–685.

Aldred, R. (2011). From community participation to organizational therapy? World Cafe and appreciative inquiry as research methods. *Community Development Journal* 46 (1): 57–71.

Anderson, L. (2011). How to . . . use the World Café concept to create an interactive learning environment. *Education for Primary Care* 22: 337–338.

Balram, S., Dragicevic, S. & Meredith, T. (2003). Achieving effectiveness in stakeholder participation using the GIS-based collaborative spatial Delphi methodology. *Journal of Environmental Assessment Policy and Management* 5 (3): 365–394.

Bergold, J. & Thomas, S. (2012). Participatory research methods: A methodological approach in motion. *Forum: Qualitative Social Research* 13 (1), Art. 30. Retrieved from http://nbn-resolving.de/urn:nbn:de:0114-fqs1201302.

Blackstock, K. L., Kelly, G. J. & Horsey, B. L. (2007). Developing and applying a framework to evaluate participatory research for sustainability. *Ecological Economics* 60 (4): 726–742.

Borg, M., Karlsson, B., Kim, H. S. & McCormack, B. (2012). Opening up for many voices in knowledge construction. *FQS* 13 (1), Art. 1. Retrieved from http://nbn-resolving.de/urn:nbn:de:0114-fqs120117.

Brown, J. (2001). The World Café: Living Knowledge through Conversations that matter. *The Systems Thinker* 12 (5): 1–5

Brown, J. & Isaacs, D. (2005). *The World Café: Shaping our futures through conversations that matter.* San Francisco: Berett-Koehler.

Bulkeley, H., Castán Broto, V., Hodson, M. & Marvin, S. (Eds.) (2011). *Cities and low carbon transitions: Routledge studies in human geography.* London: Routledge.

Callon, M. (1999). The role of lay people in the production and dissemination of scientific knowledge. *Science Technology and Society* 4 (1): 81–94.

Coenen, L. & Truffer, B. (2012). Places and spaces of sustainability transitions: Geographical contributions to an emerging research and policy field. *European Planning Studies* 20 (3): 367–374.

EU. (2010). Directive 2010/31/EU of the European Parliament and of the Council of 19 May 2010 on the energy performance of buildings. *Official Journal of the European Union*, L 153/13, 8.06.2010.

EU. (2012). Directive 2012/27/EU of the European Parliament and of the Council of 25 October 2012 on energy efficiency. *Official Journal of the European Union*, L 315/1, 14.11.2012.

Evrard, E., Chilla, T. & Schulz, C. (2014). The Delphi method in ESPON: State of the art, innovations and thoughts for future developments. *Science in support of European Territorial Development*. E. Programme. Luxembourg: ESPON. pp. 187–191.

Fastenrath, S. & Braun, B. (2016). Sustainability transition pathways in the building sector: Energy-efficient building in Freiburg (Germany). *Applied Geography*. Online first.

Fouché, C. & Light, G. (2010). An invitation to dialogue: 'The World Cafe' in social work research. *Qualitative Social Work* 10 (1): 28–48.

Freytag, T., Gössling, S. & Mössner, S. (2014). Living the green city: Freiburg's Solarsiedlung between narratives and practices of sustainable urban development. *Local Environment* 19(6): 644–659.

Geels, F. W. (2002). Technological transitions as evolutionary reconfiguration processes: a multi-level perspective and a case-study. *Research Policy* 31, 1257–1274.

Hansen, T. & Coenen, L. (2015). The geography of sustainability transitions: Review, synthesis and reflections on an emergent research field. *Environmental Innovation and Societal Transitions* 17: 92–109.

Hessels, L. K. & van Lente, H. (2008). Re-thinking new knowledge production: A literature review and a research agenda. *Research Policy* 37 (4): 740–760.

IPCC. (2014). *Chapter 9: Buildings*. Final draft report of the Working Group III contribution to the IPCC 5th Assessment Report 'Climate Change 2014: Mitigation of Climate Change'.

Jorgenson, J. & Steier, F. (2013). Frames, framing, and designed conversational processes: Lessons from the World Cafe. *The Journal of Applied Behavioral Science* 49 (3): 388–405.

Kindon, S. (2010). *Participatory action research: Qualitative research methods in human geography*. Edited by Hay, I. Oxford: Oxford University Press. pp. 259–277.

Kindon, S., Pain, R. & Kesby, M. (2007). *Participatory action research approaches and methods: Connecting people, participation and place*. London: Routledge.

Landeta, J., Barrutia, J. & Lertxundi, A. (2011). Hybrid Delphi: A methodology to facilitate contribution from experts in professional contexts. *Technological Forecasting and Social Change* 78 (9): 1629–1641.

Martin, S. (2010). Co-production of social research: Strategies for engaged scholarship. *Public Money & Management* 30 (4): 211–218.

Mason, K., Brown, G. & Pickerill, J. (2013). Epistemologies of participation, or, what do critical human geographers know that's of any use? *Antipode* 45 (2): 252–255.

McCann, E. (2011). Urban policy mobilities and global circuits of knowledge: Toward a research agenda. *Annals of the Association of American Geographers* 101 (1): 107–130.

Murphy, J. T. (2015). Human geography and socio-technical transition studies: Promising intersections. *Environmental Innovation and Societal Transitions* 17: 73–91.

Pain, R. (2004). Social geography: Participatory research. *Progress in Human Geography* 28 (5): 652–663.

Peck, J. & Theodore, N. (2010). Mobilizing policy: Models, methods, and mutations. *Geoforum* 41 (2):169–174.

Preller, B., Affolderbach, J., Schulz, C., Fastenrath, S. & Braun, B. (2017). Interactive Knowledge Generation in Urban Green Building Transitions. *The Professional Geographer* 69 (2): 212–224.

Prewitt, V. (2011). Working in the Café: Lessons in group dialogue. *The Learning Organization* 18 (3): 189–202.

Princen, T. (2003). Principles for sustainability: From cooperation and efficiency to sufficiency. *Global Environmental Politics* 3 (1): 33–50.

Rohracher, H. & Späth, P. (2013). The interplay of urban energy policy and socio-technical transitions: The eco-cities of graz and freiburg in retrospect. *Urban Studies* 51 (7): 1415–1431.

Rowe, G. & Wright, G. (2011). The Delphi Technique: Past, present, and future prospects – Introduction to the special issue. *Technological Forecasting and Social Change* 78 (9): 1487–1490.

Sachs, W. (1999). *Planet dialectics: Explorations in environment and development.* London: Zed Books.

Samadi, S., Gröne, M.-C., Schneidewind, U., Luhmann, H.-J., Venjakob, J. & Best, B. (2016). Sufficiency in energy scenario studies: Taking the potential benefits of lifestyle changes into account. *Technological Forecasting and Social Change* – in print.

Schneidewind, U. & Zahrnt, A. (2014). *The politics of sufficiency.* Munich: Oekom.

Schulz, C. & Preller, B. (2016). 'Keeping up with the pace of green building: Service provision in a highly dynamic sector', in Jones, A., Ström, P., Hermelin, B. & Rusten, G. (Eds.) *Services and the green economy.* London: Palgrave Macmillan UK. pp. 269–296.

Sheridan, K., Adams-Eaton, F., Trimble, A., Renton, A. & Bertotti, M. (2010). Community engagement using World Café. *Groupwork* 20 (3): 32–50.

Tan, S. & Brown, J. (2005). The World Cafe in Singapore: Creating a learning culture through dialogue. *The Journal of Applied Behavioral Science* 41 (1): 83–90.

Truffer, B. & Coenen, L. (2012). Environmental innovation and sustainability transitions in regional studies. *Regional Studies* 46 (1): 1–21.

UNEP. (2011). *Towards a Green Economy: Pathways to Sustainable Development and Poverty Eradication.* Retrieved from www.unep.org/greeneconomy.

Vargas-Moreno, J. C. (2008). Spatial Delphi: Geo-collaboration and participatory GIS in design and planning. *2008 Specialist Meeting – Spatial Concepts in GIS and Design.* Santa Barbara. Retrieved from http://ncgia.ucsb.edu/projects/scdg/docs/position/Vargas-Moreno-position-paper.pdf.

TheWorldCafé. (2008). *Café to go! A quick reference guide for putting conversations to work.* Retrieved from http://www.theworldcave.com/wp-content/uploads/2015/07/Café-To-Go-Revised.pdf.

13 Democratising renewable energy production

A Luxembourgish perspective

Kristina Hondrila, Simon Norcross, Paulina Golinska-Dawson, Vladimir Broz, Aydeli Rios and Jules Muller

---

Box 13.1   Students for citizens' energy: objectives, concepts and methods

During the academic year 2015–2016, our peer group set out to investigate how the challenge of democratising and developing renewable energy production in Luxembourg could be met.

This was done in the framework of the certificate in "Sustainable Development and Social Innovation" of the University of Luxembourg, of which peer group work is a central component, as it offers "a co-designed and systematic process of participatory enquiry that engages diverse perspectives from a wide range of scientific expertise, professions, interests, and experiences in a transformative learning process" (König, 2015).The main learning outcome is intended to be collaborative co-creation of shared knowledge on complex problems of sustainability.

The peer group consisted of five mature students supported by an experienced mentor, all of different nationalities.The peer group members, three females and two males, hold university qualifications in industrial engineering, philosophy, European studies, international relations, psychology, mathematics, economics and business administration.They range in age from 38 to 63 and originate from Germany/Denmark, Italy, Mexico, Poland and the United Kingdom.

The local mentor, the Luxembourgish president of one of the energy cooperatives analysed, provided guidance and invaluable knowledge of the renewable energy scene in his country.

The starting hypothesis of the peer group project was that the energy transition not only offers the opportunity of greater involvement of citizens in energy production, but actually makes it necessary. Technological transition needs to be accompanied by social innovation if wider systemic change – that is, the replacement of fossil fuel by renewable energy sources – is to be successful.

Adopting a normative stance, the peer group identified the Dutch theory of socio-technical transition presented during the certificate course as a suitable conceptual framework.The multi-level perspective offered by the theory allowed the peer group to move from the case-study of two (of the three) existing energy cooperatives in Luxembourg, EquiEnerCoop and TM EnerCoop, to the examination of conditions for the spreading of citizens' participation and, finally, to the development of ideas for even more 'democratic' forms of citizens' energy production and consumption, for which the group formulated concrete policy recommendations intended to lead to a paradigm shift in the energy market and beyond.

Accordingly, the peer group approached the topic from the perspective of social innovation, pursuing the question in which way and under which conditions energy cooperatives could become agents of a sustainable transition in Luxembourg or, to put it differently, drivers of democratisation and change at a transformative scale.

The challenge: renewable citizens' energy

In this chapter, we define 'energy democracy' and provide arguments for citizens' energy. Energy from renewable sources (i.e. wind, solar, hydro, biomass and geothermal energy) has become a cornerstone in policy strategies for sustainable development, as a substitute for fossil fuels and, thus, a pre-condition for the transition into a low-carbon era. The increasing use of renewable energy is one of the principal means by which governments in Europe and elsewhere seek to counteract climate change, the depletion of natural resources (e.g. fossil fuels) and pollution, to mention but a few of the environmental challenges that threaten the very survival of our societies, economic systems and planet. Under the European Union's 2020 policy framework, Luxembourg has committed to increase the share of electricity generated from renewable energy sources from the current 4.5% to 11% by 2020.

Many scholars and stakeholders in Europe agree that a significant increase in renewable energy generation will require not only technological change, but also a paradigm shift in the way the energy market is organised. Energy systems are a critical infrastructure, vital for the functioning of our societies and economies. In all countries a few big players (typically state-owned monopolist structures, many of which have subsequently been turned into large commercial corporations) have traditionally dominated the market, producing and distributing fossil fuel–based or nuclear energy from giant power plants via centralised unidirectional grids that, in their turn, have also been operated by state-owned or privatised monopolists. Moreover, European countries have been highly dependent on primary energy imports of oil, natural gas and coal from non-democratic countries (such as Russia, Saudi Arabia, the United Arab Emirates and Iraq[1]). To sum up, in many ways the energy market represents an interest- and profit-driven network, with close links between policy and business and a high degree of import dependence. The role of citizens is limited to that of passive consumer (and voter). It is this context that explains the significance of 'energy democracy'.

In a more general sense, the term can be used to promote the ideas of:

- Ecological sustainability: using 'home-grown' renewable energy sources;
- Macroeconomic independence from fossil fuel–exporting countries;

- Market diversity and less market dominance by large commercial enterprises;
- Consumer protection: affordable and transparent prices and free choice of energy provider.

The importance of participatory planning and an acceptance of renewable energy installations with and by citizens are also often identified as issues. Wind parks, biomass plants and solar panels are mostly situated close to local communities, often in rural areas where they are likely to be perceived as a disturbance. This clearly calls for a new planning culture and participatory decision making at a local level.

However, of even greater concern in our context is the more specific sense of the term, where energy democratisation implies the idea of renewable energy as a 'common good' that should be managed by the citizens themselves at the local level:

> From the perspective of social justice, more attention therefore must be paid to the way in which decentralised renewable energy sources are managed. In a world where energy is scarce, these sources of energy will mean income for the operators. Citizens and users therefore have every interest in keeping this local energy production in their own hands as much as possible. Governments too have every interest in anchoring decentralised renewable energy with the users as much as possible so that the added value of the production also benefits society.
> (REScoop, 2015, p. 60)

In this sense, defining characteristics of energy democratisation are:

- Decentralisation and regional value creation;
- Financial participation in and co-ownership of renewable energy installations by citizens.

This last point brings us to the concept of 'citizens' energy' as defined by Leuphana University Lüneburg and the institute trend:research, being energy owned – and controlled – by citizens either individually (private households or farms) or collectively (e.g. by cooperatives, in which citizens hold at least 50% of the voting rights) (TRLU, 2013, p. 15).

It is on the collective ownership and exploitation of renewable energy that our peer group (Box 13.1) has concentrated, as cooperatives represent a new form of social (and commercial) organisation (Box 13.2) that could lead to the move away "from a centralised, oligopolistic energy system to one that is decentralised and above all democratically controlled and operated" (REScoop, 2015, p. 69).

Box 13.2   Cooperatives and energy transition

According to REScoop.eu, the European federation of groups and cooperatives of citizens for renewable energy and energy efficiency, in 2016 there were some 2,400 such cooperatives in Europe, mostly in Western Europe, and particularly in Germany, Denmark and Austria. Of these, about half, representing 650,000 citizens, are members of the federation. Its members have invested 2 billion EUR in electricity generation equipment, with a total capacity of 1 GW. The importance of the role of citizens in increasing the share of renewable energy, and as drivers of the energy transition, is demonstrated by the case of Germany. Research by trend:research and Leuphana University on the German 'Energiewende' ("Definition und Marktanalyse von Bürgerenergie in Deutschland", October 2013) showed that in 2012, total renewable energy generation capacity in Germany was some 72.9 GW. Of this, 41% was owned by institutional investors, 12.5% by energy companies and 46% by citizens. Within this 46%, 25% of the total was in private ownership, and citizens' energy cooperatives accounted for some 9% (c. 6.7 GW) (TRLU, 2013; RESCoop, 2016).

One way of looking at energy cooperatives is as social enterprises. Applying the approach of the European Social Enterprise Research Network (EMES) to social enterprise, one can define the following main aspects of energy cooperatives (Defourny & Nyssens, 2012, pp. 8–9):

- Economic and entrepreneurial: producing and selling (renewable) energy from collectively financed and owned installations, mainly based on voluntary work, limited profit distribution among the members;
- Social (and environmental): an explicit aim of benefitting the community and environment; surpluses are reinvested in the community or new projects;
- Participatory governance: equality of members and collective decision making, each member has one vote.

Based on these characteristics, social enterprises are deemed to have a special role to play in social innovation, as "the value they create is necessarily shared value, at once economic and social" (EC, 2013, p. 7). Many scholars emphasise the potential of social entrepreneurs to introduce new services, methods of production and forms of organisation. As Geoff Mulgan puts it, "social entrepreneurship sits within a broader context of social change", typically drawing "on the often invisible fecundity of tens of thousands of individuals and small groups who spot needs and innovate solutions" (Mulgan, 2006, pp. 75–76).

The vision of citizens and communities 'taking over' the energy market, getting together to collectively produce, share and consume their own energy is put forward by a number of scholars, perhaps most prominently by Jeremy Rifkin:

> In the coming era, hundreds of millions of people will produce their own green energy in their homes, offices, and factories and share it with each other in an 'energy Internet' [. . .]. The democratisation of energy will bring with it a fundamental reordering of human relationships, impacting the very way we conduct business, govern society, educate our children, and engage in civic life.
> (Rifkin, 2011, p. 2)

Rifkin believes that, in the future, most people will be prosumers, installing solar panels on their roofs to generate and to consume their own electricity and selling surpluses to others. This "would democratise the production and distribution of energy by creating millions of mini energy entrepreneurs", leading to a new Collaborative Age and the Third Industrial Revolution (Rifkin, 2011, p. 48).

Although Rifkin's vision is not uncontested, he is not alone in believing that decentralisation in the energy sector will give rise to "fundamental shifts in social organisation that affect all areas of people's lives" with far-reaching implications for all spheres of society, such as politics, economics and culture (Gross & Mautz, 2015, pp. 8–9).

The energy source most accessible to citizens is solar energy (ranking before wind energy), due to the relatively affordable and continuously decreasing cost of investment, high availability and relatively easy rooftop installation (COM/2015/080, European Commission, p. 15). These factors have contributed to the fact that, according to the Ren 21 multi-stakeholder network, global solar energy generation has more than trebled over the past four years. In 2015 alone it grew by 32.6% according to the BP Statistical Review of World Energy (BP, 2016, p. 5).

In the same way, in Luxembourg, the three existing energy cooperatives have made solar energy their main business. Leaving the conceptual level, we shall now describe the peer group's approach and methodology before turning to the case studies.

The methods used

The group adapted the Dutch theory of socio-technical transition (Geels, 2002; Verbong & Geels, 2007) to the energy sector and the wider societal context in Luxembourg, related to the three levels of social organisation and innovation identified by the theory (see Figure 13.1 and Box 13.3):

1   Micro-level: energy cooperatives and other local and social initiatives (or "niches");
2   Meso-level: spreading of initiatives resulting in wider changes in the energy market and social organisation throughout society ("patchwork of regimes");

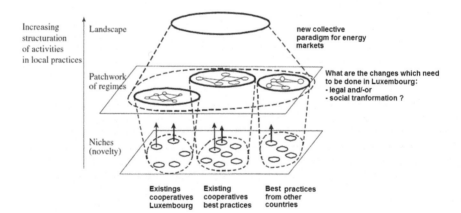

Figure 13.1 The social-technical transition model for provision of renewable energy in Luxembourg

Source: Based on Geels, 2002

3   Macro-level: paradigm-shift penetrating the entire fabric of society (or "landscape"), marking the transition to a new era.

Starting at the micro-level, the group chose observational studies as the methodology best suited to analysing the two energy cooperatives and the policy framework: conducting interviews with executives of the cooperatives (presidents and
board members) as well as regulators, attending meetings and conferences of energy
stakeholders in Luxembourg and carrying out on-site visits of solar installations.
The case studies showed that energy democratisation in Luxembourg has begun,
but that many challenges remain to be met in order for the cooperatives to become
more than isolated initiatives (or "niches"), for citizens' participation to become
more widespread and for a paradigm shift to take place.

The methodology is presented in Figure 13.2.

For benchmarking purposes, on-site visits to renewable energy projects and
interviews with a cooperative and architects were also held in Germany. They
showed that democratisation and citizens' participation could be taken yet another
step further by moving towards the 'prosumer model', by which citizens consume
their own renewable energy directly.

These findings were presented at the General Assembly of EquiEnerCoop and
were published on an energy blog.[2]

In order better to understand the current EU policy framework, enabling and
constraining legal factors affecting any expansion/scaling of the prosumer model
and potential upcoming policy initiatives, the peer group members further attended
a public hearing in Brussels on energy prosumers as the possible new "rising stars
of the [EU] Energy Union".[3]

**Box 13.3   A multi-level perspective on socio-technical
transitions**

The benchmark for our studies was the work of Verbong and Geels (2007,
2010), who developed a theory of transition looking at socio-technical transformations in a multi-level perspective (MLP) in the context of an analysis of the
Dutch electricity market and its transition towards sustainability. They considered social networks as well as new renewable energy options, structural changes
in the existing electricity market and changes in rules and regulations which
support or block transition. The MLP considers interactions between niche
innovations and existing regime developments in a broader environment (Verbong & Geels, 2007). Sustainable transition on the energy market could happen
only if and when innovations introduced by different actors in "niches" (novel
renewable energy projects) link up and are reinforced by any developments in
the existing regime and broader socio-technological environment (macro level).
The socio-technical regime consists of three dimensions (Geels, 2002):

- Network of actors and social groups;
- Formal, normative and cognitive rules that guide the activities of actors
  (laws, rules, standards and regulations, guiding principles, relationships);
- Material and technical elements; these include physical resources, such as
  grid, generation plants and renewable energy installations.

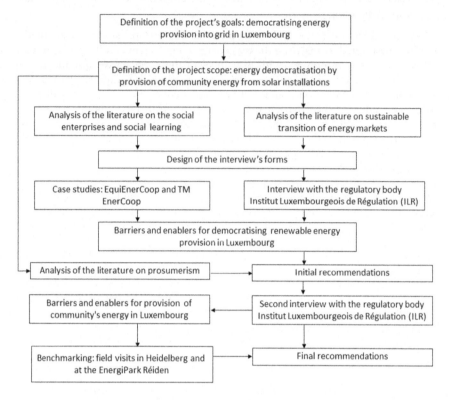

Figure 13.2 Peer group methodology

Insights: energy cooperatives as pioneers of social innovation in Luxembourg

In the towns of Esch-sur-Alzette and Junglinster, citizens, local policy makers and civil society organisations have come together to put into place a social business model that is innovative in the Luxembourgish context, with the aim of collectively producing renewable energy (Box 13.4).

Box 13.4   Facts and figures about EquiEnerCoop and TM EnerCoop

Common features of the cooperatives EquiEnerCoop (founded in 2012 in Junglinster) and TM EnerCoop (founded in 2013 in Esch-sur-Alzette):

- Objectives: social, environmental, political and, to a lesser degree, financial;

- Membership: 85 to 150, mainly local, members co-owning PV installations;
- Community context: emerging from pre-existing community activities and (transition) movements with local political support;
- Business model: producing and selling solar energy from publicly owned roofs (125,000 kWh/year and 26,000 kWh/year, respectively) to energy suppliers at feed-in tariffs (FITs of 32 and 21 cents/kWh, respectively), based on volunteering and limited outsourcing of work;
- Leadership: dedicated board members, combining multiple competences and investing significant amounts of time.

Our case studies have shown that both of these energy cooperatives have been successful in establishing their entrepreneurial activities by 'recruiting' a sufficient number of members from their communities (and beyond), mainly from the well-off middle class, willing to invest money in solar panels. They have also installed democratic and participatory governance structures (one member/one vote). Both have operated with low barriers of entry, with membership being tied to the minimum purchase of one share at the respective prices of 25 EUR (EquiEnerCoop) and 100 EUR (TM EnerCoop). Although their membership is mainly composed of local inhabitants, TM EnerCoop has also recruited non-profit organisations as members (such as Greenpeace) as well as people living outside their municipality.

Their main objectives have been to increase citizen and community involvement in renewable energy production, to increase environmental awareness, to help their municipalities reduce CO_2 emissions and to provide their members with a stable return on investment.

By seeking to replace energy from fossil fuels by renewable energy, the main motivation of the founders can be described as achieving "consistency", as defined by Schweizer-Ries and Samadi et al. (see Table 13.1).

Being pioneers, both cooperatives have had to innovate solutions and handle numerous problems relating to the technical installations (hardware, grid connection, fire protection), unexpected costs and financial management (grid connection fees, bank accounts) and administrative procedures (statutes, authorisation to use the roofs of public buildings, insurance issues, obtaining feed-in tariffs).

The key success factors for solving these issues have been the support and involvement of local policy makers, a competent and committed leadership and a sense of community and shared purpose among the members. Their ability to organise themselves, handle problems, launch their projects and mobilise significant resources within short time spans seems remarkable.

They are both economically viable, as they receive feed-in tariffs, guaranteed for 15 years, for selling electricity to energy suppliers. The feed-in tariffs are above electricity market prices (currently approximately 18 cents per kwh) and more than cover the costs of hardware and maintenance in the medium term. The cooperatives thus expect to break even after 11 to 12 years. Any profits are either returned to the members or re-invested in more joint projects serving the community.

To sum up, the cooperatives do indeed provide successful models for energy democratisation that are, in principle, transferable to other municipalities and projects in Luxembourg. Indeed, new cooperatives are currently being founded in Luxembourg. And yet, they are still 'niche experimenters'.

Their growth prospects in the current setup are limited. Their main weakness is dependence on the time and competences of a handful of board members and, thus, their limited capacity to embark on new and larger projects. Moreover, their business model is under threat, as feed-in tariffs are expected to fall below market prices in the coming years. Therefore, for new projects and after the expiration of the feed-in tariffs for the current projects, a new business model will probably have to be developed that might also involve e.g. more outsourcing.

It could also be argued that they have not yet taken social entrepreneurship, decentralisation and democratisation as far as they might. Although the cooperatives have decentralised energy production, creating local revenue by putting energy in the hands of citizens, their business model still relies on selling electricity to big energy supplying companies and has not yet managed to reach other social strata of society.

The peer group therefore recommends that cooperatives like EquiEnerCoop and TM EnerCoop scale up their activities by extending them to selling energy directly to consumers and to their members (direct marketing) and to 'turn prosumer' by embarking on projects based on collective 'in-house' consumption of self-produced energy, including by tenants in apartment blocks. By so doing, the energy cooperatives would become more independent of energy companies (as main buyers of their energy), take local value creation and social inclusion another step further and move towards autonomous community energy management, thereby 'revolutionising' the energy market. If coupled with strategies for greater sufficiency – here understood as changing consumption patterns and lifestyle (see Table 13.1) – the communities could not only reduce energy consumption, but would also achieve a better balance between energy supply (from self-production) and their own energy needs and demand (self-consumption). In this way, the cooperatives would move closer to self-sufficiency/autonomy.

Additional insights: collective prosumer projects and community grids

Prosumerism can be said to be the most democratic and decentralised form of energy provision and consumption. It is a model in which electricity is produced, owned, consumed and potentially sold and bought by the same person/entity or group of people located in the same building or in immediate geographic proximity, with no or very limited grid use for self-produced and self-consumed energy. The model dissolves the traditional separation between active energy producer and passive consumer and might well drive one of the major future trends on the energy market.

Advantages of prosumer models often cited are:

- Empowering consumers on the energy markets;
- Encouraging new, more efficient, energy consumption patterns (or "sufficiency" as defined in Table 13.1), because consumers become more aware of their energy use when shifting to producing and consuming their own energy;
- Balancing energy consumption in the grid (potentially reducing energy usage peaks);

Table 13.1 Comparison of motivations between the cases studies based on three different approaches to energy

Approach	EquiEnerCoop	TM EnerCoop	Nussloch project	Heidelberg Village project
Efficiency: Rational use of energy – using less energy to do the same job/achieve the same result[1]	✓	✓	✓✓	✓✓✓
Consistency: Consumption of energy from renewable sources in replacement of energy from fossil fuels	✓✓✓	✓✓✓	✓✓✓	✓✓✓
Sufficiency: A change of social practice leading to a self-imposed restriction of energy consumption, independent of improved efficiency[2]	✓✓	✓✓✓	✓	✓✓✓

Source: Schweizer-Ries, 2008 and Samadi et al., 2016

[1]Typically based on technical improvements – more energy-efficient equipment and machinery, better heat insulation, etc.

[2]In general, arising from a conviction that this is the right thing to do. For many observers, the most obvious form of sufficiency may be in relation to the behaviour of domestic energy consumers, but the concept is equally valid for larger-scale consumers such as industries and transport systems.

Figure 13.3 Renewable energy cooperative in Heidelberg
Source: www.heidelberger-energiegenossenschaft.de

- Reducing costs, both with regard to energy costs and exposure to price increases on the part of the prosumer as well as for public energy infrastructure;
- A new approach for companies, cooperatives and other organisations to save costs and generate revenues.

The Renewables GSR, Ren 21, identifies "the use of solar PV for self-consumption in residential, commercial and industrial sectors" as one of the new emerging business models in Europe as a reaction to decreasing or inexistent feed-in tariffs (Ren, 2016, p. 61).

Although most prosumers in Europe, albeit still few in number, are private households, farms or single small or medium-sized businesses or institutions, wider systemic change in the energy market and beyond may be expected to emerge from the spreading of collective prosumer models for multi-family residential buildings, such as apartment blocks, and from largely self-sustaining energy communities.

The peer group went to Heidelberg in Germany to visit examples of collective prosumer models. There, a local energy cooperative (Heidelberger Energiegenossenschaft) has teamed up with a housing cooperative on a residential project, "Neue Heimat Nussloch", to install and operate rooftop solar panels, producing up to 370,000 kWh per year (broadly corresponding to the electricity consumption of about 100 households). Tenants have the opportunity of becoming clients, buying electricity from the cooperative below the market price with a 20-year price guarantee. This is possible because the cooperative does not have to pay certain taxes and fees.[4] Moreover, tenants can themselves become members of the cooperative and invest in solar installations. Surplus energy is fed into the grid (with feed-in tariffs); similarly, shortages are covered by grid supply. To realise the project, the cooperative had to install its own meters in the buildings and work closely with a (green) energy provider as well as the grid operator.

Another housing project, 'Heidelberg Village', comprising 162 apartments, is currently being implemented by an architecture firm. As one core element, solar panels will be installed on the roof and facades. Any electricity not directly used will heat water or be stored in a battery. The firm is considering setting up a cooperative to facilitate direct marketing to the tenants as well as co-ownership. The energy

Figure 13.4 Designing built environment for sustainability communities: Heidelberg Village

Source: www.heidelberg-village.de by Frey Architekten

installations are one element of a broader concept of sustainable and socially inclusive housing, which can be said to include attempts to change social (community) practices, including the energy consumption patterns of the inhabitants.

Furthermore, the architects are currently working on a project to build a green district, with a Smart Green Tower at its core. Also having a solar panel façade, the tower (a residential and office building) will not only generate solar energy, but also supply it to neighbouring buildings and be connected to a large-capacity storage unit that will enable it to provide balanced loads to neighbouring districts through a smart grid.[5]

The project is thus based on the vision of local energy management, including the operation of a community grid designed for the collective sharing and storage of self-produced energy. Certain scholars and stakeholders believe that the setting up of grids suitable for a "two-way, new, decentralised energy landscape" (YEEPS, 2012, p. 9) constitutes one of the biggest public policy and technological challenges of our time.

As ambitious as they may be, these projects point towards the possibility of a future in which renewable energy installations are integral elements of all buildings (just like water pipes and electricity cables are today) and in which citizens and communities are the main agents in the energy market. Strategies for greater efficiency and consistency, as defined earlier, seem to be only the beginning of potentially deeper changes in the energy market. These would be brought about by altered individual and collective consumption patterns aiming for sufficiency and greater collective independence and self-determination via self-production, self-consumption and self-management of energy by communities. However, for this to happen, regulatory regimes require significant adaptation.

Policy recommendations and conclusion

In this section, we will briefly outline current EU policy developments, providing the framework for national policies, as a basis for formulating recommendations for policy makers and regulators in Luxembourg on how to promote citizens' energy and, more specifically, collective prosumerism.

In its Energy Union strategy, the European Commission encourages consumers to take "full ownership of the energy transition, to benefit from new technologies to reduce their bills and participate actively in the market, while ensuring protection for the vulnerable ones" (COM/2015/080). Moreover, in a communication on "delivering a new deal for energy consumers" it recognises that decentralised generation and storage options can "enable consumers to become their own suppliers and manager for (a part of) their energy needs" (COM/2015/339). In an accompanying document, it also describes best practices of "energy self-production".

However, although no official definition exists as yet, the examples given suggest that the European Commission has a rather narrow understanding of prosumerism, not including community prosumer structures. Partly for this reason, the European Economic and Social Committee has, since January 2015, published several studies and opinions. The EU consultative body argues for a broad definition of prosumerism, giving up a notion that presupposes the energy supplier and consumer necessarily to be the same entity, in order to allow tenants, for example, actively to participate in cooperative energy schemes. A definition should also cover community entities (cooperatives, local companies or authorities or similar) supplying "their electricity directly to consumers in the local area, without going via the energy market or distributors" (EESC, 2016, pp. 10–11).

During a public hearing organised by the EESC in June 2016, most stake-holders present agreed that the foundations for a prosumer revolution (i.e. an appropriate regulatory framework at national level dedicated to individual and collective prosumers' securing a level playing field, as well as incentives for self-consumption models) do not yet exist (Petrick, 2016, p. 10). As the EESC points out, there are currently "no signs of a consistently-implemented government strategy aimed at promoting civic energy", let alone community prosumer mod-els (EESC, 2015, p. 2).

This situation, indeed, applies to Luxembourg. Although measures have been put in place to encourage energy cooperatives, important obstacles and barriers to further growth of citizens' energy in Luxembourg remain:

- Limitation of feed-in tariffs to utilities no larger than 200 kW;
- Limitation of cooperatives having to be set up by at least 10 'physical' persons (with a view to strengthening multi-stakeholder networks, it would make sense to allow non-profit organisations and public structures – for example, munici-palities and local councils – to participate in the founding of cooperatives);
- Cooperatives cannot receive subsidies for major investments (whereas other legal entities can; however, a new law is currently under preparation);
- Non-transparent and sometimes high once-off grid connection fees (amount-ing to an unexpected 40,000 EUR in the case of EquiEnerCoop, for example);
- Little information or support provided by the National Energy Agency to sup-port citizens with the administrative requirements of setting up or running a cooperative;
- No information provided by the National Energy Agency on prosumer energy installations, with or without battery storage;
- Grid operator CREOS opposes prosumer installations for small and medium-sized businesses and publicly owned buildings, citing a "lack of regulation".

With regard to prosumerism, there have been some improvements due to a new tariff methodology for settling grid fees initiated by the national regulating body (Institut Luxembourgeois de Régulation, ILR).[6] It clarifies that grid fees do not have to be paid on self-produced and self-consumed electricity. Self-consumed energy is also exempted from VAT and other fees, except from the "taxe d'électricité".

However, as of January 2017, prosumers will have to pay a "grid capacity charge". The exact amount, however (currently under negotiation), is likely to depend on factors such as the availability of a storage system (which would mean that prosum-ers will feed less or no surplus energy into the grid and take out less energy from the grid at times of peak consumption).

A grey zone remains regarding collective prosumer models, where solar panel owners and consumers are not identical (not the same legal entity). This would, for example, be the case when a cooperative (or other joint legal structure of the inhabitants) produces solar energy and supplies it to residents of the building on which the panels are installed (even if the residents are members of the coopera-tive). Under EU law, all consumers have the right freely to choose their supplier, to have grid access and to be invoiced based on their actual energy consumption. Therefore, in this case, there seem to be only three main solutions:

- The cooperative becomes an energy supplier, marketing the energy directly to the inhabitants (which would then be clients of the cooperative), with all the legal, consumer protection and accounting obligations applying to energy suppliers in the market (Heidelberg case);

- Self-consumption is limited to the collective parts of the building (staircase, etc.); or
- PV panels are divided up among the inhabitants, each household owning (or renting) a part of the solar installations and receiving its energy directly and exclusively from its own panels.

These options show how complicated and unclear the current legal situation is – and how little likely it is that collective prosumerism will become widespread under the current regime.

Another factor impeding the emergence of prosumerism in Luxembourg is the fact that there is currently no financial incentive for citizens to become prosumers, due to relatively low electricity prices and higher feed-in tariffs. Clearly, prosumer models only become economically interesting once it is cheaper to consume your own energy rather than to buy it from the grid. However, as mentioned earlier, this may indeed become the situation in a few years, when feed-in tariffs may fall below market prices.

Therefore, alas, there are currently neither single-household nor collective prosumers in Luxembourg according to official statistics.[7] The peer group has only heard of two potential collective prosumer projects that are currently being planned by two of the energy cooperatives (EquiEnerCoop and Energy Revolt).

For this reason, the peer group concluded that the three existing energy cooperatives in Luxembourg are still only "niche experimenters". A real field of citizens' energy, or "patchwork of regimes", has yet to emerge. Luxembourg – like all other European countries – still has a long way to go for a paradigm shift to take place in the energy market (and beyond), to move towards a decentralised and democratic energy transition.

A lot of advocacy work needs to be done by civic and public stakeholders to overcome lock-ins in the current system that continues to be dominated by (the interests of) large energy suppliers and monopolist grid operators that do not, currently, seem willing to become partners in paving the way for the transition.

In addition to adapting the regulatory framework, there are, however, some things that policy-makers can do:

- Introduce a self-consumption premium as a financial incentive for prosumerism;
- Provide subsidies for specific consultancy services, storage systems (e.g. batteries) or grid connections for small, medium and large-scale prosumer pilot projects;
- Require public authorities, property development and urban planning agencies to consider installing renewable energy devices for any larger housing projects and innovating solutions (including new business models) to enable residents, including tenants, to co-own installations and consume energy from their rooftops.

The potential for renewable energy production already exists in Luxembourg, but the challenge is to scale it up to a broader population and also target urban areas, where many people are tenants with varying socio-economic backgrounds.

In the view of the peer group, apartment blocks and large housing projects, in particular, offer considerable potential for developing collective energy production and self-consumption, as demonstrated by projects in Germany and in other countries. So far, this potential is under-exploited in Luxembourg.

However, achieving transformational changes at the macro-level in Luxembourg may well depend on such projects, in addition to adequate financial incentives and a propitious regulatory framework.

We would like to conclude by stating our conviction that the future belongs to decentralised forms of energy production, consumption and storage, involving local communities, public authorities, housing owners, tenants and developers, architects, new businesses and cooperatives, to take the energy revolution forward.

Acknowledgements

We would like to thank Luis de Sousa (TM EnerCoop), Andreas Gissler (Heidelberger Energiegenossenschaft), Qiaozhi Meng and Juergen Heller (Frey Architekten) for the time they spent with us to explain their projects and provide feedback.

Questions for comprehension and reflection

1   Are you aware of any citizens' energy project or social enterprise in your region or country?

  • If so, how do they relate to the definitions given in this chapter?
  • If not, what might be preventing the development of such projects?

2   Could you consider becoming a prosumer yourself or teaming up with others in your neighbourhood, at your place of study or at work to launch a prosumer project? If so, how would you proceed?

3   Imagine a future in which each building or district were to produce its own energy. How would things be different from today?

Notes

1 See Eurostat statistics on energy production and imports, http://ec.europa.eu/eurostat/statistics-explained/index.php/Energy_production_and_imports

2 See www.equienercoop.lu (General Assembly on 1 June 2016) as well as Kristina Hondrila, Simon Norcross, Paulina Golinska-Dawson, Vladimir Broz, Aydeli Rios and Jules Muller. *Bürger-Energiewende & Eigen- und Direktverbrauch von Solarstrom*, www.freyarchitekten.com/pr/rising-stars-of-the-energy-union-energy-prosumers-in-europe/

3 Public hearing: "Rising Stars of the Energy Union? Energy Prosumers in Europe" organised by the European Economic and Social Committee on 28 June 2016, www.eesc.europa.eu/?i=portal.en.events-and-activities-rising-stars

4 On the self-produced electricity, the cooperative pays no grid fees and only the levy to finance the German energy transition ("EEG-Umlage").

5 Architekten, F. *Smart Grid*, www.freyarchitekten.com

6 See Règlement E16/12/ILR, 13 avril 2016: www.legilux.public.lu/leg/a/archives/2016/0091/index.html

7 See ILR statistics for 2015: www.ilr.public.lu/electricite/statistiques/releve_detaille_ilr/2015/Statistiques-imp-exp-prod-2015-01-10.pdf

References

BP. (2016). *BP statistical review of world energy*. June 2016.

COM/2015/080, European Commission. (2015). *A framework strategy for a resilient energy union with a forward-looking climate change policy.*

COM/2015/339, European Commission. (2015). *Working paper on best practices on renewable energy self-consumption.*

Defourny, J. & Nyssens, M. (2012). *The EMES approach of social enterprise in a comparative perspective.* Liege: EMES Working Paper.

EC - European Commission. (2013). *Guide to social innovation.* Retrieved from http://s3plat form.jrc.ec.europa.eu/documents/20182/84453/Guide_to_Social_Innovation.pdf

EESC. (2015). *Study of the European Economic and Social Committee on the changing future of energy: Civil society as a main player in renewable energy generation.* EESC-2014-04780-00-04-TCD-TRA

EESC. (2016). *Opinion of the European Economic and Social Committee on delivering a new deal for energy consumers.* EESC-2015-05067-00-AC-TRA

Geels, F. W. (2002). Technological transitions as evolutionary reconfiguration processes: A multi-level perspective and a case-study. *Research Policy* 31 (8–9): 1257–1274.

Gross, M. & Mautz, R. (2015). *Renewable energies.* London and New York: Routledge.

König, A. (2015). Towards systemic change: On the co-creation and evaluation of a study programme in transformative sustainability science with stakeholders in Luxembourg. *Current Opinion in Environmental Sustainability* 16: 89–98. doi:10.1016/j.cosust.2015.08.006

Mulgan, G. (2006). 'Cultivating the other invisible hand of social entrepreneurship', in Nicholls, A. (Ed.) *Social entrepreneurship: New models for sustainable social change.* Oxford: Oxford University Press: 74–96.

Petrick, K. (2016). *Topping the potential of prosumers drivers and policy options (re-prosumers).* IEA-RETD. Retrieved from www.eesc.europa.eu/?i=portal.en.events-and-activities-rising-stars-presentations.39840

Ren. (2016). *Renewables 2016 global status report.* Retrieved from www.ren21.net/status-of-renewables/global-status-report/

Rescoop. (2015). *The energy transition to energy democracy.* Retrieved from http://rescoop.eu/energy-transition-energy-democracy

Rescoop. (2016). *RESCoop facts&figures.* Retrieved from https://rescoop.eu/facts-figures-0 (Accessed on 12/07/2016)

Rifkin, J. (2011). *The Third Industrial Revolution: How lateral power is transforming energy, the economy and the world.* New York: Palgrave MacMillan.

Samadi, S., Gröne, M.-C., Schneidewind, Uwe, Luhmann H.-J., Venjakob, J. & Best, B. (2016). Sufficiency in energy scenario studies: Taking the potential benefits of lifestyle changes into account. *Technological Forecasting and Social Change.* Available online 3 October 2016.

Schweizer-Ries, P. (2008). Energy sustainable communities: Environmental psychological investigations. *Energy Policy* 36 (11): 4126–4135.

TRLU. (2013). *trend:research & Leuphana Universität Lüneburg. Definition und Marktanalyse von Bürgerenergie in Deutschland.* Initiative „Die Wende – Energie in Bürgerhand", *Agentur für Erneuerbare Energien.*

Verbong, G. P. & Geels, F. W. (2007). The ongoing energy transition: Lessons from a socio-technical, multi-level analysis of the Dutch electricity system (1960–2004). *Energy Policy* 35 (2): 1025–1037.

Verbong, G. P. & Geels, F. W. (2010). Exploring sustainability transitions in the electricity sector with socio-technical pathways. *Technological Forecasting and Social Change* 77 (8): 1214–1221.

YEEPS Young European Energy Professionals. (2012). *Imagine energy in 2050.* Retrieved from www.yeeps.eu

14 Community-based monitoring for improved water governance

A case study in Holbox Island, Quintana Roo State, Mexico

Kim Chi Tran and Ariane König

Community-based monitoring of water quality in a coastal zone under development on Holbox Island

For effective adaptive governance in the face of scarce environmental resources, new forms of social coordination are required that are effective in translating salient theory and empirical research findings for developing coping mechanisms while allowing room for some trial and error (Dietz et al., 2003; Ostroem, 1990). This chapter focuses on collaborative scientific inquiry for sustainability as one kind of social learning process that has the potential for transforming human–environment interactions. It can be argued that the key to such transformative scientific inquiry is to rethink how new actionable knowledge is co-created in collaborative processes. Physical, technological and social infrastructures for supporting the institutionalization of such community science-based self-governance processes are a key for their effectiveness (Ison et al., 2007). Understanding diverse facets of challenges from distinct perspectives of different experts and stakeholders who engage in transformative social learning from each other is a key to a better understanding of complexity (Medema et al., 2014). Monitoring and learning about changes in the environment or resources and related social practices and structures over time have been established as central elements to help communities in developing effective self-governance approaches. The question of stakeholder and public participation in science thus becomes central. Moreover, monitoring is expensive, so in particular in developing countries with scarce public resources, the work with citizen volunteers or projects embedding monitoring and combining it for the purpose of science education in schools made sense even before an increasing number of researchers became interested in public participation in science for improved governance.

Accordingly, this chapter explores whether more sustainable water governance relying on transformative social learning may be fostered with community based monitoring in a developing country. The approach in this chapter is to critically analyse a pioneering project in this area instituting community based monitoring of a water area in the coastal zone of Holbox Island in the northeast of the Yucatan Peninsula in Mexico between 1999 and 2004. The main objective of this project was to bring science from the "sophisticated" level to the local community level in order to get their participation in an environmental conservation project. The term used in the late 1990s was "community-based science" (Tran, 1999). Merits and limitations of the project are critically discussed. The retrospective analysis of this project over a decade after its completion offered in this chapter highlights design requisites for such projects in terms of designing robust institutional spaces that can withstand time in a developing country where modern communication facilities did not exist. The chapter also highlights the potential of new mobile technologies and tools for citizen and civic science for future design of such projects across multiple scales of social organisation.

Towards this objective, the chapter first introduces a framework for the analysis of initiatives for the public participation in research. The next section of this chapter provides an overview on the inputs, activities and methods used in this project of building a community that is progressively engaged in collaborative scientific inquiry over time. Subsequent sections analyse the outputs and outcomes in terms of whether actionable knowledge and evidence for learning was produced. The concluding sections analyse the merits and limitations of the project's design and implementation, and distill out design requisites for future projects in the area of civic science for improved water governance.

A framework for the analysis of initiatives for the public participation in scientific research

Diverse forms of public participation in scientific research through community-based monitoring approaches have been described (Bonney et al., 2009; Shirk et al., 2012), which in some areas have been practiced for close to a century. A basic definition is that they involve multiple participants with diverse interests and forms of expertise in design and implementation of monitoring. A litany of terms have been coined describing ways that groups of non-scientists participate in research or environmental monitoring: volunteer monitoring, participatory research, citizen science (Wals et al., 2014), collaborative or multi-party monitoring, civic science and community science (Haklay, 2015). Building on Shirk et al. (2012) all these activities can be grouped under the term of public participation in scientific research, but only the latter three challenge the traditional stance of science as objective knowledge situated outside of society. Today common examples include forest management and water quality management (Burgos et al., 2013; Stringer et al., 2014). Empirical evidence suggests that participatory monitoring projects can transform the relationship of ecosystems, local communities and economies; reconnect people to the landscape and to each other; achieve appreciation of complexity; and renegotiate what is valued in the community (Fernandez-Gimenez et al., 2008).

In this chapter, we present a case-based argument, in line with Shirk et al. (2012), that what matters most in determining outcomes and impacts of such initiatives is the quality of participation of all interested parties, including scientists, stakeholders and volunteers in the project. The quality of participation reflects on the relationships between the engaged actors in terms of the dimensions such as trust and credibility, who is given voice and agency, mutual responsiveness and notions of fairness.

Focussing on the quality of participation in such initiatives, at least three distinct categories can be distinguished that describe the quality of public participation:

1   In contributory projects, scientists just recruit volunteers to collect data for them.
2   In collaborative projects, volunteers assist scientists in research design and data collection for knowledge production towards shared goals.
3   In co-creative projects, scientists provide inputs to address issues or problems identified in a particular community of actors.

This framework, depicted in Figure 14.1, specifically allows to better understand "how scientific and public interests are negotiated for project design toward multiple, integrated goals" (Shirk et al., 2012, p. 29). This chapter will apply a framework for the analysis of public participation in research developed by Shirk et al. (2012) to the case situated in Holbox Island, Mexico. This framework, depicted in Figure 14.1, specifically allows to better understand "how scientific and public interests are negotiated for project design toward multiple, integrated goals" (Shirk et al., 2012, p. 29).

The framework serves to relate the way inputs and activities are defined, designed and allocated with resources to outputs, outcomes and impacts of the project:

> **Inputs** in this framework are, for example, interests (the hopes, desires, goals and expectations) of the scientists and other actors as they engage in the project framing phase, the phase in which issues of interest are defined, as well as the project focus and scope. (Interests of funders, management agencies and political entities are not a central consideration in this framework, although they may be important in other contexts).

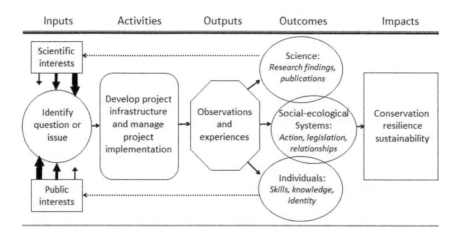

Figure 14.1 A framework for the analysis of initiatives for public participation in research
Copyright © 2012 by the author(s), see Shirk et al. (2012). The Figure is published here under a Creative Commons Attribution-NonCommercial 4.0 International License.

Activities considered in this framework include all tasks relating to the development of the project infrastructure, data gathering, storage and interpretation, training, recruitment of staff and volunteers, meetings and communication to keep all engaged and disseminate results.

Outputs and related outcomes include *individual* experiences and learning by all engaged, *scientific results* in terms of observations and measurements and *outcomes for social-ecological systems*. Outputs usually can be quantified, for example, in terms of data points in a database, or the numbers of individuals, website visits, volunteer hours, workshops and trainings (Phillips et al., 2012). Related outcomes then can include innovative research techniques and scientific publications. Qualitative analysis of outputs and outcomes in terms of meaning-making of experiences and scientific results by diverse participants are also important. Outcomes for social-ecological systems can include, for example, improved relationships and less environmental degradation, improved resource management strategies or rapid detection of and direct response to environmental problems. These, too, have been placed in a clear correlation with the quality of participation in several publications.

Impacts stated amongst the goals of such projects might include sustainable resource governance, empowerment of marginalised groups with local knowledge or an improved science–policy–practice interface. These are usually only visible over the long term and are very difficult to capture and measure.

Of particular value from this book's perspective is also the aim to understand differences in the value of the initiative itself and knowledge co-created by it from the point of view of different actors, but also how this may change for each group over time (Martello, 2004). For science, otherwise unavailable knowledge might be accessed by involving community actors, for example, by compiling large-scale data networks or depending on very localised insights (e.g. Berkes et al., 2000). Again, underlying assumptions about different types of knowledge and their legitimacy as a resource to scientific inquiry will very much shape the scientific outcomes of the project (see also Hulme, 2009, Chapter 1).

The case of Holbox Island: situation, issues and interests

The case presented in this chapter describes an approach to establish community-based environmental monitoring as a practice to allow a local community in rural Mexico to contribute to environmental management and ecosystem protection. Based on public participation, a sound development policy should reflect the needs and desires of the community. However, many development projects are undertaken in communities where the local inhabitants have little or no input to the planning and implementation of such projects. In order to ensure the active participation of the local community in a development project, the community should be involved in the project, and the ownership of the project must be established from the beginning of the project. The case study in this chapter is an environmental conservation project related to the monitoring of coastal water quality and of beach profile, carried out in Holbox Island and the adjacent Yalahau Lagoon[1] (see Figure 14.2. for a map of the area). An overview on the situation of the project, including issues and associated interests, is provided in Box 14.1.

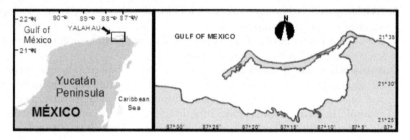

Figure 14.2 Geographic location of Holbox Island
Source: Tran et al., 2008

Box 14.1   Situation of the project, issues and interests

The site: Holbox Island and the Yalahau Lagoon are located in Quintana Roo State, northeast of the Yucatan Peninsula. The lagoon is a rich nursery place of sharks and many other living resources and resulted from the upwelling in the northeastern shelf of the Yucatan and the relative under-exploitation of resources. The lagoon is partially isolated from the ocean by a coastal fringe, well known as Holbox Island. The area of Holbox Island is approximately 80 km², of elliptic form and runs parallel to the coastline (Jiménez-Sabatini et al., 1998) (see Figure 14.2).

The island population was approximately 1,200 local habitants, but increased significantly during the tourist season. The local community in this case study included fishermen, local NGOs, shop owners, local cooperatives, local government, local high school teachers and students. There were several cooperatives in the island, namely the 'ejidos' (that is an association of landowners), fishing and tourist cooperatives. There were also some groups, registered as local non-profit organizations/NGOs, although only one was active in environmental and development issues.

At the beginning of the project (March 1999), the infrastructure of Holbox was limited: unpaved roads, no motorised vehicles except the golf cars and tricycles, small local shops, a mini supermarket, a small hotel and bungalows besides the inhabitants' dwellings (Tran et al., 2002a). There were only two land phones on the island. There was electricity but only a small percentage of the population had access to televisions. However, in the time span of the project, the urban development and tourism grew very fast. The schooling system consisted of a primary school and a junior high school. For further education, students had to leave the island for bigger towns. Therefore, the majority of the Holbox population had rather limited formal education.

The issues: There was acceleration in building construction to accommodate increased visitors. Yalahau Lagoon was declared as a bio-reserve, and as such no urban planning was programmed for the island. However, the urban development had taken place in Holbox in a rather un-planned way, creating many negative impacts to the lagoon and its surrounding ecosystems, for example, the degrading coastal water quality throughout the area (Tran et al., 2002b). The increased urbanization also resulted in a significant rise in land value, which made it increasingly difficult for local inhabitants to remain as landowners. There were conflicts of interest among different groups, including the local and state governments, in term of land use, fish catch, environmental protection and development projects.

In the late 1990s, a project to construct breakwaters around Holbox Island was planned by a governmental agency, but with no specific purpose and no public consultation. The local community of Holbox was not informed about the project. A large amount of funding from the central government was received, and in late 1999 to early 2000 several breakwaters were constructed a few hundred meters from one another on a major stretch of the beach front, causing a major change in the aesthetic aspect of the island. The poor construction of the breakwaters accelerated coastal erosion, and after a couple of years, most of the breakwaters were disconnected from the beach due to the erosion (Tran et al., 2002a; Tran, 2006).

Framing of this issue in this chapter: The project was done in collaboration between Kwansei Gakuin University (KGU), Kobe, Japan; and the Universidad Marista and Mexican National Research Center (Centro de Investigacion y de Estudios Avanzados – CINVESTAV) in Merida, Mexico. Inspiration for the project ideas and methodology came from the paper published by Kwansei Gakuin University (Tran, 1999), which reported on environmental education at high schools and local communities in developing countries to monitor the marine pollution, including coastal water quality. The project started as part of the field work for university students from the School of Natural Resource Management of Universidad Marista and part of their societal contribution to the local community. However, in the year 2000, CINVESTAV received a grant from the World Bank to build up a database of coastal water quality and beach profile (coastal erosion). The researchers proposed that the Holbox project could be used as the platform to involve the local community in data generation through a citizen science project. This would be a cost-effective way to generate a database.

Building an engaged community: inputs, activities and methodology

The project was interdisciplinary in nature, consisting of the scientific, social and educational components. The scientific aspect was on coastal water quality monitoring, where data related to water quality, such as temperature, salinity, dissolved oxygen, pH and nutrients (nitrate and phosphate) were collected. The social aspect was on how to sustainably involve the local community in the project (meaning in the long-term time scale) in order to generate a database of coastal water quality and beach profile. The educational aspect related to non-formal teaching of science to the local community on environmental related topics, namely coastal water quality and coastal erosion. The project involved multiple stakeholders with diverse interests, of which some were in conflict. The approach here was bottom-up, in which the community was empowered and played an active role in sustainable development.

The local communities are the internal stakeholders. They affect and are directly affected by the issues, which in this project are coastal water quality and coastal erosion. The interrelations among stakeholders and stake holding process are presented in Figure 14.3. The researchers are stakeholders along with the local community and the local government, although their involvement is very different. They are external, but play an important role in transformational change, such as being facilitators and mediators through their impartiality and the trust they receive from the community. To a certain extent, donors are also stakeholders, as their role is to provide financial support to the project. The state government and central government are stakeholders because their policy-making responsibility brings impacts to a remote community.

Note should be taken that some parts of the scientific process could not involve the local community. For example, in 2000 to 2002 the researchers from the team

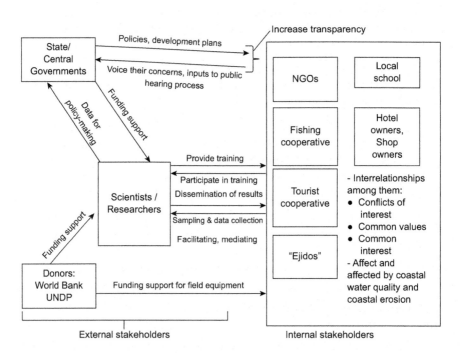

Figure 14.3 Stakeholders in water quality in Holbox Island and their relationships in the project

carried out more sophisticated monitoring, such as the data flow technique, to obtain the higher spatial variability in hydrology (Tran et al., 2002c). Other than that, the researchers visited Holbox on regular basis, at a minimum of twice a year in 1999 to 2004; each visit lasted 2 to 10 days. Besides the collection of scientific data, the visits served also for regular contacts, advice and support to community on issues of water quality monitoring, the development plan of the island and advising on policy for the local community.

The project activities consisted of (i) data collection in the dry and rainy seasons to establish a baseline of coastal water quality (published in Tran et al., 2002b); (ii) training the community in sampling and data collection; (iii) establishment of the infrastructure and support for the community to participate in the data collection; (iv) surveys to learn about the local community's perception, understanding and attitudes related to coastal pollution in order to design a relevant and effective environmental education for them; and (v) community meetings to consult the publics on various environmental and development issues as well as to disseminate the results of the project. Participating scientists included researchers (both natural and social scientists) and university students from Universidad Marista.

The progress of the project and its activities are summarised in Table 14.1.

Table 14.1 shows that involvement of selected volunteers from the community from the very beginning of the project, in particular the planning stage, helped to confirm the relevance of the project activities – for example, the sampling stations in the Yalahau Lagoon. Whilst the first few volunteers could be readily recruited, it proved difficult to engage a larger share of the local community at the beginning of the project. For example, only one person came to the first community meeting and four persons came to the second meeting. The community participation increased after the project "champions" were identified, because they were seriously involved in the project. This included the scientific training for data collection and acting as messengers to their community.

Achievements of the Holbox project in 1999 to 2004: outputs and outcomes

The baseline of the water quality monitoring at the beginning of the project has been published (see Tran et al., 2002b, 2002c). The sampling at all 43 sampling stations on a regular basis could be very costly, time consuming and not necessary because the remote sampling stations in the lagoon did not receive the impacts of development. However, data should be gathered on a regular basis at some sampling stations, particularly those which were located around the town and had shown signs of contamination or undergoing changes due to the development process. The decision on the sampling stations was taken in agreement between the scientists and the community.

On the social aspect, the results obtained in 2004 showed that Holbox people were active in the coastal water quality monitoring program. From March 2003, the community took charge of sampling for coastal water quality and beach profile monitoring on a monthly basis. Data generated by the local people were sufficiently reliable, as has been recognised in other case studies (Deutsch et al., 2007; Burgos et al., 2013).

Table 14.1 Community monitoring of water quality on Holbox Island: project overview

Field work in Holbox	August 1999	November 1999	March 2000	August 2000	Other visits (2000–2002)	Survey 2003	March 2004
Objectives and inputs	Data collection of coastal water quality in the rainy season. Invite local community in the project. Train local people on sampling and data collection	Study public perception of coastal water quality issues and local community participation in the project	Data collection of coastal water quality in the dry season. Present data of previous trips to local community. Train more local people on sampling and data collection	Community meetings. Community taking lead for the monitoring. Study possibility of regular water quality monitoring as a practical part of their curriculum	Obtain the higher spatial variability in hydrology. Workshops to train the community on coastal water quality and beach profile monitoring; regular meetings for contacts, advisory and supports to community.	Assess level of understanding and commitments of the community compared to the beginning of the project in 1999.	Presentation of results of the survey in September 2003. Discussion of the project progress since 1999 and its continuation since the project leader terminated her involvement in April 2004.
Activities and methods	Community consultation meetings. Sampling and data collection of coastal water and sediments. Scientific training for local participants.	Questionnaires (Tran et al., 2002a for details)	Community dissemination meetings. Sampling and data collection of coastal water and sediments. Scientific training for local participants.	Community meetings. UNDP grant to set up the infrastructure for community participation	Scientific research using Data Flow technique. Community meetings. Training to the community on instrumentation for data collection.	Pretest questionnaires, done in March 2003 with 30 interviewees. The main survey done in August 2003 with 246 interviewees.	Community meetings
Who was engaged?	2 researchers, 10 students from Universidad Marista, and 4 local persons.	2 researchers and 10 students from Universidad Marista.	2 researchers, 10 students from Universidad Marista, 1 science teacher and 2 students from the local junior high school.	2 researchers, 6 more local participants, school headmaster and teachers.	8 researchers and 15 local people from the community	5 researchers and 10 local people from the community	3 researchers and the local community (approx. 25)

community meetings	biologist second meeting; 3 fishermen and the biologist of the first meeting	cooperatives (land owners, fishing, and tourist)	(NGOs, cooperatives, local government, school teachers, local medico doctor, hotel owners, shop keepers, and fishermen)	(NGOs, cooperatives, local government, school teachers, local medico doctor, hotel owners, shop keepers, and fishermen).	(NGOs, cooperatives, local government, school teachers, local medico doctor, hotel owners, shop keepers, and fishermen).	(NGOs, cooperatives, local government, school teachers, local medico doctor, hotel owners, shop keepers, and fishermen).	(NGOs, cooperatives, local government, school teachers, local medical doctor, hotel owners, shop keepers, and fishermen).
Main outputs, outcomes and impacts	Relevance of the proposed sampling stations confirmed by the local community. Baseline of coastal water quality for the rainy season; 4 local persons trained.	Interests of local community in the data of coastal water quality. Baseline coastal water quality for the dry season. 3 more local persons trained	Interests and commitments from local community to join the monitoring program. Survey data for designing environmental awareness campaign and local community participation	Grant awarded from UNDP; commitments from the local community	Generation of database on coastal water quality and beach profile by the community. Trust and open communications between researchers and local community. Social platform for open discussions and conflict resolution with the facilitation of the researchers.	Compared to 1999, level of understanding and commitments of the community improved (Tran, 2006).	The way forward for the project with community monitoring of coastal water quality and beach profiles on a regular basis and CINVESTAV's support as advisory and laboratory supports when needed
Comments	The local community's perception on coastal water quality corresponded well with the results from the scientific component (Tran et al., 2002a).	More local people attended the meeting compared to August 1999, thanks to the survey in November 1999 and to the project "messengers"				The surveys were done in the form of risk communication study based on the previous questionnaires of 1999 and on information collected from a community meeting in August 2002	

Similar to several research projects that have the goal to foster social learning for integrated management (Ison et al., 2007; Steyart & Jiggins, 2007), our research group recognised the complexity of environmental issues, such as coastal water quality and coastal erosion, and also that these issues could not be solved by scientific research alone. Of course, scientific research plays a very important role in generating data. However, the scientific data must be used effectively in both the top-down management, such as formulating policies, and the bottom-up approach by providing awareness to the local community affected by the environmental issues. Then they understand the problems and are empowered by the newly acquired knowledge to voice their concerns and to participate in the problem-solving process.

Through the span of the project, we observe that the social dynamic had changed. At the beginning of our project in 1999, the survey results showed that people did not perceive coastal water pollution as a problem in their island because the issue was not perceived visually. According to their understanding, the cause of coastal water pollution was mainly perceived as a natural phenomenon; therefore, they were not responsible for the causes and solutions. The citizen science program brought to the island by our project involved the local community directly in the project activities, such as sampling, data collection and the learning process through community meetings for the dissemination of results. The people of Holbox began to understand the difference between visual pollution such as plastic and garbage on the beach and in the water, and invisible chemical pollution, which was more serious for biodiversity and public health. They also understood that they themselves had contributed to the causes of the problems and could participate in finding solutions. In the case of coastal erosion, the visual perception was very clear as the obvious sight of beach reduction; therefore, it was easier to get the community to act on that issue.

The survey in 2003 showed much change in the Holbox people's perception. Individual behaviours had changed through awareness and internalization of the causes and effects. Although they became more proactive in problem solving, many referred to the lack of rules and regulations from the government both in preventing and in solving problems. On the other hand, they did not fully trust the government in various development proposals (Tran, 2006). The citizen science played an important role here as it provided the citizens with the right understanding of the issues, their causes and effects. Hence they could actively voice their concerns and participate in the policy-making process, such as in a public hearing.

The community also actively participated in problem solving. One particular case in 2003 was the fish kill that occurred in an area of Yalahau Lagoon. The community went there immediately with the field instruments to collect data such as temperature, pH and dissolved oxygen, and they collected water and sediment samples to bring to the CINVESTAV laboratory in Merida. The timely data and samples collected by the community were very helpful for the researchers to find out the causes of fish kill, which was due to the toxins from a harmful algal bloom in the area. This had been aggravated by the poor water circulation during the dry season. Although the cause of the problem in this case was a natural phenomenon, the community also learnt that excessive nutrients from some anthropogenic activities, such as dumping of organic garbage in the coastal water, could contribute to the harmful algal bloom.

Another success story of the program is that the community collected donations from the local habitants and the government to buy and install many garbage bins all over the island. Along the beach, there were signposts asking people not to throw the motor oil and cleaning chemicals from their boats into the seawater. Through observations and communications with the people in Holbox, the project team was informed that the boat cleaning wastes were no longer being thrown into the sea. This is similar to what was reported by Steyart and Jiggins (2007), that knowing is constructed in action, that is, that it arises within the act of constructing an issue and its solutions and that creating new relationships between stakeholder groups can lead to transformative change at multiple scales.

Through community meetings at each visit, the researchers and community had developed trust and open communications. From the community, representatives were from various groups such as local government representatives, "ejidos" (association of landowners), fishing cooperatives, an environmental NGO and tourist cooperatives. Scientific results were presented at each meeting, creating environmental awareness and understanding of cause and effect of various environmental issues in the community. There were conflicts of interest among the groups, not so much related to coastal water quality but more to the beach profile. In this case, the "ejidos" wanted to sell land to foreigners for construction close to the waterline, whereas the local community did not want too much development. The project provided a platform for open discussions and conflict resolution with the facilitation of the researchers. One concrete example was that coastal erosion could be observed visually and through the results of the beach profile studies. One cause of the erosion was the construction of breakwaters before the year 2000 by the government without community consultation. Another cause was the proximity of building constructions to the water line without respecting the safety limits. Understanding the causes and effects helped the community in questioning and influencing the development projects and negotiating better ways of development based on consensus.

Insights gained on the design of community monitoring projects

Merits of the project

A major contributor to the success of the Holbox case study was that a majority of the local habitants were born and lived in the island. They had a strong identity as Holbox inhabitants and were proud of their island. A case study in another place, with a different situation from that at Holbox, such as where a majority of people are migrants or seasonal workers, would provide a different insight into the problems of involving a local community in a development project in a sustainable way.

In general, the Holbox habitants have rather low levels of formal education compared to people in developed countries. However, as emphasized earlier, their commitments were the key factor of success. They were committed in terms of their time and sometimes their financial resources to the project. In developed countries, it seems that people are busier with work, commuting, hobbies, recreation, etc., whereas life in Holbox was much less hectic. The priorities in the daily activities and how to use time are very different between citizens of a developed country and people living in the rural or small islands of a developing country. These points should be taken into account when a citizen science project is designed.

Citizen science projects are not only or primarily about teaching science to the community. It is a continuous process of teaching and learning for both the researchers and the community. There are things that scientists need to learn from the community, for example, the relevance of their project to the community, what kind of problems exists in the community, where the problems are located, etc. Therefore, it is very important to get the community involved actively from the beginning of the project.

As identified in Table 14.1, a success factor in a project was the "champions" who played a crucial role as catalysts and messengers in the community. These persons should be identified as soon as possible at the beginning of the project.

The role of the researchers is equally important. Co-creative projects in which research scientists, policy and decision makers and diverse user groups and engaged citizens work together in framing the research questions and creating and evaluating a knowledge base, rather than researchers just "using" citizen volunteers to collect data for them, are most likely to have a positive impact for change (Pahl-Wostl et al., 2008; Armitage et al., 2009).

The scientists, as outsiders in the community, can also play the roles of facilitator and mediator when there are conflicts of interest arising from different stakeholder groups. They should be patient, good listeners and impartial, thereby creating trust in the community. With concrete scientific data, they can provide their objective and impartial advice on the issues and solutions. For example, the conflicts in land uses between the "ejidos" and the communities were mediated through the scientific data that all land use activities should respect the safety limit. The commitments of the researchers, the trust and acceptance from the community to the researchers are undoubtedly as important as the commitment of the community to the project. They are the key success factors for a citizen science project.

The participation of the school in the project would be very important because water quality monitoring on the regular basis could be boring, time consuming and difficult for one group of people in the community to carry out the tasks over a long period. A successful case study of school participation is an example of field trips from Jakarta to Pulau Seribu (Indonesia) organised for high school students from the Jakarta International School to study the causes and effects of pollution on the quality of seawater and on the coral reef. This work had the objectives of teaching students the basic techniques for marine pollution monitoring and collecting data (Tran, 1999). This contributes to the heightening of the students' awareness in environmental issues and getting them involved in environmental conservation activities.

In sum, this pioneering project of establishing a community-based monitoring approach for a better understanding and action upon determinants of water quality in Yucatan, Mexico, when analysed with hindsight over a decade after its completion, offers many valuable insights for the design of civic science projects in collaboration with rural communities in developing countries:

- Valuable to all engaged;
- Institutionalisation is key and requires thought;
- Design of robust institutional and social spaces that are adaptable but also resist time;
- Technological change offers huge new opportunities.

A citizen science project is a good process in empowering citizens through scientific knowledge and public participation to solve environmental or development issues that affect their lives. If the project is well planned, it also contributes to database generation in a time- and cost-effective way. Although it takes time and patience to build up trust, commitment and openness from the citizens, the rewards are worthwhile. Empowered communities can voice their concerns and influence the policy or decision-making process in appropriate ways, which can contribute to the sustainable development at the local level. This project offers valuable insights on the production of actionable knowledge.

Limitations

As analysed earlier, one success factor for this kind of project is the frequent contacts between the researchers and the local community. The latter could participate in the data collection, but they need the feedback from the researchers on the meaning of the data, which is a form of encouragement and recognition of their efforts. The researchers are the "glue" holding the various stakeholders together. However, the challenge is that not all researchers are prepared to work closely with the community. Some would prefer to collect data and publish them rather than spend time in the community. Therefore, the weakness of the project can be seen as the dependence of the community on the researchers and the "champions" for its success.

One way of overcoming this fragility is to institutionalise the project into the system that carries out the activities on a regular basis, for example, as part of the school curriculum. In Holbox, there was discussion about making the project activities part of the curriculum of the local school, and there was high level of enthusiasm from all the teachers and the schoolmaster for this. However, the actual realization of this on a long-term basis did not happen because the necessary support and commitment from the government did not materialise.

In a recent contact with the community in Holbox, the author learnt that the project fell apart in about 2009, five years after the project leader (i.e. the author) had left the project. The main reasons for the failure were:

- The lack of communication and follow-up from the researchers, caused by the personnel changes in the research team from CINVESTAV and among the involved people from the community;
- The original level of commitment and interest were lacking from the new research team members.

The work on civic science of this project was not well cited by the scientific community who work on this topic. The main reason could be that publications from this work were published as coastal and ocean management rather than citizen science or a project on water governance. This theme was developed several years after the project became inactive.

Outlook

This case illustrated how civic science can serve to empower citizens through scientific knowledge and public participation to solve environmental or development issues which affect their lives. It also contributes to database generation in a time- and cost-effective way if the project is well planned.

Our case study shows that continuing efforts in capacity building and environmental education for the local community are important. That provides them with the knowledge and skills necessary to make appropriate choices for the preservation and development of their locality (Holbox) and the surrounding ecosystems. In the long term, the local community would not only be able to contribute directly to ongoing conservation projects, but would also be in a better position to participate with or influence the government agencies in the decision-making process.

Two main lessons were learnt in this project. First, the trust and openness between the researchers and the community are essential for the success of the project, as observed in 1999 to 2004. This trust and openness could be obtained from the frequent contacts between the researchers and the community on a face-to-face basis, from which some lasting friendships and mutual supports were formed. The second, contradictory, lesson learnt is that the long-term survival of a project that depends on personalities and goodwill is very vulnerable if strong institutional supports are absent, as observed after 2004.

Future research and design requisites for transformative science projects

Technological innovation provides new avenues for improved networking of such projects, locally and globally, to hopefully contribute in making them more socially robust and enduring. A new surge of monitoring projects relying on participatory sensing technologies with mobile devices has been driven by the exponentially accelerating innovation in cheap sensor technologies, mobile computing tools, networking applications and data aggregation and processing tools (Silvertown, 2009; Cohn, 2008; Haklay, 2013). Smart phones carried by millions of people can combine an array of sensors with rich user interfaces and store and send time- and location-tagged data. There is thus a shift to using phones as "networked mobile personal measurement instruments", allowing for greater engagement in caring for the environment (Paulos et al., 2009). Flexible open-source software tools that are easily adapted for diverse monitoring purposes for combining sensor-derived data, photographs and input of subjective data on environmental quality allow usage by citizens and researchers with little computing knowledge (e.g. Stevens et al., 2014).

However, this newly emerging societal trend of accountability and measurement regimes for sustainable development can be considered one emerging primary organizing principle for organizations and society. There are, however, two sides to this coin. On one hand, in reflexive modernity information communication technologies (ICT) are attributed key roles in structuring society and the actions of individual agents by providing new forms of accountability that can affect success of particular individuals and organizations. On the other hand, the risk of loss of individual freedoms from measurement regimes that shape expectations to meet specific standards is acute (see also Chapter 1 and Chapter 2). Top-down imposed, ICT-driven measurement regimes can thus be socially coercive; community decisions on what to measure and be accountable for are preferable; social learning processes and institutional platforms for this need be developed.

In the EU, several larger-scale citizen science projects were financed under FP7. One example is a citizen science observatory for water issues that at present focusses on the risk of flooding (BeWater project, 2013). Citizen observatories also feature in several calls in the Horizon 2020 framework programme, and the development of a platform for the coordination between diverse citizen science projects is supported by the EU Digital Unit of the European Commission. Some of these projects are developed in the co-creation logic and share similar goals with transformative sustainability science. For example, the EveryAware project engages citizens in a techno-social process to enhance collective understanding of air pollution and how better adapt and manage it.

Moreover, in practice it has been noted that with increasing ease of data collection with mobile networked infrastructures, there is also an increasing number of projects in which resulting information finds only limited use (Silvertown, 2009; Cohn, 2008). Some guidelines have been formulated, and data reliability is becoming less of an issue as mobile technologies and easy Internet access allow for redundancy in data contributions. Also in the area of water monitoring it has been noted that *the form of data and knowledge produced* play a large role in how and by whom they are used (Kim et al., 2011). Conrad and Hilchey (2011), in their 10-year review of citizen science and community-based environmental monitoring projects, identified several research gaps. In particular, these were (1) the use and usability of monitoring data by decision makers and (2) the barriers to sharing data between scientists, policy makers and decision makers and users and how they might be overcome. The design of future projects will need to consider and address these needs.

Acknowledgements

The authors would like to acknowledge the contributions and commitments of many people who made the project a success: Mr. Eloy Gil, Dr. David Valdes, Ms. Elizabeth Real, Dr. Jorge Euan, Dr. Jorge Herrera and Dr. Luis Capurro from CIN-VESTAV-Merida, Ms. Mizue Ohe from Kwansei Gakuin University, Ms. Maria Luisa Isla, Prof. Juan Carlos Sejos, Dr. Alejandro Flores from University Marista) and the Holbox community (especially Norma Betancourt, Alberto Coral and Tino).

Questions for comprehension and reflection

1 Consider the local features that are more and less favourable to the establishment of sustainable community involvement in a development project, contrasting remote communities in developing countries to urban areas of developed countries.

2 Develop an analysis in table format based on the framework presented in the second section in this chapter to relate inputs and activities to the outputs and outcomes for this project. Discuss the most salient costs and benefits and their distributions.

Note

1 The beach profile study was not in the project at the beginning, but because coastal erosion had been a serious issue for Holbox, this subject was added in by the interests of the donors and the local community of Holbox.

References

Armitage, D. R., Plummer, R., Berkes, F., Arthur, R. I., Charles, A. T., Davidson-Hunt, I. J., Diduck, A.P., Doubleday, N.C., Johnson, D.S., Marschke, M., McConney, P., Pinkerton, E.W. & Wollenberg, E. K. (2009). Adaptive co-management for social – ecological complexity. *Frontiers in Ecology and the Environment* 7 (2): 95–102. doi:10.1890/070089

Berkes, F., Colding, J. & Folke, C. (2000). Rediscovery of traditional ecological knowledge as adaptive management. *Ecological Applications* 10 (5): 1251–1262. doi:10.2307/2641280

BeWater Project. (2013). *Society adapting to global change.* Retrieved from www.bewater project.eu/

Bonney, R., Cooper, C. B., Dickinson, J., Kelling, S., Phillips, T., Rosenberg, K.V. & Shirk, J. (2009). Citizen science: A developing tool for expanding science knowledge and scientific literacy. *BioScience* 59 (11): 977–984. doi:10.1525/bio.2009.59.11.9

Burgos, A., Páez, R., Carmona, E. & Rivas, H. (2013). A systems approach to modeling community-based environmental monitoring: A case of participatory water quality monitoring in rural Mexico. *Environmental Monitoring and Assessment* 185 (12): 10297–10316. doi:10.1007/s10661–013–3333-x

Cohn, J. P. (2008). Citizen science: Can volunteers do real research? *Bioscience* 58 (3): 192–197. doi:10.1641/B580303

Conrad, C. C. & Hilchey, K. G. (2011). A review of citizen science and community-based environmental monitoring: Issues and opportunities. *Environmental Monitoring and Assessment* 176 (1–4): 273–291. doi:10.1007/s10661-010-1582-5

Deutsch, C., Sarmiento, J. L., Sigman, D. M., Gruber, N. & Dunne, J. P. (2007). Spatial coupling of nitrogen inputs and losses in the ocean. *Nature* 445: 163–167. doi:10.1038/nature05392

Dietz, T., Ostrom, E. & Stern, P. C. (2003). The struggle to govern the commons. *Science* 302 (5652): 1907–1912. doi:10.1126/science.1091015

Fernandez-Gimenez, M., Ballard, H.L. & Sturtevant, V.E (2008). Adaptive management and social learning in collaborative and community-based monitoring: A study of five community-based forestry organizations in the Western USA. *Ecology and Society* 13 (2): 4–19.

Haklay, M. (2013). 'Citizen science and volunteered geographic information: Overview and typology of participation', in Sui, D. Z., Elwood, S. & Goodchild, M. F. (Eds.) *Crowdsourcing geographic knowledge: Volunteered geographic information (VGI) in theory and practice.* Berlin: Springer. pp. 105–122. doi:10.1007/978-94-007-4587-2_7

Haklay, M. (2015). *Citizen science and policy: A European perspective.* Washington, DC: Woodrow Wilson Center for International Scholars.

Hulme, M. (2009). *Why we disagree about climate change.* Cambridge: Cambridge University Press.

Ison, R., Roling, N. & Watson, D. (2007). Challenges to science and society in the sustainable management and use of water: investigating the role of social learning. *Environmental Science & Policy* 10 (6): 499–511. doi:10.1016/j.envsci.2007.02.008

Jiménez-Sabatini, T., Aguilar-Salazar, F., Martínez-Aguilar, J., Figueroa-Paz, R. & Aguilar-Cardozo, C. (1998). *A fishing vision on Yalahau Lagoon in Holbox area, Quintana Roo State, Mexico.* Publication of the Federación Regional de Sociedades Cooperativas de la Industria Pesquera del Estado de Quintana Roo and Instituto Nacional de la Pesca, August 1998.

Kim, S., Robson, C., Zimmerman, T., Pierce, J. & Haber, E. (2011). Creek watch: Pairing usefulness and usability for successful Citizen Science. *Proceedings of the SIGCHI Conference on Human Factors in Computing Systems*: 2125–2134. Vancouver, BC, Canada.

Martello, M. L. (2004). Expert advice and desertification policy: Past experience and current challenges. *Global Environmental Politics* 4 (3): 85–106. doi:10.1162/1526380041748074

Medema, W., Wals, A. & Adamowski, J. (2014). Multi-loop social learning for sustainable land and water governance: Towards a research agenda on the potential of virtual learning platforms. *NJAS – Wageningen Journal of Life Sciences* 69: 23–38.

Ostroem, E. (1990). *Governing the commons: The evolution of institutions for collective action.* Cambridge: Cambridge University Press.

Pahl-Wostl, C., Mostert, E. & Tàbara, D. (2008). The growing importance of social learning in water resources management and sustainability science. *Ecology and Society* 13 (1): 24. Retrieved from www.ecologyandsociety.org/vol13/iss1/art24/

Paulos, E., Honicky, R. J. & Hooker, B. (2009). 'Citizen science: Enabling participatory urbanism', in Foth, M. (Ed.) *Handbook of research on urban informatics: The practice and promise of the real-time city*. New York: Information Science Reference. Chapter XXVII, pp. 414–436.

Phillips, T., Bonney, R. & Shirk, J. L. (2012). 'What is our impact? Toward a unified framework for evaluating outcomes in citizen science participation', in Dickinson, J. L. & Bonney, R. (Eds.) *Citizen science: Public participation in environmental research*. New York: Cornell University Press. pp. 82–95.

Shirk, J. L., Ballard, H. L, Wilderman, C. C., Phillips, T., Wiggins, A., Jordan, R., . . . Bonney, R. (2012). Public participation in scientific research: A framework for deliberate design. *Ecology and Society* 17 (2): 29. doi:10.5751/ES-04705-170229.

Silvertown, J. (2009). A new dawn for citizen science. *Trends in Ecology & Evolution* 24 (9): 467–471. doi:10.1016/j.tree.2009.03.017

Stevens, M., Vitos, M., Altenbuchner, J., Conquest, G., Lewis, J. & Haklay, M. (2014). Pervasive Computing, *IEEE* 13 (2): 20–29. doi:10.1109/MPRV.2014.37 Retrieved from http://ieeexplore.ieee.org/stamp/stamp.jsp?tp=&arnumber=6818498&isnumber=6818495

Steyart, P. & Jiggins, J. (2007). Governance of complex environmental situations through social learning: a synthesis of SLIM's lessons for research, policy and practice. *Environmental Science & Policy* 10: 575–586.

Stringer, L. C., Fleskens, L., Reed, M. S., de Vente, J. & Zengin, M. (2014). Participatory evaluation of monitoring and modeling of sustainable land management technologies in areas prone to land degradation. *Environmental Management* 54 (5): 1022–1042. doi:10.1007/s00267-013-0126-5

Tran, K. C. (1999). Community-based science for coastal-marine pollution monitoring: Toward policy. *Policy Studies* 7: 167–175.

Tran, K. C. (2006). Public perception of development issues: Public awareness can contribute to sustainable development of a small island. *Ocean and Coastal Management* 49: 367–383.

Tran, K. C., Euan, J. & Isla, M. L. (2002a). Public perception of development issues: Impact of water pollution on a small coastal community. *Ocean and Coastal Management* 45 (6–7): 405–420.

Tran, K. C., Valdes, D., Euan, J., Real, E. & Gil, E. (2002b). Status of water quality at Holbox Island, Quintana Roo State, Mexico. *Aquatic Ecosystem Health and Management* 5 (2): 173–189.

Tran, K. C., Valdes, D., Herrera, J., Euan, J., Medina-Gomez, I. & Aranda-Cirerol, N. (2002c). 'Status of coastal water quality at Holbox Island, Quintana Roo State, Mexico, in Coastal Environment- Environmental Problems in Coastal Regions IV', in Brebia, C. A. (Ed.) *Environment problems in coastal region IV: Fourth International Conference on Environmental Problems in Coastal Regions*. Boston: Wit-Press. pp. 331–340.

Tran, K. C., Valdés-Lozano, D. S., Elizabeth, R. & Zapata-Pérez, O. (2008). Variaciones del indice de calidad en laguna Yalahau, Quintana Roo, Mexico, basado en las caracteristicas del agua y sedimentos, en el periodo 1999–2002. *Ciencias de la Tierra y el Espacio* 9: 20–29. Retrieved from www.iga.cu/publicaciones/revista/assets/ctye-9-art4.pdf

Wals, A.E.J., Brody M., Dillon J. & Stevenson, R. B. (2014). Convergence between science and environmental education. *Science* 344: 583–584.

Part III

Tracking, steering and judging transformation

15 Sustainability indicators

Quality and quantity

Jerome Ravetz, Paula Hild, Olivier Thunus and Julien Bollati

The challenge of indicators

In our science-based society, we are surrounded by indicators. At every turn they tell us where we are, how we compare, where we should or should not be and whether we are moving towards or away from the desired or undesired state. Wealth and health of individuals, communities and natural systems are conveyed by indicators. Their representations include numbers, graphs, dials, letters, symbols and colours. And they vary in reliability, from the gauge that says your car's engine temperature is 'normal', to the prediction that next month's weather temperatures will be 'normal'. They are one of the main channels whereby the citizen interacts with science. How can the citizen make good decisions in the midst of this jungle of indicators? The citizen might well ask, "What do I really *know*, when I'm told that I have an Ecological Footprint of 4.1 global hectares? And then what should I do about it?"

This sort of question has now become urgent. In recent years, at an accelerating pace, evidence has accumulated that casts doubt on the reliability, even the veracity, of science in the policy process. The leading policy-relevant social science, economics, lost its reputation in the crash of 2007 to 2008 and has not regained it. Other sciences of complex systems, such as nutrition, suffer from exposures of past errors and misdeeds, such as with dietary fat and sugar. In many cases, there has been a 'magic number', an indicator, which was claimed to express a scientific truth that defined the correct policy. And then it turned out that their precision was no guarantee of their accuracy.

This development, recently even called a crisis, presents a challenge to sustainability science. The policies that governments will be required to implement for the achievement of sustainability will sometimes be difficult, painful and uncertain. If they do not enjoy the consent of the public, then they will surely fail. How this would happen cannot be known, but there can be no doubt that it can. Sustainability (and many other policies) will depend critically on public trust, and that will depend on the assessments made of scientific claims by people who are not expert. In that sense we have come into the age of post-normal science, regardless of whether this condition is welcomed or even recognised. For the 'extended peer community' now takes shape whenever there is a science-based policy in contention.

The critical analysis of the quality of indicators is therefore a core challenge for sustainability science. Unless citizens can make an informed assessment of the competing claims about science, usually expressed as indicators, the creation of policies for sustainability will be left to spasms of popular emotion. And for that assessment it will be necessary for citizens to participate in the social processes whereby indicators are designed and calculated, processes in which participants' inevitably diverse perspectives on reality and value are harmonised. Such 'actionable knowledge' is what sustainability science is about; the analysis of quality of indicators is a contribution to its development.

Indicators as knowledge

In traditional science, knowledge was accepted as being achieved by the discovery of the facts of nature. With indicators, the creative process is not so much simple discovery as design. Of course, the product is derived from science and is intended to represent what is out there. But the link to reality is not so direct; indicators cannot be shown to be simply true or false. For they are constructed first by the collection of data, itself heterogeneous and frequently uncertain, and then by its processing using methods that involve simplifying assumptions and conventions. Although they will usually have the same form as scientific statements, as guides to policy and action, they are affected by inexactness, uncertainty, even irrelevance. We shall see that discussions of quality, which in science can usually be left to the experts, are an essential part of the public engagement with indicators. In the absence of such critical dialogues, there is a strong risk that any indicator will either be misleading or simply be ignored.

Even in the processes of scientific discovery, design is present. We no longer believe that scientists discover facts like pebbles on a beach. Equipment and research strategies must be designed. Any research that uses statistics should be designed around a testable hypothesis. And when the result depends strongly on the outcome of statistical tests, we are in an intermediate situation between a discovered 'fact' and designed 'artefact'. Even at their most straightforward, statistical tests do not provide a 'true/false' answer, but only a 'probability' that is itself conditional on various assumptions, built into the design of the test. Most commonly there will be a 5% 'significance' test; results that score below that will generally be rejected as too insecure. But the 5% is itself a convention, which is more appropriate to some policy contexts (avoiding errors of excess sensitivity, thereby preventing false-positives) than to others (avoiding errors of excess selectivity, thereby preventing false-negatives). Even that standard test has been subject to strong criticism on methodological grounds (Ziliak & McCloskey, 2008); but more sophisticated tests may become difficult for users and even for practitioners to understand. As a result of the complex logic of statistical tests, it is all too easy for research to produce results that do not stand up to further scrutiny. Large areas of science are now known to be infected with substandard work, in what is known as 'the irreproducibility crisis' (Benessia et al., 2016). So if we are placing scientific results on a spectrum between 'fact' and 'artefact', statistics is by no means securely placed at the 'fact' end. We could say that indicators pick up at that point, and computer models are still further out on the artefact end (Saltelli & Funtowicz, 2014).

Some environmental indicators are constructed out of simple operations on the data that are collected. But where complex processes are involved, or where some of the necessary data are absent, the operations become more sophisticated and indirect. The processes of assembly, interpretation and manipulation of data can involve very many separate actions and judgements, so that performing an audit on the production of any single indicator may be quite difficult to achieve. As a small example of the way that errors can creep in, there is the case of Ecological Footprint calculations in Canada that needed correction after it was discovered that the *average* household income had been used instead of the *median* household income, as was intended (Anielski & Wilson, 2005). In such cases we are at the other end of the spectrum from the discovery of objective facts about the natural world, and well over into the realm of designing artefactual materials for well-founded decisions. The example also serves as a reminder that abstract definitions, if not checked against local conditions, can easily mislead. Indicators are thus a case of 'situated knowledge' whose quality depends on deep familiarity with the terrain that they are purporting to describe.

The Global Warming Potential is among the most important indicators for environment and technology policy, and it is a useful example of a 'composite' environmental indicator. These are the closest to policy issues, which deal with complex realities, but they are the furthest from simple measures. Thus the global warming potential using the 'greenhouse gas emissions' is (in an evaluation by the European Commission) 'based on sub-indicators that have no common meaningful unit of measurement and there is no obvious way of weighting these sub-indicators' (European Commission, 2016). The sub-indicators are the so-called greenhouse gases: carbon dioxide (CO_2), methane (CH_4), nitrous oxide (N_2O), hydrofluorocarbons (HFCs), perfluorocarbons (PFCs) and sulphur hexafluoride (SF_6). Converting these six gases to carbon dioxide equivalents makes it possible to compare them and to determine their individual and total contributions to global warming. The process involves science, but it also requires simplifying assumptions, for each gas has a different greenhouse effect and its own residence time in the atmosphere and therefore a different overall effect. The simple number, of so many parts per million, that is regularly reported as steadily increasing, is constructed out of all these various components. There can be a 'best' scheme for weighting these different elements, but there is no simple precise 'fact' that describes the global warming potential.

Understanding that indicators are designed rather than discovered, we can appreciate that their quality ultimately depends on how well they perform their functions. They are there to do a job, portraying useful information. Whereas a 'fact' when discovered by research is evaluated along the single dimension of likelihood (as modified by its intrinsic interest) the product of design is evaluated along several coordinate criteria. As knowledge, an indicator should be a good *proxy* for the state that it is intended to reveal, it must be sufficiently *accurate* in its depiction and it should be *feasible* for its constructors and for intended users. Of course, these criteria could be refined and enriched in any particular case. To be useful, the indicator must perform several functions. These can be listed as *science, policy* or *education*. For that last function, education, it is most important that they should be easy to comprehend and are vivid in their presentation. For *policy*, they must be accurate to the degree that their successive values can reflect trends in the underlying situation, so that it can be seen whether a policy is having the desired effect. For *science*, their structure of evidence and argument must be sufficiently clear that they can be tested, criticised and amended where necessary and used as background knowledge for subsequent research. We might find that some particular indicator scores well for one function and its associated leading criteria and doing not so well on others. Confusion can result if it assumed that good marks on one sort of quality guarantee excellence in all the others. Quality of indicators is thus a complex attribute, and we shall examine this in greater detail later in the chapter.

One other aspect of indicators will be useful in our analysis. Whatever their form (numbers, graphs, colours, etc.) they are representations of a certain reality 'out there'. They are generally not photographs providing a visually realistic picture, but they can be understood as images, conveying useful information about some relevant features. And as images, they can have a variety of effects. Thus they can (1) *reveal*, (2) *suggest*, (3) *distort* and (4) *conceal*. We can think of an x-ray as an example. By its means the more solid and dense tissues inside the body are revealed. This new information, as interpreted by scientists, suggests future studies; both medical practice and scientific understanding have been revolutionised. But we know that the shadows of varying intensities are not a true representation of the body structures, but are effectively a distortion of what is inside. And because the x-rays pass right through the softer tissues, they are effectively concealed from view. Another imaging technique will reveal them, with their own set of suggestions, while inevitably distorting and concealing something else. That is why sets of different imaging tools, each with their own strengths and limitations, are used in modern diagnostic practice. Their proper use is to provide ever more rich and powerful information for answers to diagnostic questions; rather than simply to 'show what's there'. Whether the image be an x-ray or a sustainability indicator, the principle is the same: as objects of design, they provide policy-relevant information rather than simple facts.

To conclude this discussion of the scientific status of indicators, we should consider the influence of values. There is a very influential tradition of imagining science as value free; and paradoxically this property is said to make science to be of very great value! Various schools of philosophers have lauded science for its 'objectivity' in contrast to the humanities; and in the scientific materials that the student sees, values can be said to be conspicuous by their absence. Looking at science in the round, we do see influences of values. The scientific knowledge that is discovered depends on research priorities that reflect values, and also the potential knowledge that remains undiscovered has that status because its projects had low priority, also reflecting values. As we have seen, all statistical tests have values built into their criteria of 'significance'. However, given all of that, there is a strong and justified appearance of objectivity in the materials produced by scientific research.

Similar arguments could be applied to the products of design. A building may be intended to reflect the architect's vision of the good life for his clients, but it must still keep out the rain. Products of design reflect the correspondence with reality, the quality of the process and the integrity of the creator, no less than those of research. However, beyond the most basic criteria of good operation (such as keeping out the rain), the quality of a design depends on its performance of a portfolio of functions. And these in their turn depend on the purposes of people and institutions that they satisfy. There is no ideology of value-free architecture, any more than of value-free medicine. What about indicators? Of course, indicators of social conditions inevitably embody values. There is a recent case that will become a classic, where the Planning Commission of India has set the poverty cut-off figure at an income of 26 rupees (= €.39) per day per person for village inhabitants (Mahapatra & Sethi, 2011).

When we see a scientific statement like "4.1 global hectares", there are no value-commitments to be seen on the surface. But we know that that particular indicator has been chosen over others because it does a particular job. Considering it as an image, we are less concerned with what it distorts and conceals, and more with what it reveals and suggests. That choice reflects our vision of what is truly essential to 'sustainability' among all the worthy causes; and that depends on our vision of what sort of good life is expressed in 'sustainability'. Thus, we might decide to choose instead an indicator that measures the difference between 'our' consumption and that of, say, the median citizen of the planet (the one at 50% on the prosperity ladder). A number expressing the ratio of 'our' consumption to 'theirs' would also be an indicator of sustainability, expressing a different set of perspectives and values. The choice of 'footprint' instead of 'inequality' is by no means 'wrong' or 'false' in itself, but that choice, like all others, cannot be said to be free of normative loading. The contribution of sustainability development indicators to political processes is inescapable, and their evaluation should include that dimension as well (Garnåsjordet et al., 2012).

This discussion of the value loading of indicators might appear to raise troubling issues about the politics of sustainability. If there are no objective facts on which all parties can agree, whereby reason can rule over passion and self-interest, what alternative is there to domination by the strong over the weak? Our reply starts with the historical observation that when the strong want domination over the weak, they can manage it quite well by manipulating an ostensibly objective process. For sustainability of any sort, social or environmental, there must be dialogue based on respect among different standpoints and perspectives. Then, as we show elsewhere in this book, a transformative social learning process can take place. The construction of indicators, appreciative of the need for situated knowledge and aware of the several dimensions of use, can play a fruitful and essential part in this process. We can now assess the challenge of indicators. As products of a social process of design and construction, they convey a sort of knowledge that is very different from the simple facts that their quantitative form suggests. They will increasingly, and justifiably, be subjected to critical scrutiny by an informed public. Their degree of quality may become more salient in policy debates than the quantitative information that they express. Our approach to this challenge is to analyse their content, seeing what sorts of judgements can effectively be made about them. As part of this work, we will discuss tools for accomplishing this analysis, notably the NUSAP notational system. Then in parallel we will take an important case study, that of the Ecological Footprint, following the debate about it, and also suggesting an exercise to test the appropriate precision of its quantities.

Box 15.1 The Ecological Footprint

The Ecological Footprint (accessed through www.footprintnetwork.org for Global Footprint Network) is one of the best-known environmental indicators, accessible to both specialists and citizens, for the calculation of environmental impacts on both a large-scale and personal basis. It is supported by a worldwide network of volunteers, and it was an important example of 'citizen science' before the idea was recognised. Yet it has attracted controversy in Europe, particularly in Luxembourg (geographically, the cradle of this book), and the first attempt to gain official status for it was not successful. Its history is a reminder of the complexity of endeavours to create a sustainability science; and we study it in that perspective.

The Ecological Footprint concept was formally introduced by Mathis Wackernagel and William Rees in the early 1990s (Kitzes et al., 2009). It measures human consumption of products and services from different ecosystems in terms of the amount of bioproductive land and sea area needed to supply these products and services. In other words, the Ecological Footprint calculates the total area needed to produce food, provide resources, produce energy and absorb the CO_2 emissions generated by the supply chains. As this land is nowadays distributed globally (e.g. products and services in Luxembourg are imported from all around the world), the Ecological Footprint is expressed in 'global hectares' (gha) (i.e., hectares of land with a world average productivity).

The area of land or sea available to serve a particular use is called biocapacity and represents the biosphere's ability to meet human demand for material consumption and waste disposal. In other words, the biocapacity represents the capacity of an area or ecosystem to generate an ongoing supply of resources and to absorb its CO_2 emissions. Unsustainability occurs if the Ecological Footprint (i.e. the demand on the system) exceeds its biocapacity.

The Ecological Footprint and biocapacity calculations cover six land-use types: cropland, grazing land, fishing grounds, forest land, built-up land and carbon uptake land, along with fisheries areas of sea. According to the 2010 edition of the National Footprint Accounts (NFA) developed by the Global Footprint Network (GFN), humanity demanded the resources and services of 1.5 planets in 2007. This situation, in which total demand for goods and services exceeds the available supply, is known as overshoot. On the global scale, overshoot indicates that stocks of ecological capital are depleting or that CO_2 emissions are accumulating (Ewing et al., 2010; Ewing et al., 2008).

The Luxembourg experience with the Ecological Footprint reveals two main benefits. It is obviously a good concept for communication and raising awareness. The per capita footprint and the 'planet approach' are easy to transmit and share. The power of individual footprints to illustrate the impact of different consumption habits on the resource depletion of the planet makes it a good educational tool. Second, the aggregation of a range of indicators to a single unit allows the comparison of humans' consumption to the 'biocapacity', the total available natural resources that the planet can provide per year without damage.

But like all images, the footprint can distort as well as reveal. The Luxembourg case is particularly useful in this respect. The national average or per capita consumption footprint is equal to a country's consumption footprint divided by its population. In 2008, with 14.7 gha Luxembourg had the highest per capita Ecological Footprint within the European Union (Hild, 2013). The average Ecological Footprint for 24 European countries, without taking Luxembourg into account, is about 5.4 gha/capita so Luxembourg's footprint is an extraordinary two and a half times the average! This is explained by the statistics, which show quite excessive quantities of alcohol and motor fuel sold in the Grand Duchy (Hild et al., 2010). It could appear that the Footprint was labelling Luxembourgers as quite profligate in their use of the planet's resources. In subsequent years the matter was sorted out. Cross-border workers (resident in France, Germany or Belgium), who make up a very large proportion of the Luxembourg workforce, would buy their low-tax alcohol near the workplace but consume it, unnoticed statistically, at their homes elsewhere. They and tank-tourists would similarly take advantage of the cheap motor fuel. When their contribution to the footprint was analysed, the native Luxembourgers' burden was reduced to a high but still reasonable 9.4 gha/capita. Details are available on the project website www.myfootprint.lu.

In addition to this vulnerability to error from local characteristics, the Footprint also suffers from extreme simplification. (The assumptions about the ecological system that define the Footprint accounts are clearly displayed in Ewing et al., 2008). Only two environmental impacts are considered, CO_2 emissions and land use, whereas it suggests by its name a comprehensive index of resource depletion. Other environmental impacts, such as water consumption and solid waste production, are outside its scope. Further, in the course of debate on quality it emerged that some calculation steps in the Ecological Footprint and biocapacity calculations (e.g. yield and equivalence factors) were not transparent and reproducible, but were actually proprietary information of the company that runs the project. This feature alone prevented it from being a fully accepted scientific method. These points may be responsible for the situation that, in spite of its recognised merits, the Ecological Footprint was not accepted as the only indicator for consumption or resource depletion by official institutions in Luxembourg (the statistical office and the environmental administration). Nor was it adopted by Eurostat, the EU statistical agency. A systematic study sponsored by DG Environment of the European Commission recommended that the Footprint be further developed to remedy its deficiencies and then employed only as a part of a 'basket' of related indicators (Best et al., 2008). Later we will examine the issues of quality in the Ecological Footprint more closely.

Uncertainty and quality in indicators

We are accustomed to using numbers to convey simple information. Something is so big, or so small, compared to a standard; or one attribute of a thing is greater than that of another. Numbers used in science do not have perfect (infinite) precision, for the quantities they represent have inherent sources of inaccuracy. The number that is cited in reports is just a 'best estimate' which is hoped to be the closest possible to a true value, and also to be sufficiently close so that it can be accepted and used. All that uncertainty is commonly represented by the 'significant digits' convention, as supplemented by the $\pm$ notation for 'random error'. More sophisticated representations, as the 'standard deviation' or 'variance' have been developed by statisticians. When we come to indicators, such straightforward means of characterising the information are not sufficient. We might even say that indicators are 'doubly complex'. There is the complexity of their production, particularly in the case of composite indicators; and there is also the complexity of their use, performing one or more societal functions as useful information. This feature is reflected in the articulation of the representations of quality, usually consisting of lists of attributes. There are several ways of organising this information, which we will review.

We shall see that the quality of a sustainability indicator may itself be complex, including several dimensions. Whether it is a 'good' indicator will depend on which aspect is considered crucial, and that decision will itself depend on the value choices that are in play. In the case of official statistics, the public is to some extent protected by the presence of well-established national and international organisations committed to the maintenance and improvement of quality. But many such organisations are reluctant to adopt particular compounded indicators on the grounds that in spite of their importance they lack reliability. Because in this case, as in official statistics, the difference (frequently small) between successive calculated values is of crucial importance, the challenges of reliability and, indeed meaningfulness, are even greater. This situation expresses the key paradox of sustainability science: the greater the urgency, the greater the challenges. To meet these, awareness and commitment among researchers and users are all the more necessary. In this way, the basic tools of sustainability science are co-created in a social design process, rather than being 'discovered' like traditional scientific 'facts'.

There are several more challenges in the effective use of indicators. The most serious is the need to employ weighting factors when aggregating entries of different sorts. The resulting indicator may depend very strongly on the choice of those weighting factors, and so the policies that are deduced from the indicator will also be influenced by that choice. The supposed objectivity of the numerical score may actually conceal a value commitment, or a policy preference, of those who constructed the indicator. Also, indicators that relate to the quality of situations will be based on data whose collection is very far from simple measurement. A scale of personal preferences may be effectively coded in numbers from −5 to +5, but is addition or multiplication meaningful for them? And even for more conventionally quantitative indicators and statistics, there is the paradox of differences. In many cases the real use of the numerical indicator is to assess performance in relation to a standard, or to detect trends. We are then in the position of giving great significance to differences, perhaps quite small, between numbers which may be uncertain to a significant degree. For some policies, even the second differences may be important; thus an annual 'growth' rate of 2.2% may be good news for a national economy, whereas if it is only 1.8% that is bad news. Such challenges are a strong reason why care and skill, expressed in quality assurance, are essential for the effective use of indicators.

Quality of official statistics

The ability of an indicator to effectively inform decision making does not rely only on its construction, but its actual implementation is also of paramount importance. The compilers of official statistics in the European Statistical System (ESS) have established a shared quality framework to ensure they achieve the best fit for purpose. The cornerstone of such an international common quality framework is the European Statistics Code of Practice first adopted in February 2005 and revised in September 2011. The Code of Practice is made of 15 key principles that describe the standards to comply with in terms of institutional environment in which the statistical compiler operates, statistical production processes and the output (Table 15.1).

Table 15.1 Quality criteria for European official statistics

Institutional environment	Statistical processes	Statistical output
1) Professional independence	7) Sound methodology	11) Relevance
2) Mandate for data collection	8) Appropriate statistical procedures	12) Accuracy and reliability
3) Adequacy of resources	9) Non-excessive burden on respondent	13) Timeliness and punctuality
4) Commitment to quality	10) Cost effectiveness	14) Coherence and comparability
5) Statistical confidentiality		15) Accessibility and clarity
6) Impartiality and objectivity		

Source: European Statistical System, 2011

The Quality Assurance Framework (QAF) of the ESS (Eurostat 2012) describes activities, methods and tools that facilitate the implementation of the Code of Practice (CoP). Users of European official statistics are encouraged to take advantage of the availability of 'metadata' describing their quality performance and their compliance with the above standards. The critical choice of indicators for decision-making and an awareness of their limitations is a step of paramount importance.

The set of five principles focusing on the statistical output appears to be able to provide the most valuable insight in the quality of statistics to non-specialists as well. They are:

- **Relevance**: relates to how well the statistics meet the needs of users
- **Accuracy:** relates to how well the statistics describe reality
- **Timeliness:** relates to how quickly the statistics become available
- **Accessibility and clarity:** relates to how easy to obtain and understandable are the statistics
- **Comparability:** relates to differences between geographical areas, sectoral domains or over time within the same data set
- **Coherence:** relates to the adequacy of the data to be reliably combined in different ways and for various uses.

Although in the Code of Practice comparability and coherence are grouped in a single principle, from a practical point of view, as in this spider chart, it seems a good idea to distinguish between them. A graphical representation of the scores assigned to each criterion (e.g. in a range between 1 and 5) with a spider chart, as in the

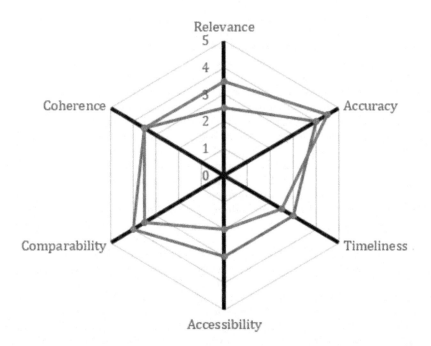

Figure 15.1 Spider chart for quality of indicators

figure, can be a useful way to summarise information and make comparisons. The objective is to achieve the best performance for all criteria; however, a number of quality principles can be seen to be in conflict with each other. For these principles, compilers try to identify and strike the best balance, but will not be able to optimise for all. This is what 'design' is all about. A typical example is the trade-off between the timeliness of a series and its accuracy. On one hand, the production of precise indicators needs a lot of raw information and time-consuming statistical procedures. On the other hand, to meet policy requirements, some indicators need to be released promptly. This is the case, for example, with unemployment statistics. The design compromise is that the statistics are published as soon as stabilized, but are then subject to revisions as additional information becomes available. Another trade-off that appears sensitive is the one between the accuracy of a statistic and its coherence. Indeed, heavily tailored statistical definitions and collection processes usually produce statistics less comparable with indicators coming from other measurement regimes. Of course, the production of inaccurate statistics is not desirable under any perspective. But for the purpose of profiling complex challenges, it can be appropriate to use a less specific indicator that can be compared with statistics coming from other knowledge systems.

Indicators as situated and actionable knowledge – relevance to scale

Indicators that are intended to inform government policies are designed around broad goals. These might seem to be independent of scale or unit of government. Welfare, of people or the environment, does not depend on how many others are included in the reckoning. In practical terms, this might imply that many indicators could be 'nested', those calculated at the local level being aggregated up through the regional, to the national and global levels. Then we would have a coherent picture of all the relevant aspects of life, capable of being unpacked at any desired level for the appropriate policy implementation.

One of the great lessons of work on sustainability is that life, and statistics, are not like that. The construction of indicators is done in very particular contexts, defined by political and administrative cultures, with considerations of priorities, resources, feasibility and public support. At the top level are national or supranational organisations, with professional staff and ongoing programmes of data collection and management. At the bottom level are local communities, defined either by political boundaries or issues of common interest. Their resources are limited in every way, and feasibility will be a critical constraint on their activities. Furthermore, because policies to be implemented at the local level will, or should, depend on local problems and needs, the indicators created to support those policies will be correspondingly local.

Awareness of these aspects of indicators can help communities to avoid failures in their attempts to implement sustainability policies, with their consequences in disappointment and disillusion. The ideals of interchangeability between indicators at any level and nesting into higher levels can be misleading. If we think of indicators not as objective facts waiting on a shelf, but as tools designed for particular functions, we avoid these pitfalls. The criteria of quality in those designs will inevitably and properly be partly local in their orientation. Imperfections in the relations among different indicator sets are the price to be paid for policies that reflect the real difference in perspective among different places and different governance levels.

This feature of indicators can be a great education in situated knowledge, the sort that is designed around particular local needs, and yet is still quite genuine.

In this way, indicators are not only situated knowledge, they are also actionable knowledge. Their criteria of quality are a blend of scientific rigour and practical efficacy. They must be robust in use, standing up both to criticism by users and to the vagaries of experience. And they must also be 'fit for function', providing information that is relevant and well designed for its use in policy processes. The evolution of an indicators project from the initial idealistic ambition to the creation of an effective tool of investigation can be painful and hazardous. If it survives contact with harsh realities of project implementation, the exercise can be transformative for all concerned.

Composite indicators and their challenges

When it comes to providing statistics to depict complex realities such as progress and well-being, designers of indicators face a special challenge. Measurement regimes usually follow the boundaries of a particular knowledge system (a coherent set of activities or judgements as recorded by some institution or recognised group), whereas complex inquiries stretching across different cognitive frameworks (such as issues related to quality or wellbeing) have broader needs of scientific information. A possible answer to these needs is provided by composite indicators.

Composite indicators or indexes are defined as "individual indicators compiled into a single index, on the basis of an underlying model of the multi-dimensional concept that is being measured" (OECD, 2004).

Composite indicators have the advantage of summarising information sourced from different cognitive frameworks. They can support the feedback between different knowledge systems and measurement regimes. Their capacity to summarise complex realities is their most appealing feature, and is the reason they are so popular within the communication with the general public.

However, the construction of a composite indicator is not a straightforward process, as the sub-indicators have no common meaningful unit of measurement and there is no obviously scientific way of weighting them (Composite Indicators Research Group, 2016). They summarise complex, multi-dimensional realities without dropping the underlying information base, they are easier to interpret than a battery of many separate indicators and they can assess progress of countries over time. They also facilitate communication with the general public (i.e. citizens, media, etc.) and promote accountability. On the other hand, they can give oversimplistic representation of complex and heterogeneous phenomena – they may invite simplistic policy conclusions. They may also send misleading policy messages if poorly constructed or misinterpreted, and they are sensitive to normalisation methods and weightings as well as missing data and outliers. Finally, they may disguise serious failings in some dimensions and increase the difficulty of identifying proper remedial action if the construction process is not transparent (OECD, 2008).

Given the importance of indicators for framing public policy, it is of great importance to appreciate their intrinsic quality features and the implications for their quality of using different types of composite indicators. A first approach used in the construction of composite indicators to overcome the different nature of the unitary measures of the sub indicators is 'normalisation'. Normalised indexes convert the different variables into unit-free sub-indicators that range between 0 and 1 and then compound them by some arithmetical procedure. A typical example of normalised indexes is the Human Development Index published by the United Nation Development Programme (UNDP, 2015). In this case, the three normalised sub-indicators are compounded by their 'geometric mean', the cube root of their product (UNDP, 2015). As for all composite indicators, the selection of aggregation model, or of weights if used, can be the subject of political dispute.

An alternative approach for building composite indicators is to transform sub-components that are not directly comparable by using a 'bridging unit' of measurement with the objective of aggregating them. Composite indicators with unitary measures have a better coherence than unit-free composite indicators. Moreover, in the process of choosing indicators the participants with scientific and statistical skills are more likely to accept indicators relying on recognised relations such as physical laws. Yet, these indicators too are susceptible to misinterpretation, as inexpert users will be less well equipped to understand the choices made in the design of the indicator than in the case of normalised indicators.

The adoption of particular conversion approaches can have implications for the interpretation of the indicator. For example, it is common to aggregate the emissions of greenhouse gases in terms of tonnes of CO_2 equivalent. However, although this indicator will be useful for studying trends in global warming, it will fail in the provision of knowledge on the impact of a changed climate on human health. In practice, these indicators enable interlinkages between different knowledge systems, but only for a very limited amount of information. In most cases, the bridging unit forces the flattening of the complexity of information to a single aspect. We have seen an example of this in the case of the Ecological Footprint. Thus, in that case as others, any single indicator will simultaneously reveal, suggest, distort and conceal. The question is then whether this is understood by participants, both as a general principle and in particular cases.

The manipulation of primary information required to produce composite indicators is usually deemed excessive by the official statistical systems. That is why composite indicators are generally not recognised as official statistics. The verdict is "composite indicators combining individual indicators that have no common meaningful unit of measurement and implying arbitrary choices for weighting the sub-indicators cannot be labelled as official statistics and should thus remain in the research or political sphere" (European Statistical System, 2011).

The European Union has an ambitious programme of development for sustainability in all its dimensions, including environmental, socio-economic and personal and cultural. For its implementation, the European Commission, operating through Eurostat, has a correspondingly ambitious programme of monitoring through indicators. This includes a broad selection of 130 indicators, organised under 10 themes, together with a complementary programme of quality assurance and development. Useful reports are European Statistical System 2011a, 2011b and OECD 2008.

It has an accessible website on which its activities are displayed (http://ec.europa.eu/eurostat/web/sdi/indicators).

Monetary units

There is a measurement unit that is particularly attractive because of its ease of communication with and interpretation by the general public. These are monetary units (i.e. currencies). For instance, it is very common to learn about socio-economic phenomena in terms of impact on the gross domestic product.

The conversion toward a monetary unit (e.g. Euro) has also the advantage of ensuring a good comparability and therefore excellence coherence. Nevertheless, despite monetary units being able achieve a formal coherence, there is still a comparability issue. As is well known, the value of currencies changes across time (inflation) and space (purchasing power). This aspect of monetary units of value has to be accounted for when considering phenomena with an intergenerational and/ or international dimension. All monetary units are subject to the 'social discount rate' – how important the future is to us – and the calculated value of an item in two generations' time will depend crucially on the value chosen for that parameter.

Nevertheless, monetary units are very special units. Although the conversion among units is in some cases relatively undisputed, as in the cases where a physical law can be applied, economics is not an exact science, and the literature is full of fragmented and competing theories. For example, market prices can be seen either as a measure of the scarcity of a resource or of its utility to economic agents who would be spending other resources in order to consume it – these attributes relate to opposite ends of the transaction. The accuracy of these indicators as measures of something real is therefore often subject to criticism. There is also the danger that the adoption of a monetary unit gives the impression that the object of measurement is just a commodity, to be bought and sold like any other. When this is applied to things with non-monetary value, as those with beauty or personal or spiritual significance, the technique is inevitably seen as political.

The flaws that can weaken indicators expressed in monetary units are often sufficiently important to raise the question whether their advantages are sufficient to justify their adoption. In this case of extreme simplification, the questions are particularly crucial: do they distort and conceal more than they reveal and suggest? As for all indicators, it is of paramount importance that users are fully aware of the quality and limitations of the indicators they embrace for their special purposes and that they are conscious of the consequences that the choice might have on their decisions.

Skills in the construction of indicators

To work towards sustainability, a community must have a plan, incorporating an assessment of where it is and where it wants to be. For this, indicators are essential. The choice and then the implementation of the indicator set are discovered to be very demanding tasks. The skills required are a combination of the scientific, administrative and social. Here we cite just two examples, one of failure on a grand scheme, the other of success on one that was carefully defined and reflectively implemented. It is a good example of transformative science in action.

Luxembourg, like most European countries, developed its own list of sustainable development indicators in order to follow the evolution of its strategy defined within the national plan of sustainable development (PNDD Luxembourg, 2010). The approach followed by the Ministry for Sustainable Development and Infrastructures was, in the first stage, to join together the various 'active forces' of the society in order to discuss, and if possible to agree, on the paradigm of a sustainable development. This group of visionaries first defined 14 trends deemed unsustainable, and on this basis they defined 18 qualitative aims. For each one of those, objectives of action (150 in all) were imagined and then specified in detailed measures for implementation. The indicators ranged from very local and detailed issues, as 'integrating residents and commuters", to rather general ones, as "elimination of global poverty". This variety is an indication of the challenges that are confronted by any attempt to plan 'sustainability' along the same lines as an engineering project.

In the second stage, each speaker round the table had the possibility of proposing the indicators that seemed justified to them. At the conclusion of the discussions, a list of 197 indicators was drawn up. This was clearly not manageable, either in terms of resources or of communication with the public. For that, a short list of 15 or 20 is effective.

The persons in charge of the project then started a new round of discussion aiming at selecting the most relevant indicators out of the total list of 197. Unfortunately, these discussions led to failure. Indeed, each participant was unwilling to sacrifice their own proposals of indicators, insisting that they had the highest priority. This position is frequently met within the framework of participatory approaches. Each participant is focused on their own sphere of activity and does not present themselves at the negotiating table with a greater desire of synthesis, but rather with the ambition to advance their own priorities.

The persons of the Ministry then took the initiative to set up a small group of experts recruited from the scientific world. The mission of this group was to achieve the selection of the 15 to 20 key indicators. The approach chosen by this group to operate this selection was very innovative. They avoided the delicate task of choosing indicators within each of the 18 qualitative aims, and then attempting in every case to justify why some particular objective would not be represented. Instead they decided to identify those indicators which were most representative of each of the five concepts of sustainability which had been defined by the preceding group. This approach had the advantage of selecting the most generally applicable indicators for the previously defined qualitative aims. The last stage of the approach defined by the Ministry was to present this selection of the indicators to the representatives of the 'active forces' of society in order for them to approve and to adopt this new list of the indicators of a sustainable development.

In spite of these creative interventions, the sustainability indicators project of Luxembourg, in its current development, failed to complete the process. The headline indicators were never compiled and diffused among the general public. The leading authorities never had a mandate from all the various participants in the project to impose their solution. The task of constructing sustainability indicators revealed its complexity and resistance to simplification, which in this case was decisive.

For those who are intending to engage in a rigorously scientific programme of construction of local indicators, the paper by Stahl and her colleagues at the U.S. Environmental Protection Agency (EPA) in Philadelphia (2011) offers valuable guidance (Stahl et al., 2011). They use a case study to demonstrate how a transparent, transdisciplinary approach to decision making enables the EPS Region III to fulfil its decision-making responsibilities while taking critical steps toward engaging in sustainability discussions. The presence of a single organization that defined the limited goals, unlike in Luxembourg, was an important component of success. But flexibility and sensitivity in the articulation of the goals was equally important.

Box 15.2   Indicators for resource allocation

The case study goals were to use information about environmental conditions to inform staff, and to accomplish fiscal resource prioritization and allocation for the federal 2010 fiscal year. They used a select group of three indicators (two of air quality and one of water) to show 1) that data are not the same as indicators, 2) that it is feasible to use disparate data in the same analysis and 3) that specific discussions about indicators can lead to transdisciplinary learning, supporting more informed decision making. When used in a transdisciplinary, iterative learning process, these indicator lessons provide a stepping stone for organisations to approach sustainability as something more than just a lofty, ethical concept. Instead, organizations can routinely and substantively address sustainability issues through a progression of individual decisions. Sustainability can be linked to decision making through a process that requires stakeholders to articulate and confront their values. In this process, an integral part of the assessment of environmental conditions is the selection and indexing of indicators and understanding what those choices imply regarding the salient issues and the affected populations. The process is lengthy and demanding, but stakeholders come away satisfied that the most appropriate data were used, the issues were fairly vetted and the values transparently applied. These also happen to be salient features of the sustainability process.

Applying NUSAP to indicators

The notational system 'NUSAP' enables the different attributes of quantitative information to be displayed in a standardized and self-explanatory way. It enables providers and users of indicators to be clear about its strengths and limits. The NUSAP system fosters an enhanced appreciation of the issue of quality in information. It thereby enables a more effective criticism of quantitative information by clients and users of all sorts, expert and lay. Its website (www.nusap.net) has a clear statement of the tasks of sustainability science and of the role of NUSAP in its implementation.

The NUSAP system is based on five categories, which generally reflect the standard practice of the matured experimental sciences. By providing a separate box, or 'field', for each aspect of the information, it enables a great flexibility in their expression. By means of NUSAP, nuances of meaning about quantities can be conveyed concisely and clearly, to a degree that is quite impossible otherwise. The name 'NUSAP' is an acronym for the categories. The first is Numeral; this will usually be an ordinary number, but when appropriate it can be a more general quantity, such as the expression 'a million' (which is not the same as the number lying between 999,999 and 1,000,001). Second comes Unit, which may be of the conventional sort, but which may also contain extra information, such as the date at which the unit is evaluated (most commonly with money). The middle category is Spread, which generalizes from the 'random error' of experiments or the 'variance' of statistics. Although Spread is usually conveyed by a number (characterised by ±, % or 'factor of') it is not an ordinary quantity. In the case of a number intended for use in policy, a useful interpretation of Spread would be a sort of 'just noticeable difference', so that trends in successive values of the number would be meaningful. Thus if changes of 1% in an indicator were significant, but changes of 0.1% were not (as is the case with macro-environmental indicators like Global Warming Potential), the Spread entry would be 1%.

On the more qualitative side of the NUSAP expression is Assessment; this provides a place for a concise expression of the salient qualitative judgements about the information. In the case of statistical tests, this might be the significance-level; in the case of numerical estimates for policy purposes, it might be the qualifier 'optimistic' or 'pessimistic'. In some experimental fields, information is given with two ± terms, of which the first is the spread, or random error, and the second is the 'systematic error' which must be estimated on the basis of the history of the measurement, and which corresponds to the Assessment. It might be thought that the 'systematic error' must always be less than the 'experimental error', or else the stated 'error bar' would be meaningless or misleading. But the 'systematic error' can be well estimated only in retrospect, and then it can give surprises (Funtowicz & Ravetz, 1990).

In the policy context, as with indicators, Assessment would usually be an expression of its strength as evidence, as established by its robustness in the face of criticism. This would give policy makers an indication of how much they could rely on it when basing an argument wholly or partly on a particular value of the number.

Finally, there is P for Pedigree. It might be surprising to imagine numbers as having pedigrees, as if they were show dogs or racehorses. But where quality is crucial, a pedigree is essential in this area as others. In the case of indicators, the pedigree does not show ancestry, but is an evaluative description of the mode of production (and where relevant, of anticipated use) of the information. Each special sort of information has its own pedigree; and research workers can quickly learn to formulate the distinctions around which a special pedigree is constructed. In the process they also gain clarity about the characteristic strengths and weaknesses of their own field.

For a sample of a Pedigree matrix, see one in Table 15.2 relating to 'emission monitoring data', coming out of a definitive Dutch study led by Jeroen van der Sluijs (2005).

The columns describe the different aspects of the information. First is the 'proxy', or how well the measured quantity relates to the thing being measured. It can be noticed that the relationship is weaker going down the column. This is standard in the Pedigree matrices relating to scientific information. Similarly, the other columns, for 'Empirical', 'Method' or 'Validation', display a weakening with descent. For any particular item of information, a box will be ticked in each column or the index number on the left will be recorded. If a gauge of the strength of the information is desired, as for the Assessment rating, then those indices can be averaged.

Everyone using this tool should be well aware that it is not a measure in any strict physical sense. It is a way of lending some discipline and reproducibility to the qualitative judgments that go into the assessment of strength of the information. Moreover, it is essential to realize that materials with a low score on Pedigree are not thereby deemed to be of low quality. The Pedigree rating shows what can be accomplished in any given scientific context. Within that context, research can be

Table 15.2 Pedigree matrix for emission monitoring data

Score	Proxy	Empirical	Method	Validation
4	Exact measure	Large sample direct measurements	Best available practice	Compared with independent measurements of same variable
3	Good fit for measure	Small sample direct measurements	Reliable method commonly accepted	Compared with independent measurements of closely related variable
2	Well correlated	Modelled / derived data	Acceptable method limited consensus on reliability	Compared with measurements not independent
1	Weak correlation	Educated guesses / rule of thumb estimate	Preliminary methods unknown reliability	Weak / indirect validation
0	Not clearly related	Crude speculation	No discernible rigor	No validation

done better or worse. In general, when research is done in ignorance of the weakness of the material, its quality will suffer. Also, when ratings are all high, the situation is closer to classical laboratory science; and when they are low, that is a sign of post-normal science.

In ordinary scientific practice, and even more in scientific education, it is universally assumed that the most important aspect of the information is at the 'hard' left-hand side, mainly the Numeral as contextualised by the Unit. All too often, even the indication of the Spread is considered as an optional extra; numbers are, after all, the bearers of true knowledge. The 'softer' categories, Assessment and Pedigree, are generally neglected or dealt with in fine-print footnotes. Even the 'harder' categories, which are usually accepted as unproblematic, may contain concealed hazards. For instance, if a quantity is known only to within, say, +50%, then this is a range of a factor of three, and it is a case of 'not even one significant digit'. How, then, is the quantity to be represented? If the differences between quantities are important, then even smaller uncertainties can vitiate an argument. For example, consider a pair of successive index numbers 6.1 ± 0.7 and 5.8 ± 0.5; if a policy maker wants to know whether there is a downward trend, what message should s/he be given? (Graphing the numbers with error bars will make the point quite clear.) In these ways, clarity of expression of uncertainty and quality in quantitative information is essential for it to be genuinely useful.

When it comes to the policy domain, or to post-normal situations in general, the balance of importance shifts. If the number at the 'hard' end is irremediably 'soft', then the qualitative information about the number, our Assessment and Pedigree, which are now called the 'metadata', can be critical in the assessment of its usefulness. After all, if a particular quantity is uncertain by +50%, then providing a three-digit precise value somewhere within that range, an uncertainty of a factor-of-three, does not have much purpose. And if the quantity, even with that uncertainty range, is strongly contested among reputable experts, then it would be unwise for policy makers to make any particular value crucial to their arguments. In NUSAP terms, in such cases, the Pedigree, not the Numeral, should be the focus of attention. Further, most numbers used in the policy context are produced by the compounding of several distinct operations. Normally, when they are passed from one specialty to another, they are (in Jeroen van der Sluijs's words) 'thrown over the fence', stripped of all the information that enables their skilled use. In such circumstances, all users would do well to adopt a NUSAP representation, with greater emphasis on the construction of an appropriate Pedigree matrix. Jeroen van der Sluijs and his colleagues have given practical examples of how this can be done by a group of experts (van der Sluijs, 2002).

Pedigree analysis of energy consumption surveys

As a class exercise in the course 'Science and Citizens Meet the Challenge of Sustainability', a Pedigree analysis was done on the report of an energy usage survey. This was the *Manual for statistics on energy consumption in households*, published by the Eurostat in 2013. The survey was organised in 2008 in response to a new awareness of the importance consumption for the strategic perspective on energy. Each member state organised the survey in their own way. The *Manual* has a summary for each survey, arranged so that the reader can easily see how it was done, and how its special challenges were met. There is also a general discussion of the methods for such surveys, so that readers can appreciate how much skill is required for obtaining worthwhile results.

The main lesson from the exercise is that finding out about energy consumption is very different indeed from just counting all the units of consumed energy. It would be prohibitively expensive to collect all the data on energy use from all the households. Therefore, some indirect methods are necessary. These might be sampling, models or a combination thereof. Data might be taken from interviews with consumers or from records (such as bills) of supply. The variety is enormous, and strong simplifications are inevitable. There are also many sources of error, not least in the responses of consumers. For all these reasons, data must be 'captured' as much as 'collected'. Perfection is an impossible, perhaps misleading, ideal. The whole exercise must be designed in terms of the degree of quality that is appropriate for the function of the outcome in policy.

In the exercise, the class engaged in a very simple, but still useful analysis of the methods of the survey. This provided a good example of the strengths, as well as the limits, of the Pedigree methodology. A mere 16 boxes are incapable of conveying all the rich variety of methods that were found in the survey reports. However, for a sketch snapshot of the situation, indicating the basic types of enquiry with their characteristic strengths and limitations, the Pedigree is very useful indeed. The Pedigree matrix that was used is shown in Table 15.3.

Students analysed the returns for three member states, Germany, Spain and the UK. The descriptions in the reports were not always easy to match with the categories in the pedigree matrix. However, overall the ratings did show the characteristic differences between the three member states. The ratings are shown in Table 15.4.

Some of the ratings, particularly the 1 for the UK Source Type (Modelling), might be surprising. This might stimulate readers to check the report to confirm

Table 15.3 Pedigree matrix for energy surveys

Value + 1/2	Source type	Frequency	Data	Validation
4	Large sample direct measurement	Annual	All measured	Independent measurements of same variable
3	Small sample direct measurement	Every 3 years	Partially measured	Independent measurements of related variable
2	Survey/Adm. register	Every 5 years	Estimated	Indirect validation (ex: on total)
1	Modelling	Once	Speculation	No validation

Table 15.4 Pedigree ratings for three national surveys

Member state	Source type	Frequency	Data	Validation
Germany	3	1	2	2
Spain	3	1	4	3
UK	1	1	4	2

the rating, and they might thereby learn something of the constraints on the collection of information that is relevant to sustainability. The students found the exercise quite illuminating.

Criticism and defence of the Ecological Footprint

The Ecological Footprint (see Box 15.1) has been the target of severe criticism, perhaps more than any other environmental indicator. This may be a result of its salience in the environmental movement and its initial scientific promise. The impact is heightened by its annual announcement of an 'Earth Overshoot Day', as if this quantity could be known with a precision of 0.3%. For a time the Ecological Footprint enjoyed an official status in Luxembourg, and so the debate over its status was especially intense there. Whatever may be the deficiencies of the Footprint, it certainly cannot be claimed that its promoters are indifferent to methodological issues. The movement's website has a full and accessible discussion of them, and the authors engage with critics, as we shall see later. Whether the Footprint is really more vulnerable to criticism than related indicators, as the Planetary Boundaries of J. Rockström and colleagues, is for us an open question. But we shall use the Footprint as the case study in our discussions.

Our first item of evidence is a brief assessment of the Footprint made by a group at Eurostat, the European statistical agency (Schaefer et al., 2006). This was the first of several evaluations, as the Footprint seemed to many at the time to provide the answer to the problem of sustainability indicators (Best, 2008). The reports are now more than a decade old, and we can expect that on many issues the authors of the Footprint will have made appropriate changes. But some of the observed weaknesses are inherent to the technique, and so these reports can still function as a useful guide to the methodological issues. The key points made in the early evaluation are:

- Non-robust policy message: changes in any inputs can modify outputs significantly.
- Heterogeneous ingredients: "the fact that a single figure is obtained does not guarantee that its interpretation is straightforward".
- Limitation of scope: the range of processes covered is very much smaller than the set of significant factors.
- Sensitivity to problems of data quality: high quality is needed for all input variables, and it is frequently not available.
- Lack of transparency of the assumptions and selections: documentation of the crucial intermediate steps is insufficient (this has been remedied to some extent; see www.footprintnetwork.org/en/index.php/GFN/page/public_data_package. Accessed 02/01/2017).
- Scientific basis of the weighting factors: many assumptions, some strongly policy laden, go into the calculation of these; documentation is insufficient.

In summary, the report says that "at the moment the selection of variables, the origin of the data and the weighting factors that are used can be perceived as being of an arbitrary nature and based on in-transparent assumptions".

Another useful assessment was made along the lines of the quality criteria adopted by the European Commission mentioned earlier. This was summarized (Lock, 2014) in a spider chart, slightly different in form from the one shown earlier.

We notice that the ratings are favourable for Relevance, at 4, but otherwise weak, as 2 for Comparability and Timeliness, but only 1 for Accuracy and Coherence and Accessibility. Reading clockwise from Relevance, that would give a Pedigree rating of 4,1,2,1,2,1. A full statement of the Pedigree would involve descriptions

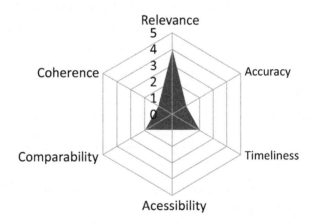

Figure 15.2 Assessment of conformity of the Ecological Footprint with Eurostat quality dimensions

Source: Lock, 2014

corresponding to the various numbers for each aspect, filling all the cells in the matrix. But for our present purposes it is sufficient to have the list of relevant aspects, with numerical ratings for each.

Recently the promoters of the Footprint have engaged in a published debate with two longstanding critics, Mario Giampietro and Andrea Saltelli (Galli et al., 2016). Ten issues were aired, with questions and answers stated briefly. In this discussion, fundamental issues were dealt with more fully than in the critical analyses of the previous decade. The Ecological Footprint thus provides an excellent example for the history of the critical study of a sustainability indicator. The Footprint's merits are recognised, as well as the sustained international effort that has gone into its construction and development. Yet its critics, themselves sympathetic to its goals, have not yet (at the time of writing of this analysis) been satisfied that its deficiencies have been adequately remedied. The different historic phases of the debate are available for analysis by students. And debate will doubtless continue, and can be instructive for all concerned.

Box 15.3 Questions for debate on the Ecological Footprint

1 What underlying research question does the Ecological Footprint address?

2 How is the research question underlying the Ecological Footprint relevant or irrelevant to policy concerns?

3 If the research question underlying the Ecological Footprint has some relevance, are there more accurate methods available for answering this particular question?

4 Given the quality and accuracy limitations of current Ecological Footprint results, is society better off without these results?

5 What are the external referents used for quantitative assessments in Ecological Footprint Accounting, and what is the accuracy of the measurement schemes used to quantify their relevant characteristics?

6 When calculating Ecological Footprint and biocapacity, is the measurement referring to the characteristics of a typology (which typology?) or to the characteristics of an instance (an instance of what)?

7 Ecological Footprint Accounting methodology generates a unique set of quantitative assessments obtained combining together measurements referring to processes taking place at different scales. This unique set of quantitative assessments is supposed to be equally useful for agents operating at different scales (e.g. farms, cities, provinces, nations, macroregions and the planet). What are the main benefits and the main limitations of this approach?

8 Ecological Footprint Accounting measures the biocapacity required for energy security as the land required to absorb the CO_2 emissions. But

then why does it not adopt the same logic for calculating the biocapacity required for food security (e.g. measuring the land required to absorb human dejections and food waste) rather than the land required to produce food?

9   Can the assessments generated by Ecological Footprint Accounting methodology define (i) whether the flows considered in the quantitative assessment are in steady-state or referring to transitional states (i.e. flows obtained by stock depletion and filling of sink); (ii) how much these flows are beneficial to the economy; and (iii) how much these flows are damaging ecological funds; why should assessments of exported (or imported) biocapacity be relevant?

10   How is the Ecological Footprint Accounting similar or different from other international metrics/indicator systems?

Source: Galli et al., 2016

Earth overshoot day: the NUSAP spread

We can integrate the discussions of this chapter by applying a NUSAP expression to 'Earth Overshoot Day'. This is the day when, according to Footprint calculations, humanity has used up a full year's worth of the planet's supply of resources. It occurs sometime in August, indicating that we are using about 1.6 times as much as our supply. In 2016, it is on August 8, or the 220th day of the year. So the NUSAP expression would start as 220 : days. What is the Spread? How precise is our knowledge of the overshoot? To be clear on that, we should look at the Pedigree. Taking the ratings from the spider chart (Figure 15.2), we have one 4, two 2's and three 1's. The mean, or average, is just under 2; or if we prefer a nonparametric statistic the median is 1.5 – not very strong at all. Either of these will do for the Assessment. Hence the NUSAP expression, lacking a number for the Spread, will appear as 220: days : S : 1.5 : 4,1,2,1,2,1.

The power of NUSAP is that it prompts us to look critically at our numbers. If we calculate Spread from the N, U end, we get something like 0.3%; that would correspond to the unit of a day, 1/365 of a year. That does seem rather excessively precise, given the low accuracy registered in the A, P end of the expression. Let us have a discussion focusing on the Unit. Given the low Pedigree rating, we might ask whether the Unit should be broader than a day, perhaps a week, a month or even a quarter-year. But if, on the other hand, a day is not too precise, then what about an hour, or even a minute to define the overshoot? The matter is important, for once that NUSAP notation has made us aware of the dangers of excess precision, we must do something about defining the precision of the Overshoot threshold.

A class might then debate what value of Spread is appropriate for Overshoot. This would involve a review of the Eurostat Pedigree in the light of the recent discussions, as supplemented by materials on the Footprint website.

To conclude with some light relief, we report on a class exercise, an Enhanced Learning Activity, involved in calculating a Personal Footprint and then playing with the program to discover the paths to ecological virtue.

Box 15.4   Exercise: the Ecological Footprint calculator

Based on a Swiss version of an Ecological Footprint calculator developed by the Global Footprint Network, a questionnaire was adapted to Luxembourg's results, enabling the calculation of individual footprints. The calculator includes 42 questions in total – 16 are taken into account in the footprint evaluation and the rest only have an awareness-raising function. The calibration of the calculator is based on an average person living in Luxembourg; thus, a typical answer corresponds to a footprint of 7.0 global hectares. The individual responses are rated either higher or lower from the mean: The 'best' consumer behaviour results in a footprint of 3.0 gha; the 'worst' consumer behaviour results in a footprint of 11.5 gha.

The student footprints can be generally divided into three groups: the group of vegan and very conscious consumers with a footprint between 3 and 4 gha; the group of consumers that try to reduce their consumption a bit in all categories, with a footprint between 5 and 7 gha, depending on the effort they are making; and the group of consumers that are often unaware of the strong impact transport has on the result, with a footprint between 8 and 11.5 gha, depending on the frequency of car use and air travel.

By playing with the calculator, some footprint-reducing impacts can be pinpointed: eat vegetarian or even vegan; do not use a car or a plane; avoid buying new clothes, furniture and electronic devices.

The calculator is available for use on the website of the Global Footprint Network at www.footprintnetwork.org/en/index.php/GFN/page/cal culators/. There are two cities for which the parameters of the calculator have been set: Calgary and Luxembourg. Because the climate and general conditions in Luxembourg are more typical of temperate-zone locations, it is recommended for classroom use.

Conclusion

In spite of their numerical form, giving the appearance of an exact science, indicators are very much a product of sustainability science as we understand it here. We have seen that each indicator is doubly complex, both in the processes of its design and construction and also in the contexts of its intended use. Assessments of the quality of an indicator will be correspondingly complex and themselves subject to debate. Examination of the meaning of the numerical expressions of indicators, such as through the NUSAP notation, will reveal implicit assumptions and also show the way to creative criticism. Trial applications of the 'Ecological Footprint' indicator and a study of the debates on its (complex) quality will be a valuable, indeed transformative, educational experience for a student whose previous experience has been restricted to dogmatic 'normal science'. They will learn that for indicators, as generally in sustainability science, there is no simple choice between 'true' and 'false'; rather, each engaged person must make their own assessment of quality, relative to their needs and growing out of a dialogue with colleagues.

Questions for comprehension and reflection

1   What is a global hectare? Does the concept reveal and suggest more than it distorts and conceals?
2   Should there be a day marking 'overshoot', or perhaps some other unit of time? If so, which unit?
3   Which is the most important dimension on the quality chart for the Footprint?
4   Does having a lower Ecological Footprint indicate that one is a better person?

References

Anielski, M. & Wilson, J. (2005). *Ecological footprints of Canadian municipalities and regions.* Report for the Canadian Federation of Canadian municipalities. Edmonton: Anielski Management Inc. Retrieved from www.anielski.com/Documents/EFA%20Report%20 FINAL%20Feb%202.pdf (Accessed 18/04/2017)

Benessia, A., Funtowicz, S., Giampietro, M., Pereira, Â. G., Ravetz, J. R., Saltelli, A., . . . van der Sluijs, J. P. (2016). *The rightful place of science: Science on the verge.* Tempe, AZ and Washington, DC: Consortium for Science, Policy, & Outcomes.

Best, A., Giljum, S., Simmons, C., Blobel, D., Hammer, M., Lewis, K., . . . Maguire, C. (2008). *Potential of the Ecological Footprint for monitoring environmental impacts from natural resource use: Analysis of the potential of the Ecological Footprint and related assessment tools for use in the EU's Thematic Strategy on the Sustainable Use of Natural Resources.* Report to the European Commission, DG Environment. Retrieved from http://ec.europa.eu/environment/natres/ pdf/footprint_summary.pdf (Accessed 30/12/2016)

Composite Indicators Research Group, JRC. (2016). *What is a composite indicator?* Retrieved from https://composite-indicators.jrc.ec.europa.eu/?q=content/what-composite-indicator (Accessed 30/12/2016)

European Commission. (2016). *Composite indicators research group.* Online dictionary. Retrieved from https://composite-indicators.jrc.ec.europa.eu/?q=content/what-composite-indicator (Accessed 17/06/2016)

European Statistical System. (2011a). *Sponsorship group on measuring progress, well-being and sustainable development.* Final Report adopted by the European Statistical System Committee, November 2011. Retrieved from http://ec.europa.eu/eurostat/ documents/42577/43503/SpG-Final-report-Progress-wellbeing-and-sustainable-deve (Accessed 18/04/2017)

European Statistical System. (2011b). *European statistics code of practice.* Luxembourg: Eurostat. Retrieved from http://ec.europa.eu/eurostat/documents/3859598/5921861/KS-32-11-955-EN.PDF/5fa1ebc6-90bb-43fa-888f-dde032471e15 (Accessed 02/01/2017)

European Statistical System. (2012). *Quality Assurance framework of the European Statistical System.* Luxembourg: Eurostat. Retrieved from http://ec.europa.eu/eurostat/docu ments/64157/4392716/qaf_2012-en.pdf/8bcff303-68da-43d9-aa7d-325a5bf7fb42. (Accessed 28/07/2017)

Ewing, B., Reed, A., Galli, A., Kitzes, J. & Wackernagel, M. (2010). *Calculation methodology for the national footprint accounts, 2010 Edition.* Oakland: Global Footprint Network.

Ewing, B., Reed, A., Rizk, S. M., Galli, A., Wackernagel, M. & Kitzes, J. (2008). *Calculation methodology for the national footprint accounts, 2008 Edition.* Oakland: Global Footprint Network.

Funtowicz, S. O. & Ravetz, J. R. (1990). *Uncertainty and quality in science for policy.* Dordrecht: Kluwer Academic Publishers.

Galli, A., Giampietro, M., Goldfinger, S., Lazarus, E., Lin, D., Saltelli, A., . . . Müller, F. (2016). Questioning the Ecological Footprint. *Ecological Indicators* 69: 224–232.

Garnåsjordet, P. A., Aslaksen, I., Giampietro, M., Funtowicz, S. & Ericson, T. (2012). Sustainable development indicators: From statistics to policy. *Environmental Policy and Governance* 22 (5): 322–336.

Hild, P. (2013). *Ecological Footprint (Update 1) – tables and figures: Luxembourg's consumption footprint from 2000 to 2008.* Unpublished. Retrieved from www.myfootprint.lu – included in the French publication: L'Empreinte écologique du Luxembourg – Edition 2013 (version française) (Accessed 19/04/2017)

Hild, P., Schmitt, B., Decoville, A., Mey, M. & Welfring, J. (2010). *The Ecological Footprint of Luxembourg: Technical report (version 4.0 – extended Scoping Study Report).* Luxembourg: CRP Henri Tudor/CRTE.

Kitzes, J. & Wackernagel, M. (2009). Answers to common questions in Ecological Footprint accounting. *Ecological Indicators* 9 (4): 812–817.

Lock, G. (2014). *Eurostat position on the Ecological Footprint.* Presentation at course Science and Citizens Meet the Challenges of Sustainability, University of Luxembourg, 22 April.

Mahapatra, M. & Sethi, N. (2011). Spend Rs 32 a day? Govt says you can't be poor. *The Times of India*, 21 September.

OECD. (2004). *Glossary of statistical terms: Composite indicator.* Retrieved from http://stats.oecd.org/glossary/detail.asp?ID=6278 (Accessed 18/04/2017)

OECD. (2008). *Handbook on constructing composite indicators: Methodology and user guide.* Retrieved from www.oecd.org/std/42495745.pdf (Accessed 18/04/2017)

Saltelli, A. & Funtowicz, S. O. (2014). When all models are wrong. *Issues in Science and Technology* Winter: 79–85.

Schaefer, F., Luksch, U., Steinbach, N., Cabeça, J. & Hanauer, J. (2006). *Ecological Footprint and biocapacity: The world's ability to regenerate resources and absorb waste in a limited time period.* Luxembourg: Office for Official Publication of the European Communities. Retrieved from http://ec.europa.eu/eurostat/documents/3888793/5835641/KS-AU-06-001-EN. PDF (Accessed 02/01/2017)

Stahl, C. H., Cimorelli, A. J., Mazzarella, C. & Jenkins, B. (2011). Toward sustainability: A case study demonstrating transdisciplinary learning through the selection and use of indicators in a decision-making process. *Integrated Environmental Assessment and Management* 7 (3): 483–498. doi:10.1002/ieam.181

UNDP. (2015). *Human Development Report 2015: Work for human development.* New York: United Nations Development Programme.

van der Sluijs, J. P., Craye, M., Funtowicz, S., Kloprogge, P., Ravetz, J. & Risbey, J. (2005). Combining quantitative and qualitative measures of uncertainty in model-based environmental assessment: The NUSAP System. *Risk Analysis* 25 (2): 481–492.

van der Sluijs, J. P., Potting, J., Risbey, J., van Vuuren, D., de Vries, B., Beusen, A., . . . Ravetz, J. (2002). *Uncertainty assessment of the IMAGE/TIMER B1 CO_2 emissions scenario, using the NUSAP method.* Report no: 410 200 104. Dutch National Research Programme on Global Air Pollution and Climate Change. Retrieved from www.nusap.net/workshop/report/finalrep.pdf#page=88 (Accessed 18/04/2017)

Ziliak, S. T. & McCloskey, D. N. (2008). *The cult of statistical significance: How the standard error costs us jobs, justice, and lives.* Ann Arbor: The University of Michigan Press.

16 Complex learning and the significance of measurement

Sebastian Manhart

The challenge: learning in a complex world

Every human being has extensive experience in dealing with complexity because every human is complex. Our difficulties in approaching complex systems, such as the interconnected circuits of nature, with multiple feedback loops resulting in non-linear relationships and time delays, are nonetheless considerable. *Homo sapiens* shows an impressive resistance to learning to embrace complexity, one of whose causes is how the neurophysiological constitution of our brain and bodies evolved (Sonnleitner, 2017). However, humankind has always been able to adjust to the changing living conditions it has been constantly faced with. Individual civilisations may have failed, but humans as a species learn. They have adapted their behaviour to the conditions and have altered these conditions through their behaviour. Thus, although contemporary human society is undeniably in danger of failing to appropriately observe and control the countless complex influences it exerts on nature and the way it depends on it, this failure is arguably not due to humankind's innate inability to deal with complexity.

This chapter presents a conceptual framework, which allows us to reflect on and communicate about the similarities and differences between the social and individual use of language, numbers and picture-based representations in transformative learning for addressing complex sustainability challenges. Different ways to represent complexity result in drawing different productive boundaries around systems of interest. Moreover, they are deemed key also for the embracing of uncertainty and ignorance, a requisite for opening reflective spaces for creativity in devising solutions and better handling surprises. From this perspective, transformative learning requires a synthesis of diverse subjective and social constructions of the world in light of different polysemiotic representations such as diagrams, in which numbers, language and pictures can be combined.

All self-conscious subjects have a variety of experiences when trying to more or less successfully control their own predetermined self-organising biological complexity on the basis of the reactions of their natural and social environment. The social attribution of subject status is tied to the capacity to perform actions that are expected by oneself and others, although the degree of expectability may vary within a cultural spectrum of 'normality'. The social environment which is presupposed through the expectation and which a person participates in producing at the same time is, in turn, an enormously complex context affecting the continually learning individual in the form of a multiplicity of feedback loops. Thus, each individual is always surrounded by multiple convoluted complexities; first, the complexity of their brain; second, that of their living body which is controlled by this brain; third, that of the person's versatile social system; and fourth, that of their all-encompassing natural environment. The life of every person as a social, self-conscious being is a permanent engagement with complex interrelations and a continual navigating through the countless natural and social structures of these interrelations. Inevitably, every human being learns to deal with complexity.

Nevertheless, our doubts about our capacity to understand complexity are completely valid. The fact that something is complex means exactly this: we cannot understand it. Following Descartes, understanding means that something is so clear and comprehensible that we figure out its system. However, we are unlikely to figure out complex interrelations. It is vital here to differentiate between complicated and complex circumstances. Knowledge about something is not to be confused with insight, revelation, clarity and evidence – in other words understanding in a sophisticated sense. We know lots of things without understanding them. For instance, we can mathematically describe numerous insights from particle physics but cannot imagine them; we register a vast number of market developments as price movements, but their causes remain obscure to us; and we can extrapolate from emotions and shifts of interest to the mundane changes in our bodies, but we do not comprehend the actual biochemical processes. At least, we can occasionally figure out complicated interrelationships, although these attempts to understand are not infrequently associated with a considerable effort. When it comes to complex systems, though, such attempts fail despite every effort. The capacity and operating principle of our consciousness are simply not sufficient. For a start, this is no reason for concern because, having been surrounded by self-organising systems in the course of evolution, we have apparently developed strategies and skills which allow us to deal with complexity without being reliant on a comprehensive understanding of this complexity. This fact, however, does not imply any certainty that this will continue to be so in the future. Considering that the future is always unknown and uncertain; sustainable development necessitates that we address the question of which requirements and strategies have enabled humankind to survive in a world of complex systems until now. An engagement with this issue will help us to better harness these experiences for the future. So what are we already capable of, living in and learning from complex circumstances? And which learning opportunities correspond to these capabilities, that is, which learning opportunities demand neither too little nor too much from human capabilities?

Transformative learning as a sustainable social process

The problem is likely to be not only complicated but also complex. However, this is no reason to forgo a systematic look. Focusing on the issue of how to successfully socially include individuals in a participatory social process of change aiming for more sustainability, one also finds, despite all the differences between the approaches introduced in this anthology, a number of significant similarities. Based on new forms of representation of complex technical and social transformation processes (Geels, 2002; Verbong & Geels, 2007/2010; Burt & van der Heijden, 2008; Wegerif, 2008), several scholars expect the global problem of sustainable development to be solved first and foremost by more appropriate systematic thinking (Checkland, 1981; Checkland & Poulter, 2006; König, 2017). The development and expansion of the necessary new social structures, practices and individual skills will rely less on theoretical-abstract teaching, but will require learning by experimentation and doing in practice. Examples in this book present a diverse set of sustainability challenges and different analytic concepts and approaches to act upon them that help to make sense of complexity in practice in diverse groups and in different realms, such as water supply (König & Tran, 2017), agriculture and the food system (Davila & Dyball, 2017) or energy-efficient home design (Hondrila et al., 2017). In these projects a number of different individual as well as organised perspectives, interests and forms of knowledge are systematically related to one another over the course of a prolonged period; this happens in the form of collaborative work involving problem framing as well as data collection, evaluation and communicative validation. Diverse approaches for developing and reflecting on shared representations of complex circumstances in group formats are developed. All these approaches are adapted to the concrete cultural and social conditions and ideally prompt individual, organisational and social-systemic processes of change that are ideally coordinated as a social learning event. Due to the complexity of the individual, social or technical-cultural variables which influence what is happening, the projects evolve very open and are loosely coupled and are therefore contextually sensitive (Weick, 1976): all settings for learning depend on chance.

All projects are regularly forced to respond to the challenge of initiating a productive dialogue between very different individuals from different social status groups with very different interests, values, perspectives and worldviews, and engaging them for a prolonged period. But it is only within the frame of an arrangement that integrates several people with multiple perspectives under the terms of a modern, democratic society that a permanent and constructive process of change can be initiated. For a person's learning process, which is linked back to society in multiple ways, the shared experience of solving a problem together with people who bring various other perspectives and semiotic approaches to the table is significant. In this process, social differences, which always exist in the form of individual value and knowledge systems, are not supplanted by 'better knowledge' but are conceived and related to one another as productive differences in social learning processes. Therefore, the projects refrain from reproducing social hierarchies: scientific and political experts are consulted on one level as laypeople; different forms of knowledge are embraced as attempting not to privilege one over the other. Any ambition of controlling the learning process by means of complexity-reducing expert knowledge should be excluded, even though they may promise security and legitimacy. Even in organised contexts, learning is an evolutionary process. The possibilities of individual control and rationality in a transformative learning process (Mezirow, 2012) and the improvement of learning through deliberate critical reflection (Mezirow, 1990) should not be overestimated here (Simon, 1979; Weick, 1995; Manhart, 2014b). Rather, the focus should be on the rich indeterminacy of open social arrangements as productive learning opportunities. Although transformative learning is less certain in terms of its results, this also promises a more creative process.

This does not mean that nothing can be said about the requirements of this process. Several central elements can be gathered from observing the co-constructive processes of individual-social learning, which include both human subjectification and the projects presented in this anthology. Social actors, like individuals or organizations, "are at the same time the creators of social systems", the producers of culture and social structures "yet created by them" (Giddens, 1991, p. 204).

Social structures, which are shared expectations, include formulated and informal rules, values or habitualised practices. Culture is the scope of achievable distinctions within. This scope is essentially structured by signs and is not only a well-ordered pattern or network, but rather a complex tangle of connectable distinctions. Knowing something is to actualize a local part, that is, to realize some connections of this tangle as sense-making. That is why knowledge is a process. We do knowledge. For us every item or object refers to another item or object. This reference is an element of a sign – an habitualised element of our perception and in the stream of consciousness which is constantly weaving countless references. A certain relation to one or between some distinctions is a sign. The process of connecting signs – for instance, as information or communication – is sense-making. The logic of inter-action within individual social tangles as well as the observation of and providing guidance for the learning processes are closely linked with the chaining rules of signs. Every perceived or only registered change in the achievability of distinctions is connected with learning. Transformative learning is a complex structural change, which includes all social levels: individual habitualised practices, perceptions and cognitions as well as interactions, organisations and the social systems. Such a great transformation needs a polysemiotic synthesis. Therefore, some notion of the sign and the associated forms of sense-making at the individual level is addressed in Figure 16.1.

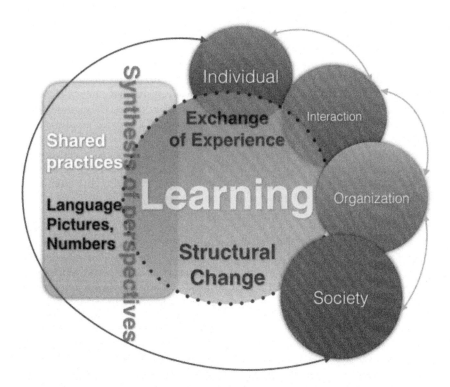

Figure 16.1 Individual learning as polysemiotic synthesis

Learning

Learning is an event dependent on chance. Again and again, people learn without intending to, and not infrequently, people intend to learn but do not learn anything or they only learn little. Learning is not a generally uniform ability independent from the learning topic. Learning successes vary from person to person, from time to time, and depend on the learning content as well as the respective circumstances. Thus, it does not make sense to speak of only one generally uniform learning ability independent from the topic. In any case, it is a number of different changes that are summarised under the heading of learning (Klein, 1987; Holzkamp, 1995; Meyer-Drawe, 2008; Bodenmann et al., 2011; Felden et al., 2012). As a result, people differ, and they differ even if they were exposed to the same influences for a prolonged period. Learning is a complex event, and it can neither be grasped by linear-mechanistic causal models, nor can it be directly influenced by the consciousness, no matter how much the person wants it or how motivated the person is. Overestimating someone's motivation and willingness to learn quickly leads to pedagogic hubris. In the likely event of the learner's failure, it also creates the impression of their individual failure (Manhart, 2014b).

Besides, people learn many things, if not most things, without them intending to learn these things, let alone noticing anything while they are learning. The only thing they do not learn is learning (Winch, 2008). This is because learning as a complex event is largely inscrutable for both an external observer and the learners themselves. Learning occurs on its own, that is, it is neuronally self-directed; for precisely this reason, it is not self-determined. When attempting to observe our own learning process, our consciousness notices other things than an external observer would. However, this does not mean that our consciousness's knowledge is more extensive or better. It is different. Our consciousness is no more able to see through the organic processes that are actually occurring than social agents can. The results of psychological and neurological research have not been able to change this. By no means does science understand what exactly is happening on an individual level when people are learning this or that. Of course, certain neuronal processes are well-known and generally tied, for example, to the improvement of both the memorizing of learned material and motor skills or to an expansion of skills and competency development. But this knowledge does not allow for a direct managing of concrete learning processes. This is unsurprising, because learning is a complex process which remains unfathomable, even if we know some of its individual details.

That's why learning and especially transformative learning is not an action. It is an emergent event that depends just as much on the learning environment we can't control as on ourselves and our attitude. We do not know how learning processes are embodied. Thus, we can learn while performing an action, but learning is not identical to this action. We can intend to acquire knowledge or a skill, but the learning process is neither this skill nor the knowledge. Actions are always limited in terms of their scope, and the same is true of the pedagogical communication that motivates these actions.

However, both the observation of the self by the consciousness as well as its observation by others, such as conversation partners and teachers, can influence the learning event. Therefore, the symbolic form of observation, such as in measurement, grading and ranking, or merely having conversations on what might have been learnt, also plays a decisive role. These are not factors causing specific learning

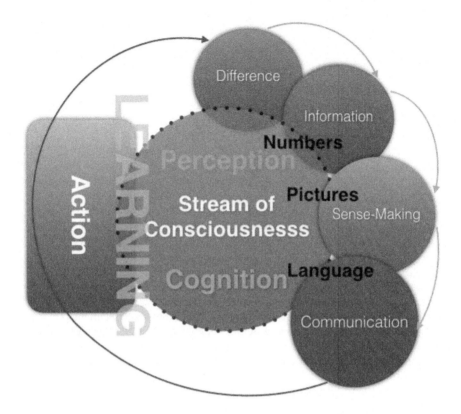

Figure 16.2 Transformative learning as a social process

processes, but elements of a predominantly tentative context control, trigger-causalities based on the arrangement of the learner's perceivable environment. This environment of the learning process does not only involve the usual suspects, nature and society, but also the consciousness itself. But in this stream of consciousness the environment at all social levels consist of nothing but signs, especially in their conventionally symbolic forms of sense-making like numbers, pictures or language, which structure the individual perception and cognition in a habitualised cultural manner. Sharing practices connected with a polysemiotic synthesis of individual perspectives can initiate transformative learning as a complex structural change of all social levels: individual, interactions, organisations and the society (Figure 16.2).

The role of signs in transformative learning

Sustainable forms of knowledge about complex systems originate from specific sign-based mechanisms of coordination that were successful during evolution. Performing and consciously changing these mechanisms allows humans to deal with complex systems in a way that facilitates sustainable co-evolutions between organic, cognitive and social systems without humans needing to understand in detail the processes and structures of these co-evolutions. To use signs, one must learn certain forms of sense-making which are established and distinguished on the basis of certain rules pertaining to the conjunction of signs. The use of these forms of sense-making, which is largely routinised, shapes both the social structures of society and one's individual perception of elements of sense and their linkage in one's stream of consciousness. Learning these forms of sense-making is a central element of socialisation; in all modern societies, this is what the systematic teaching efforts of the education system are based on. In this context, what is important is the way pupils deal with three types of signs, namely words, numbers and pictures, which serve as elements of different forms of sense-making.

Whereas expressions like words and sentences are loosely connected by means of syntactic, grammatical and stylistic rules, the relationships between numbers are strictly regulated. Furthermore, numbers are precise and their sense is principally determined by comparison with other associated numbers. In contrast, the sense of words and sentences is inherently vague. The sense of linguistic expressions alone is never unambiguous and not infrequently markedly obscure; this is even more so the case with the semantic field that reaches beyond this sense. More precise ways of phrasing can only be somewhat validated in particular social contexts (religious rituals, legal dogmas, scientific definitions). The overflowing of linguistic sense towards a strongly context-dependent semantics can hardly be prevented anyway. Therefore, semantic-free sense constructions that are purely made up of formal language are exceedingly rare. In contrast, calculations require every single element (i.e. every single number and its conjunctive sign) to be taken into account; otherwise, their sense is not achieved. Meanwhile, sentences and texts can handle re-formulations, re-phrasings or even an occasional omission when it comes to their significance, as it is impossible to give only one specific sense to linguistic expressions anyway. Calculation rules and linguistic rules exclusively govern their own sign register. Numbers lose their precise sense as soon as they are merged with language (i.e. with reference) and are understood (e.g. as communications). The vagueness that increases in the process is not only a feature of numerology; rather, a combination of forms of sense-making always induces a new richness in terms of references, which generates a more or less vague reference halo. By contrast, language is reduced in its possibilities of sense if one attempts to limit it to the precision of calculation-analogous rules of derivation in the sense of mathematical logic (Schneider, 1999). But in this way, too, a unique semantic space emerges.

Pictures, on the other hand, are visual structures of conjunction whose similarity to the thing they represent requires a synthesis of vague and precise conjunctions. The conjunctions between pictures obey very diverse rules, depending on whether one is dealing with single pictures, a series of pictures or a film. When it comes to the ways of referencing the world beyond the picture as well as the rules of connection to other pictures, there rarely exists more than a vague common sense exceeding the learned production and perception routines of time- and culture-dependent stylistics. As in the case of calculations, linguistic analogies do not contribute much to the understanding of the sense contained within the picture because there is neither syntax nor grammar for pictures and numbers that could serve as a chaining principle for their sign elements. On the other hand, mathematics and language do not operate by iconological rules. Whatever mental pictures are evoked by metaphors or other rhetorical tropes in our consciousness and no matter which associations are triggered, as forms of sense-making characterised by social connectivity, metaphors exclusively follow the rules of language – and they even do so if they supposedly suspend the latter temporarily. Metaphors are simply not pictures painted by language (Davidson, 1990), but a more (un)determinate form of language.

Understanding learning in this sign-based way, we are not talking about direct interventions into complex systems, but their context management: data and calculated results provide information about decay processes at the subatomic level, price changes about market developments, emotions convey hormonal changes to one's awareness, linguistic utterances are regarded as an expression of subjective processes, and the manipulation of the latter is in turn employed to provoke changes in the assumed causes (i.e. to provoke learning). Along these lines, specific conventional signs are used, among other things, to make it apparent that something was learned in the right or wrong way. This is accompanied by the expectation that other people's perception of this evaluative designation of specific forms of expression and the sense-making processing of the latter that takes place in the consciousness will have a positive impact on the future course and success of learning. However, what exactly happens or can happen during this learning event remains systematically vague. This is because modern pedagogical concepts can solely revolve around the context management of complex procedures, the establishment of more or less successful trigger-causalities, but not around effect-causality in relation to the learning event.

A better understanding of the role that signs play in a person's interaction with the internal and external world makes accessible the space of possibility which exists for the establishment of sustainable development paths regarding complex systems. The only thing that humans have at their disposal when it comes to the productive representation (i.e. the controlled observation and manipulation of complex individual and social systems) is signs. Signs play a fundamental role in the human constitution of consciousness, one's relation to oneself and the world and the coordination of social exchange relationships. They also structure both individual and social learning processes. Their perceivable forms and the rules of their conjunction determine what we can and do know. We live in a world of signs, which, even in high-tech modernity, is still very much more than simply a "data-rich environment" (Bernanke & Boivin, 2003; Trichet, 2004, p. 27) – although it certainly has been such an environment for a long time. Yet we know nothing about how the world is if we do not name it.

The semiotic diversity of worlds and knowledge

Issues of semiotics (Peirce, 1986; Eco, 2000) are of particular interest if the initiation and manipulation of learning are not understood as purely individual events. Only then is it possible to exercise a relevant influence on these processes. This is because signs are specific elementary distinctions which can be found at the same time and in the same form on both sides, that is, within the individual consciousness as well as the social interaction. Consciousness and, for instance, the social process of communication markedly differ in how they connect with the same signs. As sensual information, signs collectively capture attention and thereby generate a social space of shared perceptions, precisely because each individual consciousness is divorced from the social process during the interpretation of these perceptions (cf. Whitehead, 1979, p. 342). Therefore, each perceived sign is the unit that distinguishes social unanimity and individual deviation; in other words, it is the productive unit of consensus and dissent.

This division of signs is unavoidable because both sides continue to use the sign on completely different operational bases. Not even the greatest didactic or media effort can prevent the differing ways of connection. However, certain forms of connection can be rendered more likely on both sides. Nevertheless, differences in terms of the subjective connection are there and will continue to exist even in the case of a successful external conditioning. These differences are the basis of sociality. A common exchange, the communication via and about sign objects, is only necessary because a shared understanding in the sense of a synonymity of signs cannot be ensured on the level of individuals. Thus, regarding socially transformative learning processes, it is imperative to use techniques of social exchange which are open ended and unbiased in terms of content and polysemiotic dialogic. In this way, the knowledge potential which exists in the differences of subjective processing of communications, pictures and informational data can be made available for social connections in a jointly developed representational space. In turn, socially coordinated feedback can only occur by means of specific chains of signs perceivable by everyone. Specific social and semiotic arrangements increase the chances that the forms of further subjective-internal and social processing will become paralleled, at least in some cases, although a fixed linking of the different ways of processing is neither achievable nor desirable. Thus, consensus as a shared production of sense is only verifiable on the social side of sign use; besides, this is the only side on which it is socially relevant. One's individual interpretation of what is perceived always remains subjective and therefore deviates more or less significantly from this consensus. But it is especially this difference that is productive for the initiation of transformative learning processes. Because it is precisely these difference in individual interpretation, the deviations and digressions of one's individual stream of consciousness, the gaps of the socially shared interpretations and familiar perspectives, that hold potential for creative approaches. Social consensus and individual difference are compounded in the transformative new.

One can only monitor one's own learning progress on the basis of the externally perceivable side of the inscrutable internal learning process, and one can only do so by means of specific signs. It is the perception of physical but, above all, of social reactions that the individual consciousness registers and interprets as a sign of the successful learning process. Afterwards, it continues to use these reactions for guidance. What is crucial here is that the consciousness itself does not learn. It merely registers and processes certain signs that are interpreted as observed effects of the neuronal learning process that remains unconscious. It is both the degree and the form of difference between one's own expectation and other people's reactions that is productive or unproductive for the learning process. In modern society, this function of significant feedback is also fulfilled by technical apparatuses, such as the various measuring devices whose numeric single values as well as their chaining in measuring regimes, which in turn shape expectations, are taken as guiding principles by the perceiving consciousness to further influence the invisible learning event (Mayer-Schönberger & Cukier, 2013; Zillien & Fröhlich, 2016). Individual ways of expression are therefore understood by others as a sign of the change of subject-internal abilities, skills and knowledge (i.e. as a sign of learning). Self-evidently, this learning-oriented semiotics of individual behaviour must be learned as well.

The diversity of sense-making: numbers, words, pictures

Thus, the different sign forms are not substitutable regarding the sense that can be expressed with them. Although widely assumed to be possible, the translation of sense from one sign form to another is not achievable in the same way it can be done within the same form. The sense of a sentence can more or less exactly be translated into another language, whereas the translation of its reference is more difficult. The world of numbers is completely different from the one portrayed with words or with the help of pictures. But it is also the connection between the represented sense and the world outside of these signs to which they refer (i.e. their reference) that differs. In contrast to pictures, numbers are precise in their chaining with other numbers, but they do not have any reference at all; in other words: they do not have any given external reference. The meaning-making reference is only ascribed to each number through particular practices, namely through counting and measuring (e.g. in the form of measuring units). It is only in early modern society that the chaining of signs becomes systematically decoupled from the semantic requirements of language. On the one hand, this made the inherent logic of numbers unfold in calculation methods which in turn grew increasingly complicated. On the other hand, this led to the emergence of the highly technical and formalised measurement methods which re-connect chains of numbers to the reality outside of mathematics (Manhart, 2016). Since that time, it is absolutely irrelevant what is counted or measured for the chaining rules of numbers and their relationship to take effect, be it the radioactive decay of an element, the amount of leaves of a flower or the number of correct answers in a test. Pictures, by contrast, create the reference through the claimed similarity of their form to the thing they represent, that is, the rules and forms of representation change according to the object displayed. In the case of abstract or concrete art, it is actually the viewer's way of perceiving that is the object. In turn, words and sentences are symbolic forms which bear no similarity whatsoever to the sense and the reference they convey both through their vague coupling with other words and sentences and their more or less routinised, yet exceedingly solid conventional references to non-linguistic things. However, language competence means that one not only associates a certain sense with words and sentences, but always also specific references, which are solidified through processes of familiarisation. In speech acts, this leads to constant interferences between sense and reference and to deviations in individuals' interpretations of these speech acts. The attempt to achieve a shared linguistic representation can therefore only be an interminable communication process; in other words, the attempt generates more communication.

In many years of socialisation and organised education, society and every subject socialised within it have learned to handle productively the considerable differences between sign chaining rules in the form of separated forms of sense-making that have already been institutionalised and organisationally used multiple times. Literature, fine arts or science are cultivated as autonomous forms of sense possessing their own inherent logic; they mostly focus on the chaining rules of specific sign registers and their sense. All human subjects are, within their own self-relation, not only reliant upon the ability to process different sign logics in a parallel-synthetic manner in their stream of consciousness, but also upon the ability to refer sense to sense. It is only within the frame of these polysemiotic methods that reference emerges between the sign-specific forms of sense-making. Indeed, all members of society learn, in time- and milieu-specific ways, to use language, pictures and number-based comparisons, as well as calculations, for the purpose of governing their internal selves and their bodies without having to understand to any extent how this relationship between the sign-based consciousness and the unconscious neuronal processing of signs actually works.

The separation of communication and information processing

It is important to keep in mind this subjective ignorance when talking about what the three different forms of sense-making, which establish themselves on the basis of their differences concerning their regulation of signs, evidently mean. Due to their largely autonomised dynamic, they are constitutive for modern society. Unsurprisingly, language is extremely closely linked with communication. In communication, language is translated into a more determinate, self-reflexive form; this is in contrast to other physical expressions that are also part of the communicative process, such as postures and gestures. Written and spoken language is basically always read in view of its communicative sense. In contrast, when it comes to numbers, it is solely their informational content that is of interest. As a rule, chains of numbers and calculations are not regarded as communications. Thus, the numeric result of a measurement or calculation is not understood as a message, but only as the result of a counting operation or measuring operation; the latter has an indexing function. A subject socialised in modern society expects numbers to contain a specific informational content imparting that the numbers represent this numerical relationship and not any other one, but the subject does not expect numbers to contain any other communicative sense potentially attributable to the author, whether the latter is an individual or an organisation. Chains of numbers establish the strictly regulated form of sense-making that purely consists of information processing. It is only in the act of counting and measuring that numbers are associated with non-numbers (e.g. objects) and that they are given an additional reference. Ideally, this process involves technically strict coupling. The numeric result is then understood as information in the sense of registering a given thing (i.e. it is understood as an index). As a rule, the modern way of dealing with numeric measurement results no longer involves any attempts to communicatively understand the intentions ascribed to the author.

Pictures, too, are a form of sense-making containing its own logic; in this form of sense-making, a spatiotemporal connecting structure is visualised as a simultaneous perceived event. Depending on the respective cultural context and the technique of the recording system, people understand pictures (e.g. paintings) as communicative references to sense or, as in the case of x-ray images, only as purely informative patterns. In the medicinal context, an x-ray image is only seen as an index (i.e. as the effect of a represented thing) and is insofar reduced to its sheer informational value. Neither language nor mathematics, neither the focus on the communicative nor on the informative side of sense, can replace the unique performative power of the iconic form of sense-making, as we can experience watching a film at the cinema or looking at a sculpture in a museum. Besides, the organisation of modern sciences, its successes and its problems are closely linked not only to the forms of sense dominating research and representation, but also to their monologic use in the frame of theoretical and methodological programmes of different disciplines. In interdisciplinary collaboration, which is increasingly promoted, diagrams and multimedia simulations assume a particular significance. Adding more sign registers and integrating them in order to produce an independent form of sense-making is only necessary because a translation of sense – if possible at all (Quine, 1960, pp. 52ff., 70ff., 92ff.; Quine, 1974, p. 98ff) – can only take place within the logic of one sign form, that is, from one language into another, from one numeric calculation into another mathematical formulation or from one pictorial representation to another iconic form. Therefore, diagrams as well as other polysemiotic representations cannot be categorised as any of the previously discussed monosemiotic forms of sense-making. They are discernibly more than that. Being syntheses of different sign forms, they generate an independent semantic space which is extraordinarily important for the productive handling of complexity and especially for the productive work on the separation of scientific theory and non-scientific practice (Table 16.1).

Information and communication sciences

To be able to better assess the non-trivial consequences of the semiotic approach chosen by this chapter, it seems useful to explain, once again, the concept of the separation of sign forms, using the example of the classification of sciences. From a semiotic perspective, the traditional classification into natural sciences and humanities, or the one into social sciences and cultural sciences, can be replaced by the differentiation between information and communication sciences. The central tendencies of the modern development of science have been the regulation of research methods and the conflation of terms into theories by means of a specialisation in specific sign registers. The extreme points of the continuum of disciplinary fields are marked by the exclusively number-based orientation towards information and the purely language-based orientation towards communication. The distinction of university disciplines on the basis of separate theory–method paradigms, which is still dominating the scientific discourse, becomes considerably less important from a semiotic perspective (Manhart, 2011; Manhart, 2014a). The analytical structuring of the science continuum should rather be based on the generative principle which determines both scientific research and the representation of knowledge: the dominant logic of signs and sense-making.

As long as research and representation predominantly employ information processing through either measurement-based numbers or language-based communication, divergent perspectives and different scientific worlds will emerge. What differs then is what is negotiated as a scientific fact and how a scientifically legitimised consensus can be reached. What differ are both the forms of research and the legitimate ways of representing results within these forms of research. Disciplines such as physics, biochemistry or engineering distribute their results and predominantly control the representation of these results within their theories on the basis of non-communicative practices of number chaining. The legitimate production of numeric results is made dependent on highly standardised measurement methods, no matter what other intentions the involved researchers may have. By contrast, in the communication sciences, the methodically governed communication (i.e. the observation with the aid of conceptually and theoretically regulated jargon) is the research process itself. The communication sciences predominantly deal with linguistic utterances and texts about other texts and linguistic utterances. Thus, the empirical evidence of the communication sciences is, for the most part, nothing but communication. One of the results of this development is that most inter- and transdisciplinary projects arise from the attempt to bring together research and theoretical approaches which are exclusively information and communication based, respectively, in order to solve overarching, 'practical' societal issues. This is because 'practical issues' are frequently complex and always require a polysemiotic representation. This means that neither the monosemiotic analysis by means of numbers nor language alone can fully grasp the connective structures of these issues. This is also true of the practical problem of sustainable development.

However, the difference is not only found in the respective sign-specific process of object or world constitution, but primarily involves the function that these different ways of representation fulfil for society and the individuals operating within it. It makes a huge difference whether an academic discipline examines the reference of a given phenomenon, that is, the effects of research on individuals and society – in other words, whether the communicative sense of a scientific fact is the focus – or whether the discipline's epistemological interest lies in the registration and regulatory conjunction of numeric data. The research of all communication sciences aims to analyse what scientific phenomena mean for us; therefore, all communication sciences are language based. The research of information sciences, by contrast, focuses on the registration of 'data' generated by means of measuring regimes; thus, what is dominant here is the form of information processing. Number-based information sciences are not concerned with what academic insights mean for human ways of sense-making. The reference of data only plays a secondary role – for instance, when it comes to possible practical applications of research results or their popularisation, which is no longer part of academia and which tellingly resorts to the pictorial and linguistic representation of these results. But the designation of a fact as scientific is exclusively based on the consistency check of measurement and calculation results by means of mathematical-statistical methods. In the frame of further academic processing of the data, mathematical modellings are used and the precise probability of a phenomenon's occurrence is calculated; thus, the possible vagueness of results is determined with mathematical precision (measuring tolerance, error intervals, etc.) by the discipline itself. Mathematical vagueness, then, has nothing to do with the vagueness of the semantics inherent in every communication, the mutual ambiguity of the reference-making purpose of an utterance and the diffuseness of the multiple linguistic reference chains. The vaguenesses of communication are vague vaguenesses because they correlate diffuse sense references and linguistically vague regulations. The vaguenesses of physics, climate research or economic game theories are precise vaguenesses because, as statistically gathered probability values, they assume a numerically precise form, no matter how false or correct this precision may prove to be in the future. Every numeric error is precise as well.

Between the extremes of pure information and communication sciences, there are numerous intermediate forms of sciences, such as psychology or educational science. The research and theories of theses sciences rely more or less equally upon both communications and measurements and numeric values. Thus, these sciences are predominantly polysemiotic regarding the analysis and representation of their results. These sciences' polysemiotic basis of research and representations does not play a decisive role for the conventional classification of sciences, such as the differentiation between humanities and (natural) sciences or the three-part division of cultural, social and natural sciences. Rather, the special status of psychology, educational science and engineering is normally legitimised through the assumption that these disciplines spend more time dealing with practical or technical issues. Therefore, they do not appear to be as 'theoretical' as, for example, 'pure' mathematics or language philosophy.

The alleged difference between theory and practice is a rather diffuse expression of the underlying difference between monosemiotic and polysemiotic academic research and representation. The existence of sciences whose research and representational practices integrate several sign registers for the production of sense not only points to the similarity of all the sciences occupying the continuum between information and communication sciences, but it is also an expression of the complex constitution

Table 16.1 Conventional forms and relations of signs

	Numbers	*Words/Sentences*	*Pictures*
Relation	Strictly connected	Loosely connected	Loosely connected
Certainty of sense-making	Precise and determined by comparison	Inherently vague semantic fields	Synthesis of vague and precise conjunctions
Reference	Without	Habitual	Similarity
Regulations	Universal rules of counting and calculation	Diversity of specific syntactic, grammatical and stylistic rules	Diversity of individual and cultural styles
Typical practice	Counting, measuring	Speaking, writing	Drawing, filming
Typical product	Key figure ranking	Article, book	Painting, movie
Favourite form of sense-making in modern society	Information	Communication	Information communication
Polysemiotic Synthesis	Diagrams, scenarios, simulations, games		

of a world which cannot be represented by only one sign register. Therefore, even the 'hard' natural sciences, which are typical forms of number-based information sciences (aside from mathematics), have not entirely abandoned the use of language-based, communication-related terminology. Measurements regularly produce semantic constructs (i.e. entities) which are mainly reformulated by means of language, such as energy, dark matter, intelligence or sustainability, which then inevitably generate a communicative surplus of sense through being received by other researchers and non-academic experts (politicians, managers, environmental activists, etc.). Because it goes beyond the results of measurement and calculation methods, a precise understanding of this communicative surplus of sense is and always remains contentious.

Complementarity of forms of sense-making and polysemiotic synthesis

To mediate between the numeric data and the linguistic expression of its reference, pictures and pictorial comparisons, diagrams and, increasingly, multimedia representations such as simulations are regularly used in many domains, for instance, in transdisciplinary approaches and in the popularisation of research results. In the traditional understanding, pictures, diagrams or simulations serve to master the problem of translating between informational and communicative senses (Table 16.1). But different forms of sense-making cannot replace one another; their sense is precisely not translatable into one another, but they are positioned complementary to each other. With regard to specific problems rooted in the separate usage of different monosemiotic forms of sense-making, a complementary parallelism of the different semiotic ways of representation expressing the same sense can be achieved. Such an achievement is accomplished in successful interdisciplinary research associations.

Nevertheless, other new opportunities of knowledge acquisition and reference generation, of understanding and learning, can be created through the connection of different sign regimes. This is the direction taken by 'transdisciplinary concepts'. For the semiotic approach chosen in this article, it is crucial that diagrams, simulations and other multimedia representations are not pictures. Rather, they are polysemiotic forms of sense-making which combine several sign registers and thereby create new semantic spaces in the process of perception and sense-making. In the case of traditional diagrams, they contain pictorial elements combined with words and numbers. During the perception and cognitive processing of signs, the rules of connection of the respective elementary sign form are transcended and a new regulation is established as a way of making sense; this is a polysemiotic correlation of sense. In this way, a new space of possibility for significant constructions of the world is opened up; this space generates forms of insight, understanding and learning that are impossible to create by means of the separate usage of language, pictures or numbers and their complementary parallelisation. The controlled use of these representations generates a polysemiotic sense which may not put us in a position to overcome the semiotic difference between academic theory and concrete practice, but at least it allows us to use this difference productively in a way that encourages learning and reference-making.

These forms of representation can only fulfil their special function of generating knowledge and reference for us because they correspond to the polysemiotic synthesis of the stream of consciousness, in which several sign registers (words, pictures, numbers, sounds, smells, colours, emotions, etc.) always appear as one simultaneous correlation of sense. The fact that specific registers dominate this process, like language or pictures dominate the thinking process, can be tied back to cultural influences on selective ways of perceiving and processing which are learned through constantly dealing with socially customary forms of representing the world, for example, in texts, films or diagrams. In reality, numerous other sign forms and forms of sense-making are always also more or less present and relevant in every conscious processing context. This simultaneousness of forms of sense-making which are overlapping and appearing in more or less prominent ways is the rule for our stream of consciousness. It is also the reason for the fact that the stream of consciousness is permanently subjectively significant. It is this polysemiotic stream of consciousness through which humans' self-relation and their interaction with the complex social and natural environment is constituted.

Thinking specific things is learned to be tantamount to distinguishing and chaining signs. The specific characteristics this chaining assumes in each case in relation to the consciousness and only half-processed or unconsciously processed signs depends upon social-cultural preconditions which become established in the mind of every individual as familiar pathways during socialisation. In practically all social interactions, several sign forms are used at the same time. Depending on time and place, different cultural patterns of sign use have an effect on the social exchange. For instance, these patterns may prefer oral forms of language-based communication to picture-based interactions and may eschew numbers almost entirely; this has been the case in many societies of the past.

Characteristic of modern society is the forced distinction between different forms of sense-making based on the chaining of only one specific sign register; for example, the monosemiotic mental calculation is based on the strict connectivity logic of numbers. This can be established in the consciousness – if at all – only with a considerable individual as well as social learning effort. On the part of society, this effort is put in by organisations which actively regulate the review of certain sign forms and forms of sense-making. This is one of the reasons for the worldwide expansion of a school model relying on regulation and familiarisation. In several years of instructional education, the same recording and notation systems are always deployed as learning materials; most of all, these systems expose pupils to monosemiotic chains of signs and their formations of sense. In particular, what is acquired by pupils, for example, in the tradition of schooling based exclusively on books, is the ability to concentrate on specific chaining rules concerning sign and sense. Learning the rules of written language and those of numbers, pupils achieve a focus that simultaneously tries to suppress other connective possibilities and associations in the consciousness, or at least tries to control these associations' potential interference with the learning process. In the past, this has proved to be an enormously successful approach. To solve the problems of modern society, however, it is necessary to also cultivate more systematically (i.e. consciously) those abilities that allow us to re-establish a synthetic polysemiotic correlation of sense between individual sign logics. In this way, we are enabled to better understand the way we deal with complex systems and creativity is promoted. This does precisely not mean that we should renounce the insights of monologically specialised forms of representation of the world; in other words: that we should regress to a state before the successes of academic specialisation, language-based sharpening of terminology or strictly economic accounting. That would mean a considerable loss of possible insights and ways of organising.

The separation of chains of signs regarded as legitimate in research and representational practice has an advantage that can hardly be overestimated: Due to it, we can systematically develop the full potential of certain sign registers, such as the calculational chaining of numbers and their generation via increasingly complicated measurement methods, in the course of unlocking and representing the world. In contrast, this approach also has a disadvantage that should not be underestimated: in the evolutionarily advancing research process, we commit ourselves to the internal logic and the space of possibility of only one sign form. Other potential constructions of the world and other ways of formulating issues are then systematically neglected, if not delegitimised. Besides, we may perceive the monosemiotic sense constructions of the modern world as increasingly meaningless. In view of contemporary issues that are pressing today, we will no longer be able to afford such a reciprocal constriction of perspectives and possibilities.

This has systematic reasons. The manipulation of complex interrelations such as climate change, which are beyond holistic understanding by the human consciousness, can only succeed through the synthesis of different perspectives, that is, on the basis of monosemiotic-complementary forms of sense-making and their polysemiotic representation. Thus, this approach precisely does not exclude information- or communication-based specialised research, but it points to the imperative of a polysemiotic synthesis which needs to be developed in a participatory manner and which needs to generate a new, independent-significant sense if individuals and organisations are to draw complexity-appropriate conclusions for the context management of transformative learning. This manipulation can only be successful if the conditions of the learners themselves are taken into account; in other words, learning can only be influenced in correspondence to the specific capabilities of persons and organisations to change individually or in a coordinated way. It is a question of exposing the quasi-natural polysemiotic synthesis, which is already constantly taking place in the stream of consciousness, to a more comprehensive reflexive observation; this is achieved by means of organised feedback loops to modern techniques of multimedia polysemiosis. In this way, the creative potential that lies in the entanglement of the evolutionarily open learning event and the different regulations of sign forms and forms of sense-making can be used differently (i.e. ultimately in a more comprehensive manner). Improving our understanding of our capabilities and restrictions in dealing with complex systems is therefore by no means less important than the detailed analysis of the complex systems itself. Prompting and influencing learning is such a form of dealing with complex systems. This is because the learning event of individuals, organisations or even of society as a whole is a complex process which defies classic causal analysis and can therefore not be directively controlled.

Measuring as an important pedagogical practice

Society and, with it, the modern individual react to this uncertain situation of being dependent upon inaccessible dynamics with a search for reliable criteria which allow them to observe and control more precisely, if not the learning event itself, then at least the results of the learning event. As learning is a complex interconnection of processes, its performative surface is the only place where its progress can be subdivided through symbolic markings. In this way, stages of already acquired knowledge and skills can be identified as countable progressions, compared to others and encouraged. Therefore, the inscrutable internal operations – the learning event – are routinely symbolically subdivided on the basis of externally perceivable interim results and presented as this result-producing sequence of actions; in addition, increasingly often, they are also evaluated (i.e. numerically fixed). Here, symbolic techniques of systematising comparisons play a crucial role. Number-based measuring in particular has long become a central pedagogical practice (Manhart, 2016), which is not only used in organisations for the observation of learners by others, but is also adopted by numerous individuals to deliberately observe themselves. In the context of the establishment of this form of observation, what emerges is the increasingly nuanced development of pedagogical-economic evaluation practices and the installation of long-running and increasingly densely interconnected measuring regimes focusing on the control of learning progress and performance; these are deliberately employed as context-managing practices of changing people. Measuring is a standardised form of observation which changes subjects in a particular way. Because the subjects expect the results of a prospective measurement and the further repetition of these results, they orient themselves, in the course of their development, to time-stable measuring regimes and the numeric results generated within these regimes. Thus, the processes of change are actually induced in subjects by the measuring process; the latter also influences these processes' results, which it claims to be objectively counting. The reflexive use of this form leads to the common practice of pedagogical measuring, which ranges from school grades to the quantified self-movement activity trackers to bankers' bonuses, and which always encompasses more than merely a sequence of numbers. Measuring is a dynamising, reference-making case of polysemiosis.

Thus, the manipulation of learning processes can be achieved by observing, changing and using sign-based interaction formats. Individually as well as socially, far-reaching learning processes, which are the goal of transformative learning of sustainability, require the transformation of consciousness-based, organisational and systemic ways of processing. The difficulty lies in achieving a largely similar, parallel development of individual, organisational and societal forms of sense-making across different status groups. This can only be achieved by using different sign formats in a coordinated manner. To successfully use the effects of different sign forms and forms of sense-making on learning, we must develop social interaction formats which allow for different sign forms to be arranged as learning opportunities in an adequate, socially coordinated and polysemiotic way. The dialogue formats, which are typically language based (Wegerif, 2008; Burt & van der Heijden, 2008), should therefore be expanded to equally include the use of number-based, pictorial and other sign formats which appeal to the great variety of human ways of sense-making. This polysemiotic quality that is ascribed to 'practical' approaches, often without thinking, is also the basis of the persistent fascination with and expectation towards the latter. The assumption that practical approaches provide a copious amount of possibilities is often the reason why they are preferred to the 'mere theory' when it comes to choosing a learning environment. Not infrequently, this is correct.

The approaches to sustainability challenges presented in the chapters of this book can also be conceived as dialogic arrangements which are socially conceived in different ways and which facilitate sustainable processes of change through their grappling with practical issues. Complementing previous approaches, new multimedia technologies in the form of diverse presentation tools, simulations and even the gamification of the process of how to approach problems now provide additional new possibilities of manipulating and reflexively controlling the polysemioses conducive to learning. Being polysemiotic instruments of working on concrete projects, simulations and scenarios serve to initiate shared practices open to the different perspectives and world constructions not only of individuals but also of social agents, be it organisations of science, politics or economy. Different perspectives are correlated to one another as social co-constructions of the world through the shared use of different sign forms and forms of sense-making without supplanting within individuals the differing ways of processing and the reference of the forms of sense-making. Rather, this variety of sign registers and forms of sense-making sustains this diversity of ways of processing and simultaneously propels the shared generation of experience in social exchanges. This is because the generation of experience takes place under the conditions of the consciousness and the social systems which both share regarding the polysemiotic synthesis of references. Only in this way can the complex dynamic of transformative learning be triggered as an expansion of the frame of thought and the organisational forms of society, and be influenced in the direction of sustainability.

Conclusion: productive boundaries for transformative learning

In sum, learning as an event will remain inscrutable. However, if we connect it to the integrated variety and richness of sign forms and forms of sense within semiotically and socially de-hierarchical social arrangements, learning can be encouraged more easily. The polysemiotic synthesis of information, pictures, communications and practical actions, which is induced in technologically supported formats, follows different rules and offers other opportunities for creative transformation than the traditional formats of teaching, organised schooling or language-centred dialogue, which are often specialised in terms of semiotics and media. Consisting in reflexive-intentional context management of an unintentional-inscrutable event, these new forms are and enable education (Manhart, 2008). But these polysemiotic arrangements, too, are subject to the risk of failing. From a scientific perspective, nothing else is possible. And yet, if we are interested in individuals evolving into autonomous subjects, we should take a positive stance on learning being and remaining an automatic, largely inaccessible, complex event; political-moral considerations oblige us to do so. A directive causal control of the learning event would simply be counterproductive. This is because dealing competently with one's own internal inscrutability and responding to the learning processes' potential failures and successes in autonomous, socially appropriate and, in the best case, creative ways is exactly what successful individual education means. It is this form of unavailable, autonomous individuality that democratic societies essentially rely upon.

If we want to increase the chance of a transformative learning process being initiated which spans across different people and social structures, the semiotic and media forms of social arrangements should approach the polysemiotic diversity of the individual stream of consciousness. In light of the complexity of the task of sustainable development, every semiotically uniform conception of the problem is reductionistic and not sufficiently complex. The traditional difference between theory and practice is incessantly reproduced both in the form of monosemiotic forms of sense-making and in their complementary parallelism. Naturally, monosemiotic formations of sense continue to be necessary for the generation of special and detailed knowledge, but they are not sufficient to initiate and influence productive social change which can by no means be effected by specialised experts alone. A participatory problem management, which equally benefits from other social agents' perspectives and knowledge whose semiotics differ from one's own, requires a manner of representation which is polysemiotically dense and rich, and which uses different perception and sign registers in a parallel way. This manner of representation establishes a form of observation which productively abolishes the distinction between theory and practice and whose meaningful contents are collectively worked out, assessed and changed in a shared polysemiotic space of experience based on the differences between the subjective perspectives.

Only by synthetically mixing up diverse perspectives can transformative learning be achieved with regard to sustainable development. This transformative learning requires a continuous negotiation of subjective and social constructions of the world in light of different polysemiotic representations. The polysemiotic synthesis of number-based information processing, pictorial representation, linguistic communication and other forms of sense-making aims for a mode of processing which has its own inherent logic, markedly goes beyond the traditional language-based dialogue formats and makes full use of the technical possibilities of multimedia representational formats. A polysemiotic-multiperspective simulation of our natural and social environment which is created through social exchange does justice not only to the diversity of this environment but also to the synthetic functioning of the human consciousness. In this way, a successful change may not be guaranteed. But this approach provides the opportunity to take into account and make use of the human competence in dealing with complexity, which emerged during evolution. We can use this competence to take a sustainable development path for our society without needing to understand everything in its complexity.

Questions for comprehension and reflection

• Why is transformative learning for sustainable development not simply an individual process?
• What are relevant 'learning environments' for transformative learning? Why is it important to understand the physical functioning of ourselves, as well as surrounding environmental and social structures and practices? How are all these 'environments' influenced by signs?
• How can symbols structure transformative learning processes and their observation, considering that they appear as structuring elements of social interactions as well as of individual stream of consciousness?
• How has each of these categories of signs enabled scientific and technical achievements?
• Why are polysemiotic representations helpful in the face of complexity? How do they help to initiate and guide transformative learning?

References

Bernanke, B. S. & Boivin, J. (2003). Monetary policy in a data-rich environment. *Journal of Monetary Economics* 50 (3): 525–546. doi:10.1016/S0304-3932(03)00024-2

Bodenmann, G., Perrez, M. & Schär, M. (2011). *Klassische Lerntheorien: Grundlagen und Anwendungen in Erziehung und Psychotherapie*. Bern: Huber.

Burt, G. & van der Heijden, K. (2008). Towards a framework to understand purpose in Futures Studies: The role of Vickers' Appreciative System. *Technological Forecasting and Social Change* 75 (8): 1109–1127.

Checkland, P. (1981). *Systems thinking, systems practice*. Chichester: John Wiley & Sons Ltd.

Checkland, P. & Poulter, J. (2006). *Learning for action: A short definitive account of soft systems methodology and its use for practitioners, teachers and students*. Chichester: Wiley.

Davidson, D. (1990). 'Was Metaphern bedeuten', in Henrich, D., Luhmann, N., Herborth, F., Davidson, D. & Schulte, J. (Eds.) *Wahrheit und Interpretation*. Frankfurt/Main: Suhrkamp. pp. 343–371.

Davila, F. & Dyball, R. (2017). 'Food systems and human ecology: An overview', in König, A. (Ed.) *Sustainability science: Key issues in connecting learning, research, and practice*. Abingdon: Routledge. Chapter 10, pp. 183–210.

Eco, U. (2000). *Kant und das Schnabeltier*. Munich: Carl Hanser Verlag.

Felden, von H., Hof, C. & Schmidt-Laufen, S. (Eds.) (2012). *Erwachsenenbildung und Lernen*. Baltmannsweiler: Schneider Verlag Hohengehren GmbH.

Geels, F.W. (2002). Technological transitions as evolutionary reconfiguration process: A multi-level perspective and a case-study. *Research Policy* 31 (8–9): 1257–1274.

Giddens, A. (1991). 'Structuration theory: Past, present and future', in Bryant, C.G.A. & Jary, D. (Eds.) *Giddens' theory of structuration. A critical appreciation*. London and New York: Routledge. pp. 201–221.

Holzkamp, K. (1995). *Lernen: Subjektwissenschaftliche Grundlegung*. Frankfurt/Main: Campus Verlag GmbH.

Hondrila, K., Norcross, S., Golinska-Dawson, P., Broz, V., Rios, A. & Muller, J. (2017). 'Democratising renewable energy production – A Luxembourgish perspective', in König, A. (Ed.) *Sustainability science: Key issues in connecting learning, research, and practice*. Abingdon: Routledge. Chapter 13, pp. 234–249.

Klein, S. B. (1987). *Learning: Principles and applications*. New York: McGraw-Hill.

König, A. (Ed.) (2017). *Sustainability science: Key issues in connecting learning, research, and practice.* Abingdon: Routledge.

König, A. & Tran, K. C. (2017). 'Community-based monitoring of water quality in a coastal zone under development – A case study in Holbox Island, Quintana Roo State, Mexico', in König, A. (Ed.) *Sustainability science: Key issues in connecting learning, research, and practice.* Abingdon: Routledge. Chapter 14, pp. 250–268.

Manhart, S. (2008). 'Im Begriffsgeflecht. Zur Entstehung der Bildungssemantik um 1800 zwischen Selbstorganisation, Leben, Mensch und Markt', in Thompson, C. & Weiß, G. (Eds.) *Bildende Widerstände-widerständige Bildung: Blickwechsel zwischen Pädagogik und Philosophie.* Bielefeld: Transcript Verlag. pp. 171–193.

Manhart, S. (2011). *In den Feldern des Wissens: Die Entstehung von Fach und disziplinärer Semantik in den Geschichts- und Staatswissenschaften (1780–1860).* Würzburg: Königshausen & Neumann.

Manhart, S. (2014a). 'Organisiertes veralten – Veraltete Organisation? Zur Stabilität der Universität in den Neuerungsdynamiken von Wissenschaft und Reformen', in Weber, S. M., Göhlich, M., Schröer, A. & Schwarz, H. (Eds.) *Organisation und das Neue: Beiträge der Kommission Organisationspädagogik.* Wiesbaden: Springer VS. pp. 259–269.

Manhart, S. (2014b). Anerkennung durch Lernen: Folgen einer begrifflichen Umstellung. *Der pädagogische Blick* 22 (1): 19–32.

Manhart, S. (2016). 'Pädagogisches Messen: Messen als Organisationsform pädagogischer Praxis', in Göhlich, M., Schröer, A., Weber, S. M. & Schwarz, H. (Eds.) *Organisation und Theorie.* Wiesbaden: Beiträge der Kommission Organisationspädagogik. pp. 53–61.

Mayer-Schönberger, V. & Cukier, K. (2013). *Big data: A revolution that will transform how we live, work and think.* Boston and New York: Houghton Mifflin Harcourt.

Meyer-Drawe, K. (2008). *Diskurse des Lernens.* Munich: Verlag.

Mezirow, J. (1990). *Fostering critical reflection in adulthood: A guide to transformative and emancipatory learning.* San Francisco: Jossey-Bass.

Mezirow, J. (2012). 'Learning to think like an adult: Core concepts of transformation theory', in Taylor, E. W. & Cranton, P. (Eds.) *The handbook of transformative learning: Theory, research and practice.* San Francisco: Jossey-Bass. pp. 73–95.

Peirce, C. S. (1986). *Semiotische Schriften, Bd. I-III.* Edited and translated by Kloesel, C. & Pape, H. Frankfurt/Main: Suhrkamp.

Quine, W. V. O. (1960). *Word and object.* Cambridge, MA: The MIT Press.

Quine, W. V. O. (1974). *The roots of reference.* Chicago: Open Court Publishing Company.

Schneider, H. J. (1999). *Phantasie und Kalkül: Über die Polarität von Handlung und Struktur in der Sprache.* Frankfurt/Main: Suhrkamp.

Simon, H. A. (1979). Rational decision making in business organizations. *American Economic Review* 69 (4): 493–513.

Sonnleitner, P. (2017). 'Cognitive pittfalls in dealing with sustainability', in König, A. (Ed.) *Sustainability science: Key issues in connecting learning, research, and practice.* Abingdon: Routledge. Chapter 4, pp. 82–95.

Trichet, J. C. (2004). 'The ECB's use of statistics and other information for monetary policy', in OECD (Ed.) *Statistics, knowledge and policy: Key indicators to inform decision makings.* Paris: OECD Publishing. pp. 20–28.

Verbong, G. P. & Geels, F. W. (2007). The ongoing energy transition: Lessons from a socio-technical, multilevel analysis of the Dutch electricity system (1960–2004). *Energy Policy* 35 (2): 1025–1037.

Verbong, G. P. & Geels, F. W. (2010). Exploring sustainability transitions in the electricity sector with socio-technical pathways. *Technological Forecasting and Social Change* 77 (8): 1214–1221.

Wegerif, R. (2008). Dialogic or dialectic? The significance of ontological assumptions in research on educational dialogue. *British Educational Research Journal* 34 (3): 347–361.

Weick, K. E. (1976). Educational organizations as loosely coupled systems. *Administrative Sciences Quarterly* 21 (1): 1–19.

Weick, K. E. (1995). *Sensemaking in organizations.* London: Sage Publications Ltd.

Whitehead, A. N. (1979). *Prozeß und Realität: Entwurf einer Kosmologie.* Frankfurt/Main: Ontos Verlag.

Winch, C. (2008). Learning how to learn: A critique. *Journal of Philosophy of Education* 42 (3–4): 649–666.

Zillien, N. & Fröhlich, G. (2016). 'Reflexive Verwissenschaftlichung: Eine empirische Analyse der digitalen Selbstvermessung', in Mämecke, T., Passoth, J.-H. & Wehner, J. (Eds.) *Bedeutende Daten: Modelle, Verfahren und P-raxis der Vermessung und Verdatung im Netz.* Wiesbaden: Springer VS. pp. 1–16.

17 Uncertainty as a key to sustainability economics

Jerome Ravetz

The challenge of sustainability economics

Our present social-technical order is unsustainable in many ways. If a new sustainable order is to be achieved, there will need to be a science of the management of resources. But the present version of that science, economics, is deeply rooted in the present order in all aspects ranging from basic assumptions to detailed techniques. The challenge of creating a new, appropriate sustainability economics is urgent. Box 17.1 describes how the author engages with the science and research stance. There is a place from which to start – that is the matured and successful field of Ecological Economics. There we find an appreciation of social and environmental values that are excluded almost by definition from conventional economics, the limitations of which are revealed most clearly in its difficulties in the management of uncertainty. In this respect conventional economics has failed dramatically; nearly a century after the work of J.M. Keynes, the field is notoriously unable to predict an imminent crash. Ecological Economics has the inherent flexibility enabling it to meet this challenge, but it can be helped by enhanced systems thinking, including awareness. We will develop this theme in this chapter.

Box 17.1 How the author engages in/with the science and research stance

> **The research question:** How can economics, traditionally so closely involved with the unsustainable scientific-technical-social order, become part of transformative sustainability science?
>
> **Main object of research:** To analyse the assumptions underlying economics and to explore their contradictions, with a view to articulating a version that is appropriate for the present challenges.
>
> **Disciplines:** The author has extensive experience in studying the social problems of contemporary science, with a special focus on those that use quantitative methods. He has also contributed to the philosophy of economics, especially the management of uncertainty.
>
> **Methods:** The chapter is largely descriptive, analytical and historical.

Main beliefs about the role of science in society: Economics has always been oriented to policy. Integrity is not achieved by 'objectivity', but by openness to critical scrutiny and willingness to learn.

Views on sustainability: This is not a state that can be defined in advance, but a process of mutual education for social and personal transformation.

Why a new economics is needed

First, we recognise that the civilisation that has brought us to where we are, that we might call modern Europe or modern capitalism, has shaped and has been shaped by the dominant science of economics, based on individualism and wealth (Ravetz, 1994). Alternative approaches to conceiving and organising our economic affairs have been marginal, obsolete or unsuccessful. Even the one successful 'socialist' economy (China) has replaced command-and-control mechanisms by markets in many crucial areas. However, the economic system is far from perfect in its operation; three areas are particularly problematic. First, the instabilities in the financial system have not been resolved; disastrous collapses still occur, without any prospect of a regained control and stability. Then, the tasks of protecting the social and environmental aspects of welfare are not integrated at all well into the present market-driven system. Human suffering and environmental degradation do not count as deficits in 'gross national product'. And, most serious, the worldview that is assumed in mainstream economics is one of isolated individuals (persons or firms) concerned only to maximise their 'utility'. Anything relating to broader questions of meaning and value is strictly inconceivable in that framework; in market terms they are 'externalities'. The question facing us is whether these are only accidental imperfections in a practice and its associated science, which will be remedied in good time, or whether some radical reform is necessary. In the context of transformative sustainability science, we will discuss what such a reform could look like, and how it could be implemented.

In Box 17.2 we review a classic text of economic theory, the essay by the Victorian scholar Stanley Jevons. It marks the transition of economics thinking away from what had been called 'political economy', which considered all the 'factors of production' along with the relevant operations of the social and political order. In the place of all that complexity Jevons substituted a severely simplified picture of isolated individuals who are concerned to maximise their 'utility'. In this he was following the nineteenth-century British tradition of 'utilitarianism', which was the main ideology of those who reformed its antiquated social institutions. He was quite clear that this 'utility' had nothing to do with the higher values that motivate people; this was left to other, separate sciences. We reproduce his discussion here (in a convenient length) so as to show the plausibility and coherence of this vision; Jevons was a great thinker, who was fully aware of all those deeper issues and discussed them in the course of his argument. Appreciating the strength of these ideas along with their limitations, we can better equip ourselves as we work towards a vision that we believe to be better attuned to the needs of a theory of a sustainable economic order.

Box 17.2 Jevons on political economy

1.1

The science of Political Economy rests upon a few notions of an apparently simple character. Utility, wealth, value, commodity, labour, land, capital, are the elements of the subject; and whoever has a thorough comprehension of their nature must possess or be soon able to acquire a knowledge of the whole science. As almost every economical writer has remarked, it is in treating the simple elements that we require the most care and precision, since the least error of conception must vitiate all our deductions. Accordingly, I have devoted the following pages to an investigation of the conditions and relations of the abovenamed notions.

1.2

Repeated reflection and inquiry have led me to the somewhat novel opinion, that value depends entirely upon utility. . . .

1.34

The utilitarian theory holds, that all forces influencing the mind of man are pleasures and pains; and Paley went so far as to say that all pleasures and pains are of one kind only. Mr. Bain has carried out this view to its complete extent, saying, "No amount of complication is ever able to disguise the general fact, that our voluntary activity is moved by only two great classes of stimulants; either a pleasure or a pain, present or remote, must lurk in every situation that drives us into action." . . . Motives and feelings are certainly of the same kind to the extent that we are able to weigh them against each other; but they are, nevertheless, almost incomparable in power and authority.

1.35

My present purpose is accomplished in pointing out this hierarchy of feeling, and assigning a proper place to the pleasures and pains with which the Economist deals. It is the lowest rank of feelings which we here treat. The calculus of utility aims at supplying the ordinary wants of man at the least cost of labour. Each labourer, in the absence of other motives, is supposed to devote his energy to the accumulation of wealth. A higher calculus of moral right and wrong would be needed to show how he may best employ that wealth for the good of others as well as himself. But when that higher calculus gives no prohibition, we need the lower calculus to gain us the utmost good in matters of moral indifference. There is no rule of morals to forbid our making two blades of grass grow instead of one, if, by the wise expenditure of labour, we can do so.

Source: Jevons, 1879, Chapter 1

Alternative visions: Maslow and Schumacher

Jevons's vision of the isolated utility-maximising '*Homo economicus*' has been very powerful. It has led to our domination by what has been called 'the dismal science', whose essential feature is the exclusion of a human community from its picture of reality. We can ask whether such a world could be sustained, or even could it be worth sustaining in that form (Marglin, 2010). If economic theory is to be integrated into all the full richness of human activity, we need a simple scheme, an 'heuristic', to locate it in that context.

This has been provided by the visionary American psychologist, Abraham Maslow. His 'hierarchy of needs' puts the material aspect at the foundation, but reminds us that the higher 'needs' are there to give it meaning. The simple 'pyramid' image (Figure 17.1) has been enormously influential, both in the mainstream business world and the various 'alternatives'. Of course the scheme has been much criticised (see Wikipedia). In particular, we could question whether it is necessary for people to have their 'lower' needs satisfied before their 'higher' ones can be realised. Expressing it crudely, do people really need to become rich before they can be ethical, only then able to enjoy an affluent life in the realms of esteem and self-actualisation? If so, then there is no effective argument against utilitarianism. The influence of the higher values also works the other way: although it was not noticed by the founders of 'utility' economics, we have now learned that the ordinary activities described by economics need to have an ethical component if they are to be conducted well. Otherwise, they will fall prey to corruption and then fail in many ways. The infamous 'Gordon Gekko' principle, "Greed is good" (the ultimate expression of individualised utility), was not merely reprehensible as it stood; it also contributed to the universal cheating that was

Figure 17.1 The Maslow pyramid
Source: McLeod, 2007

Box 17.3   Schumacher

(From Small Is Beautiful)

To get to the crux of the matter, we do well to ask why it is that all these terms – pollution, environment, ecology etc. – have so suddenly come into prominence. After all, we have had an industrial system for quite some time, yet only five or ten years ago these words were virtually unknown. Is this a sudden fad, a silly fashion, or perhaps a sudden failure of nerve?

The explanation is not difficult to find.

The next four or five years are likely to see more industrial production, taking the world as a whole, than all of mankind accomplished up to 1945. In other words, so recently that most of us have hardly yet become conscious of it – there has been a unique quantitative jump in industrial production.

Partly as a cause and also as an effect, there has also been a unique qualitative jump. Our scientists and technologists have learned to compound substances unknown to nature; against many of them, nature is virtually defenceless.

Because they have no natural enemies, they tend to accumulate, and the long-term consequences of this accumulation are in many cases known to be extremely dangerous, and in other cases totally unpredictable.

In other words, the changes of the last twenty-five years, both in the quantity and in the quality of man's industrial processes, have produced an entirely new situation – a situation resulting not from our failures but from what we thought were our greatest successes. And this has come so suddenly that we hardly noticed the fact that we were very rapidly using up a certain kind of irreplaceable capital asset, namely the tolerance margins which benign nature always provides.

And what is my case? Simply that our most important task is to get off our present collision course. And who is there to tackle such a task? I think every one of us, whether old or young, powerful or powerless, rich or poor, influential or uninfluential.

To give only three preliminary examples: in agriculture and horticulture, we can interest ourselves in the perfection of production methods which are biologically sound, build up soil fertility, and produce health, beauty and permanence.

Productivity will then look after itself. In industry, we can interest ourselves in the evolution of small-scale technology, relatively non-violent technology, 'technology with a human face', so that people have a chance to enjoy themselves while they are working, instead of working solely for their pay packet and hoping, usually forlornly, for enjoyment solely during their leisure time. In industry. Again – and, surely, industry is the pace-setter of modern life – we can interest ourselves in new forms of partnership between management and men, even forms of common ownership.

Source: Schumacher, 1973

largely responsible for the 2008 financial crash. An alternative, more elaborated scheme has been proposed by the Chilean visionary economist Manfred Max-Neef. It is based on experience with South American peasants rather than North American graduates. It is non-hierarchical, with a matrix showing the interaction of nine basic, perhaps universal 'needs' with four 'existential categories' and five 'satisfiers' (some false). It has been influential in connection with the movement for 'human-scale development'.

With all these caveats, which could form a useful basis for discussion, we can use Maslow's hierarchy as a convenient first guide to our enriched picture of economic activity, beyond that of Jevons and his successors. And it serves as a reminder of why such an enriched picture is necessary. But its focus is on individual psychology rather than a societal economic activity, so for us it serves better for inspiration than for a detailed guide to practice.

A first attempt to bring Maslow's ideas down from the heavens, as it were, was made by E.F. Schumacher, the author of *Small Is Beautiful*. He was a professional economist with a very practical day job. But he was committed to bringing an ethical, indeed a spiritual, dimension to economic theory; he called it 'Buddhist economics'. Its subtitle is 'A study of economics as if people mattered', implying that for conventional economics, people don't and can't matter, even if those economists personally might care about them. He linked the two themes of the dignity of work and respect for the environment, arguing that the two concerns cannot be separated. His explanation of his position, in Box 17.3, contains a warning about the unintended environmental consequences of our powerful technology, which relates to our sustainability as a species in our increasingly artificial environment.

But even Schumacher's work is rather on the inspirational side. To be effective in practice, economics must have an articulated theory whose concepts can be related to practice through reliable routines of data capture and processing. This challenge has been met by several developments. 'Ecological Economics' is the most systematic and coherent, but there are some bridges from the mainstream to an environmental awareness that are worthy of mention.

The first enrichment: Ecological Economics

As the main organising science for modern life, economics has recently faced a rival: ecology. Both words have the same Greek root, *oikos*, or household, but historically they have come to mean very different things. 'Economics' in one form or another goes back to the eighteenth century and was defined around wealth, as in Adam Smith's classic book *The Wealth of Nations*. 'Ecology' was coined by the philosophically minded German biologist Ernst Haeckel in the mid-nineteenth century. It has a broad family of meanings, stretching from reference to a natural science with a systematic approach in environmental studies, over to an idealistic way of life, including non-violent activism. Quite recently, a group of scholars set about to integrate the two historic approaches to economy, bringing environmental awareness to economics and the management of economic issues to ecology. Thus was born 'Ecological Economics'. We regard it as the foundation element of transformative sustainability economics; there is nothing in its teaching that is alien to this new discipline. But as we come onto

Table 17.1 Comparison of economic models

	Current economic model	*Ecological economic model*
Basic worldview	**Mechanistic, static, atomistic, individual tastes and preferences** staken as given and the dominant force. The resource base viewed as essentially limitless due to technical progress and infinite substitutability.	**Dynamic, systems, evolutionary human preferences, understanding, technology and organization** co-evolve to reflect broad ecological opportunities and constraints. Humans are responsible for understanding their role in the larger systems and managing it sustainably.
Time frame	**Short:** 50 yrs max, 1–4 yrs usually.	**Multi-scale:** Days to eons, multi-scale synthesis.
Primary policy goal	**More:** Economic growth in the conventional sense, as measured by GDP. The assumption is that growth will ultimately allow the solution of all other problems. More is always better.	**Better:** Focus must shift from merely growth to 'development' in the real sense of improvement in sustainable human well-being, recognizing that growth has significant negative by-products.
Primary micro goal	**Max profits (firms) Max utility (individuals)** All agents following micro-goals lead to the macro-goal being fulfilled. External costs and benefits given lip service but usually ignored.	**Must be adjusted to reflect system goals** Social organization and cultural institutions at higher levels of space/time hierarchy ameliorate conflicts produced by myopic pursuit of micro-goals at lower levels.
Primary measure of progress	GDP.	Index of Sustainable Economic Welfare (ISEW), Genuine Progress Indicator (GPI) or other improved measures of real welfare.
Space frame	**Local to international:** Framework invariant at increasing spatial scale, basic units change from individuals to firms to countries.	**Local to global:** Hierarchy of scales.
Species frame	**Human only:** Plants and animals only rarely included for contributory value.	**Whole ecosystem:** Including humans. Acknowledges interconnections between humans and rest of nature.
Distribution/ poverty	Given lip service, but relegated to 'politics' and a 'trickledown' policy: a rising tide lifts all boats.	A primary concern, because it directly affects quality of life and social capital and is often exacerbated by growth: a too rapidly rising tide only lifts yachts, while swamping small boats.
Economic efficiency/ allocation	The primary concern, but generally including only marketed goods and services (GDP) and market institutions.	A primary concern, but including both market and nonmarket goods and services, and effects. Emphasis on the need to incorporate the value of natural and social capital to achieve true allocative efficiency.

	Current economic model	*Ecological economic model*
Property rights	Emphasis on private property and conventional markets.	Emphasis on a balance of property rights regimes appropriate to the nature and scale of the system and a linking of rights with responsibilities. Includes larger role for common-property institutions.
Role of government	Government intervention to be minimized and replaced with private and market institutions.	Government plays a central role, including new functions as referee, facilitator and broker in a new suite of common-asset institutions.
Principles of governance	Laissez-faire market capitalism.	Lisbon principles of sustainable governance.
Assumptions about technical progress	Very optimistic.	Prudently sceptical.
Academic stance	Disciplinary monistic, focus on mathematical tools.	Transdisciplinary pluralistic, focus on problems.

Source: Costanza et al., 2014, pp. 58–59

the scene later, we 'stand on the shoulders' of the founders of Ecological Economics and assume their definition of *what* it is about. Then we can incorporate these other ideas, deriving from Maslow, Schumacher and others, into our vision of *how* it is to be applied.

The transition from mainstream to Ecological Economics is well illustrated by a table published in the foundational textbook on the subject (Costanza et al., 2014). In that book we find an illuminating historical background, conceptual framework and studies of institutions, instruments and policies. For those wanting to apply economic science to sustainability problems, it is the essential resource. Table 17.1 lays out the terms of a critical dialogue with the assumptions of the mainstream economics paradigm, clearly stated here. Anyone who has been raised up in that worldview should be able to engage, if they wish, with the ecological-economics critique of inherited ways of thinking. It is true that Ecological Economics has generally made only modest inroads into the standard university economics curriculum. But given the criticism to which that curriculum has been subjected, even by its students, the absence of Ecological Economics may reflect more on the deficiencies of the mainstream than on itself (Earle et al., 2016).

The project of Ecological Economics has matured continuously since its first announcement. Recently its leaders have articulated a clear vision that includes sustainability at its core. This appeared in a succinct programmatic statement delivered to the United Nations 'Rio +20 Conference' in 2013. An extract from the executive summary is in Box 17.4.

Box 17.4 The Ecological Economics vision

Our material economy is embedded in society, which is embedded in our ecological life-support system, and that we cannot understand or manage our economy without understanding the whole, interconnected system; and growth and development are not always linked and that true development must be defined in terms of the improvement of sustainable well-being (SWB), not merely improvement in material consumption; and a healthy balance must be struck among thriving natural, human, social and cultural assets, and adequate and well-functioning produced or built assets. We refer to these assets as "capital" in the sense of a stock or accumulation or heritage – a patrimony received from the past and contributing to the welfare of the present and future. Clearly our use of the term "capital" is much broader than that associated with capitalism.

...

Balancing and investing in all the dimensions of our wealth to achieve sustainable well-being requires that:

we live within planetary boundaries – within the capacity of our finite planet to provide the resources needed for this and future generations;
that these resources are distributed fairly within this generation, between generations, and between humans and other species; and that
we use these finite resources as efficiently as possible to produce sustainable human well-being, recognizing its dependence on the well-being of the rest of nature.

Source: Costanza et al., 2013

The precision challenge to Ecological Economics

Transformation is achieved in response to challenges. The normal institutional reaction to 'uncomfortable knowledge' (Rayner, 2012) is to neutralize it somehow. This may bring temporary relief, but it prevents learning. To illustrate the bridge from ecological to sustainability economics, we will consider one particular problem: the management of uncertainty. This has been an issue in conventional economics for a long time, but has mainly been managed by neglect, either benign or otherwise (Ravetz, 1994)

In the abstracted image of economic activity that Jevons created, there was no mention of uncertainty. Utilities and probabilities were assumed to be capable of totally precise measurement. Only on the first occasion, when economics got it disastrously wrong, in the crash and Depression of 1929 onwards, could an engagement with ignorance and uncertainty get a hearing. This was due to one of the great polymaths of the century, John Maynard Keynes. He had previously been much concerned with the analysis of uncertainty, and the harsh conditions of the Depression provided an occasion for a public critique of the comforting assumptions of conventional economics. In his radical economic analysis, Keynes moved away from what we now call micro-economics in the tradition of Jevons and focused on the macro scale. He justified this by observing how economic agents do not, and cannot, predict the future with any reliability. However, there are regularities in large-scale behaviour, sometimes quite harmful, but sometimes capable of modification by government policies. It was around these that he built his theories and recommendations for practice.

The classic quote about the work of his contemporaries from his *General Theory* is

> Too large a proportion of recent "mathematical" economics are mere concoctions, as imprecise as the initial assumptions they rest on, which allow the author to lose sight of the complexities and interdependencies of the real world in a maze of pretentious and unhelpful symbols.
>
> (Keynes, 1936, Book 5, Chapter 21)

Although many of Keynes's policy prescriptions have been adopted, sadly his insights about mathematics have been seriously neglected. His criticisms, as in the quote noted earlier, are as relevant today as they were eighty years ago. He has by no means been ignored, either as an economic theorist or as a philosophical critic. But his vision of an economics where uncertainty is managed and mathematics is well crafted is still to be achieved and is more necessary than ever. And we will see that Ecological Economics still has the same challenges of uncertainty as the conventional sort.

In spite of the radical philosophical differences between the 'ecological' and 'mainstream' perspectives, there are already important areas of overlap in practice. These deal mainly with the metrics of economic activity, focused on the real-world 'macro' level. A review of papers in a journal of Ecological Economics will show that the authors are immersed in practical issues, using quantitative data and mathematical analysis to solve the relevant scientific problems. Although the discipline is largely free of the sorts of abstruse exercises that Keynes condemned, its practitioners generally need numbers to relate their concepts to reality. We can actually witness a sort of convergence between the two sorts of endeavours. One builds a bridge outwards, extending the standard macroeconomic indicators to include the social and ecological dimensions, whereas the other works inwards, using a modified concept of capital ('natural capital') to describe the predicament of our unsustainable economy in a politically effective way. The one has been promoted by a distinguished critical economist, Joseph Stiglitz, and the other by a founder of Ecological Economics, Robert Costanza. In both cases, the proponents are acutely aware of the methodological issues that they confront, although we find that they still have challenges on the crucial issue of the representation of uncertain information.

Macro-economics is a relatively young branch of the discipline. It was created in the 1930s in response to urgent policy concerns and from the outset was conscious of methodological issues. Its traditional special focus has been on gross domestic product (GDP) and its variants. This is now the standard measure of success of an economy, and its rate and direction of change are publicly scrutinised as anxiously as the pulse of a feverish patient. Yet even in the most responsible circles there is a widespread concern that GDP actually (in the terms of our analysis later) distorts and conceals more than it reveals and suggests about the real state of economic life, to say nothing of the state of society and the environment. To remedy the latter defects, the science of econometrics has been extended to the creation of indicators in those more complex fields. Of course, the traps and pitfalls in categorization and data collection are even more menacing. The recent magisterial study by Stiglitz,

Sen and Fitoussi (2010) provides an exhaustive study of the sources of error and distortion in the statistical techniques that form the basis of the key indicators. One could imagine this volume being intended to serve as the methodological companion to an eventual handbook of indicators, which does not yet exist. Indeed, the paucity of numbers in their discussion is quite striking; only a few examples are provided, and even these raise unnoticed methodological questions.

Thus in Stiglitz et al. (2010, p. 37) there is a brief table (1.2), giving 'Percentage change at annual rate, 1995–2006', for three nations, for 'Adjusted disposable income plus housework' and the same plus 'leisure', with entries for 'Total' and 'Per consumption unit'. There are a dozen entries, and the numbers are expressed to two digits – a single percentage plus one decimal place. This is a fairly standard sort of national economic-social statistic. It is a sign of the authors' sensitivity to uncertainty issues that only two significant digits are used.

But very surprisingly, the leading comment is " The imprecision associated with the above estimates should be reiterated here. These are orders of magnitude at best and should not be over-interpreted" (Stiglitz et al., 2010, p.54). Now if this statistic is 'orders of magnitude at best', then what is the meaning, if any, of the second digit in the expression, or even of the rest. We are not told whether this statistic, along with another accompanied by the same caution ((Stiglitz et al., 2010, p.54), Table 13, p. 135), is exceptionally imprecise. If so, we need to know why; and if not, then what are we to make of the great mass of economic and social statistics, whose composition is discussed in a general way (without numbers) in the report? For most of such statistics, as they are published, a mere two places of precision is indeed an exception; it is only natural to ask: What proportion of these statistics are as precise as their representations, and what proportion are 'orders of magnitude at best'?

The authors are certainly aware of issues of systematic uncertainty. There are penetrating, although brief, discussions of two issues of uncertainty: one about the future (Siglitz et al., 2010, p. 74) and the other about the influence of normative considerations on accounting conventions (Siglitz et al., 2010, p. 185). But on the issue of precision of representation, which in a traditional empirical science is crucial for competent communication of results, the readers must make do with the two examples cited.

Making the bridge from the other, ecological, end is the study of 'natural capital' and 'ecosystem services' led by Robert Costanza. This gained notoriety when its first publication announced a global figure of some $33 trillion per year (with an error bar of +50%) for the services (Costanza, 1997). These "represent the benefits [that] human populations derive, directly or indirectly, from ecosystem functions". These

Table 17.2 Household income in real terms. Percentage change at annual rate, 1995–2006

	France	United States	Finland
Adjusted disposable income plus housework			
Total	1.9%	2.9%	2.0%
Per consumption unit	1.1%	1.7%	1.6%
Adjusted disposable income plus housework and leisure			
Total	1.4%	2.3%	1.4%
Per consumption unit	0.7%	1.0%	0.9%

Source: Stiglitz et al., 2010

latter "refer variously to the habitat, biological or system properties or processes of ecosystems". 'Natural capital' may be considered as the stocks of natural assets, whose flows enable the services. Criticisms of the valuation project include the obvious methodological issues of the reduction of complex, uncertain, value-loaded situations to a single monetary metric, as was discussed in Chapter 15 on indicators. In addition, there is a political dimension to the critique, namely that it implicitly but inevitably absorbs the worldview of the marketplace into our conception of the environment. It does appear that some who apply these techniques really do believe that environmental values can and should be as 'fungible' as those of a more conventional sort, so that ecosystem health and beauty should be tradeable just like the fictitious bets and debts that collapsed in 2008 (Monbiot, 2014). Costanza has been fully aware of such critiques since the start of his work; he has also been aware that sometimes the refusal to engage on the monetary evaluation of an environmental good is effectively a surrender to commercial greed. The uses and limitations of the 'ecosystem services' concept in agricultural transitions is discussed in Chapter 9.

Unlike Stiglitz and his colleagues, Costanza is not shy about showing his numbers. With that, he reveals methodological issues which are quite similar to those that challenged Stiglitz and his colleagues. In a definitive paper (Costanza et al., 2014a) expounding and defending the natural capital concept, there is a table of sample values of various environmental parameters. The areas of the different sorts of land or water surface are given to the nearest million hectares (with some entries listed with two 0's, indicating an awareness of uncertainty), but then the (necessarily highly artefactual) 'Unit values' (in 2007$/ha/yr) are listed with up to five or even six significant digits! Such precision is uncommon in ordinary commercial transactions, and it is hard to see how it could be justified in this context.

Such excessive precision, or indeed pseudo-precision, is not a trivial matter. A high precision in expression implies a high accuracy in knowledge, and this is misleading. It is easy to see that the management of uncertainty in such tabulated information is an extremely difficult task, for which there is as yet no standard solution. If we round off the large numbers with strings of 0's, then in any aggregation the small ones would be swallowed up. And how much to round off particular entries? Attempting to clean up the precision on such a table leads to a host of dilemmas. As a consequence of this unmet challenge, and fostered by the underlying philosophy, excessive precision is universal in statistical practice.

This anomaly provides us with an important clue. When even the leaders in their fields do not meet a challenge that is obvious on reflection, we can be sure that the failing is not personal, but rather is the result of contradictions that lie deep in their system of ideas. We can recall that for Jevons, the model of a science is what we can now call pre-Einsteinian physics. That is, it seemed that there was no limit to the precision with which natural science could produce measurements and predictions. But as the twentieth century wore on, particularly after Heisenberg's announcement of the Uncertainty Principle, the limits of precision of measurement were realised. However, the quantitative social sciences stayed in their Victorian past, in spite of the arguments of J.M. Keynes. Precision was associated with the objectivity of knowledge, and it also assumed the simplicity of the world being studied. Mainstream economics notoriously assumes a simple world of isolated utility-maximisers, and it is not clear to what extent Ecological Economics has so far embraced the world of deep complexity.

Our formulation uses the four themes of complexity, contingency, contradiction and ignorance/uncertainty. Some of this is implicit in the NUSAP notational

Table 17.3 Changes in area, unit values and aggregate global flow values from 1997 to 2011 (portion) Note: The values that have increased are in **bold**; the values that have decreased are in *italic*

Biome	Area			Unit values		
(e6 ha)			Change	2007$/ha/yr		Change
1997		2011	**2011–1997**	1997	2011	**2011–1997**
Marine	36,302	36,302	0	796	**1,368**	572
Open Ocean	33,200	33,200	0	348	**660**	312
Coastal	3,102	3,102	0	5,592	**8,944**	3,352
Estuaries	180	180	0	31,509	*28,916*	*-2,593*
Seagrass/Algae Beds	200	**234**	34	26,226	**2,916**	2,690
Coral Reefs	62	*28*	*-34*	8,384	**352,249**	343,865
Shelf	2,600	2,660	0	2,222	2,222	0
Terrestrial	15,323	15,323	0	1,109	**4,901**	3,792
Forest	4,855	*4,261*	*-594*	1,338	**3,800**	2,462
Tropical	1,900	*1,258*	*-642*	2,769	**5,382**	2,613
Temperate/Boreal	2,955	**3,003**	48	417	**3,137**	2,720
Grass/Rangelands	3,898	**4,418**	520	321	**4,166**	3,845
Wetlands	330	*188*	*-142*	20,404	**140,174**	119,770
Tidal Marsh/Mangroves	165	*128*	*-37*	13,786	**193,843**	180,057
Swamps/Floodplains	165	*60*	*-105*	27,021	*25,681*	*-1,340*
Lakes/Rivers	200	200	0	11,727	**12,512**	785
Desert	1,925	**2,159**	234	-	-	0
Tundra	743	*433*	*-310*	-	-	0
Ice/Rock	1,640	1,640	0	-	-	0
Cropland	1,400	**1,672**	272	126	**5,441**	5,441
Urban	332	**352**	20	-	**6,661**	6,661
Total	51,625	51,625	0			

Source: Costanza, 2014a

system that we introduced in Chapter 15 on indicators. Although we have not at all solved the problem of representation of uncertainty in quantitative data, we believe that this new framework will enable an effective solution to be found. The point may be expressed this way: we now know that the reality we are studying is complex. Numbers are inevitably a simplification, a reduction of complexity (Radermacher, 2006). If the process of representation does not respect this difference between object and image, it will encounter insoluble anomalies, as we have seen in the two examples earlier. We would repeat the basic error of mainstream economics in using representations that reveal and suggest less than they distort and conceal. Our information about the complex world around us would be vitiated and misleading. Our understanding of ourselves in the world would be impaired, as we would lack effective information about our interactions with it. The chances of implementing sustainability would be correspondingly reduced. The challenge, then, is to create enriched representations of quantities that are still convenient and effective in use. This will involve fostering a critical self-awareness about our beliefs and assumptions, which can be threatening and painful. But that is the way to transformation.

Towards a transformative sustainability economics

We have traced an intellectual development in economics, from the restricted conception of Jevons through to the enriched vision of Ecological Economics. What is left to add for the transformative sustainability perspective? Here we will sketch what a transformative sustainability economics would be like, using key ideas from the chapters of this volume. In the conclusion (Chapter 18), we will discuss the heuristic principles that have, implicitly or explicitly, guided the analyses of the various chapters. Here we will show how they can be employed in a description of transformative sustainability economics, and leave a fuller discussion for then.

To see how the perspective of 'transformative sustainability economics' differs from that of 'Ecological Economics', we can imagine a sequence of enrichments. Starting with Jevons, mainstream economics imagines isolated individuals acting in full knowledge of the reasons for, and the consequences of, their actions. With ecological awareness, the elements of the theoretical framework include layers of context, as society and environment. Consideration of values, ethics and long-term consequences, excluded from Jevons's vision, are integrated into the analysis. With transformative sustainability science, our theoretical framework includes ourselves as individuals engaged on social learning and personal growth. For this, it embeds us in a world of deep complexity, of situated, contingent knowledge and of uncertainty tinged with ignorance. This is, after all, only a recognition of the way things are in the present age. When we speak of the 'Anthropocene' we are referring to eventual drastic changes in our environment, made by ourselves, whose consequences are unknowable.

For centuries we have accepted Descartes' belief in ourselves as 'masters and possessors of nature'; but now we begin to wonder whether we are instead just sorcerer's apprentices. Although science will be an essential element of our meeting of this challenge, we have seen that the science framed first by Descartes will not be sufficient. The belief that there can be a precise 'blueprint for survival', with which the radical ecological movement was launched in 1972, now seems antique. Yet intelligent decisions for sustainability will require management of resources of every sort, and so an appropriate science of economics will be essential. And it will be one that involves ourselves, learning, reflecting and being transformed by the experience while employing quantitative data as appropriate.

Ecological Economics has made the first bridge between the visions of Maslow and Schumacher and the practical world of decisions and calculations. We have seen in Chapter 5 that effective systems analyses need not always be caught in the trap of excessively precise numbers; sometimes, a + or − may be a totally adequate degree of quantification. But for more detailed studies, numbers are essential. Their uncertainties can be managed effectively when they are employed within a framework that embraces complexity. For that perspective ensures that quantities are seen in a context that precludes the illusion of perfection or completeness. Everything is contextual and situated, including the operations whereby the quantities are produced. Such an analysis has been produced by Mario Giampietro and his colleagues in a variety of studies. We give a very brief example of his analysis, showing how the systems perspective works for him (Box 17.5).

Box 17.5   Giampietro energy grammar

This chapter illustrates the integrated accounting for exosomatic energy flows across different scales of analysis bridging the internal and the external view on the societal energy metabolism. The energy grammar deals with: 1) the presence of an internal autocatalytic loop of energy carriers; 2) the coexistence of two distinct energy forms, mechanical and thermal energy, whose quantities cannot be summed; 3) the requirement of biophysical production factors, including exosomatic energy, power capacity, human activity and managed land, for the conversion of exosomatic energy flows into applied power (end uses); and 4) the impossibility of calculation a simple output/input ratio (energy return on investment to assess the quality of energy sources.

Source: Giampietro et al., 2014

Very similar to the 'transformative sustainability economics' discussed here is the work of Joe Ravetz (2018) (statement of interest: he is my son). With his personal experience of the follies of mechanistic planning and the pitfalls of participatory approaches, he has developed a conceptual framework based on self-aware dialogue, which covers all aspects of the movement towards sustainability. A sample of his discussion of economics is displayed in Box 17.6.

In this book we have devoted Chapter 7 to a particular heuristic for reaching consensus in policy issues, the 'Theory U'. In the same family of methods we find the works of Adam Kahane (2017). His first success was when, using scenario techniques, he helped the opposing sides in South Africa see that for each of them their best solution was preventing a bloodbath. This laid the foundations for a genuine, if imperfect, reconciliation. Since then he has reflected on his successes and failures; his most recent work has the provocative title *Collaborating with the Enemy: How to Work with People You Don't Agree with or Like or Trust*. He recommends 'stretch collaboration', an abandonment of all the "unrealistic fantasies of harmony, certainty, and compliance, and embraces messy realities of discord, trial and error, and co-creation". On this minimalist basis, along with plenty of patience and self-scrutiny, real progress can be made.

This supplementary set of heuristic ideas brings us back to the insight of E.F. Schumacher. With all our appreciation of the scientific quality of the economic analysis of Stanley Jevons, we see clearly how limited was his vision when he separated the 'utility' studied by economics from the higher values relating to human aspirations. At the most practical level, when decisions are being made about the allocation of scarce resources, there will inevitably be some who feel themselves winners and others who feel themselves losers. Unless there is some process of real reconciliation, which normally involves a transformation of consciousness on all sides, grievances, bitterness and renewed conflict will remain. This is why for true sustainability, dialogue and reconciliation are essential elements of economics.

Box 17.6   Joe Ravetz on Economics III

Economics III is about the pathways from a myopic and unthinking system, towards something (at least a little) more responsive, intelligent and even 'wise'. To map this space of opportunity we identify three main levels of system organizations:

'Mode-I' (1.0): linear and 'clever' economics: here the economy works as a 'functional system': it follows direct instructions and responds to direct pressure & short term changes.

'Mode-II' (2.0): evolutionary and 'smart' economics: here the economy works as a 'complex adaptive system', evolving by natural selection, competition, innovation and self-organization.

'Mode-III' (3.0): co-evolutionary and 'wise' economics: here the economy works on the cognitive level, with capacity for collective intelligence, and a cycle of co-learning, co-innovation and co-creation.

At present it seems that the global economy is incredibly sophisticated, but at its core runs on '2.0' extractive type logic, of 'animal spirits' and 'winner takes all' competition. This helps to explain how such an ingenious system can generate such stupid results, with financial crises, obscene social inequalities, and reckless destruction of its ecological life-support. So, we aim for a new angle on the economy, not just as 'stuff' (i.e. 'production') pushed around by animalistic consumer desire and investor returns, but as a kind of open-mind-scape. We can begin to nurture the seeds of a society-wide collective intelligence, with the capacity to organize real resources for real human needs.

This opens the door to a mode-III economics which is not only a 'dismal science' and myopic study of *'choices on allocating scarce resources to satisfy unlimited wants' in the words of Investopedia.* This forward looking economics is more an exploration of *"choices on synergistic & collaborative co-creation & co-production".* To do this we look beyond the 'animal spirits' of 'wild west' free markets and evolutionary economics, to the 'human spirits' of collaborative markets and social institutions. We look for ways in which the economic production machine and market jungle – very ingenious, but very problematic – could co-evolve towards an economics III, one which is fit for humans.

Source: Ravetz, 2018

How it is starting to happen

Many strands of economic thinking are now pointing in similar directions. We have complexity economics and systems thinking; relational economies and experience economies; new economics and green economics; evolutionary and innovation economics; institutional analysis and economics; behavioural economics; Buddhist and mindful economics; wiki-nomics and long-tail production; conscious business and cognitive capitalism; and post-industrialist or post-capitalism. Each of these raises questions and contradictions; each in its way reflects the signals of crisis and/ or opportunity (Ravetz, 2017).

To give an idea of the rich variety of scholarly and practical activities that are already being promoted under the programme of Ecological Economics, we reproduce the agenda of a meeting of the European Society of Ecological Economics (Box 17.7). This is only one of several flourishing societies; there is no doubt that the discipline has matured and can contribute strongly to the development of the necessary transformation of knowledge and practice for sustainability.

Box 17.7   Ecological Economics programme

Today's challenges require true engagement and novel solutions from Ecological Economics. Academic and practitioner communities must enact meaningful participative and mutually empowering activities across disciplines and different knowledge systems. Ecological Economics can contribute to generating inclusive and reflective research in a number of ways: as transformative science; as advocacy for non-human beings and future generations; as advocacy for environmental and social justice; as policy science; through understanding and promotion of broadly defined well-being; and through empirical insights and real-life impacts. Scientific and governance practices should be closely linked to the explicit spatial context of ecosystems and the biosphere, taking into account the needs of the non-human world. The 12th International Conference of ESEE aims to support this reflective and responsible turn in sustainability science in general and Ecological Economics in particular.

The conference will bring together diverse sets of actors who are engaged in co-producing Ecological Economics insights and advice for responsible and creative pathways towards sustainability. We seek to open up disciplinary boundaries through collaboration and discussion with conservation biology, environmental psychology and sociology, political ecology, social anthropology (amongst others), as well as through critical engagement and mutual learning with practitioners and local community efforts that aim to realise transformation towards sustainability. Novel socio-ecological insights and dialogues aim to encourage pathways to individual, collective and institutional change by virtue of collaboration, connection and meaningful knowledge-sharing through diverse expressions of human thought.

Source: ESEE, 2017

Conclusion

The variety of perspectives on economics that we have discussed here are a reflection of the evolution of society. Jevons was a far-sighted analyst of Victorian English society, but he shared its assumptions of individualism. Keynes reacted to the turbulence of the early twentieth century. Maslow and Schumacher expressed the new consciousness that took shape in the 1960s and 1970s, when 'ecology' became popular. As that matured, a new sort of economics, the 'ecological', was created. And now, in an age of complexity and uncertainty in all spheres, we are attempting to forge a further advance in economics as in other disciplines, based on a transformation of ourselves and the world, involving awareness, compassion and humility. A critical challenge is the creation of a system of representation of quantities whereby the deep complexity of economic affairs is effectively represented. Then the enriched perspective of transformative sustainability economics will be made effective for its many practical tasks.

Questions for comprehension and reflection

1   If Jevons is wrong, could we run an economy on Maslow's lines?
2   Oscar Wilde defined a cynic as "a man who knows the price of everything and the value of nothing". How would we define someone who knows the value of everything and price of nothing?
3   Can we really put a monetary value on environmental goods without marketising them?
4   How many significant digits are appropriate for the monetary valuation of a landscape?

References

Costanza, R., Alperovitz, G., Daly, H., Farley, J., Franco, C., Jackson, T., . . . Victor, P. (2013). *Building a sustainable and desirable economy-in-society-in-nature.* Canberra: ANU Press.

Costanza, R., Cumberland, J. H., Daly, H., Goodland, R., Norgaard, R. B., Kubiszewski, I. & Franco, C. (2014). *An introduction to ecological economics.* London: CRC Press.

Costanza, R., d'Arge, R., de Groot, R., Farber, S., Grasso, M., Hannon, B., . . . van den Belt, M. (1997). The value of the World's ecosystem services and natural capital. *Nature* 387: 253–260.

Costanza, R., de Groot, R., Sutton, P., van der Ploeg, S., Anderson, S. J., Kubiszewski, I., . . . Turner, R. K. (2014a). Changes in the global value of ecosystem services. *Global Environmental Change* 26: 152–158.

Earle, J., Moran, C. & Ward-Perkins, Z. (2016). *The econocracy: The perils of leaving economics to the experts.* Manchester: Manchester University Press.

ESEE. (2017). European society for ecological economics. *Ecological economics in action: Building a reflective and inclusive community.* 12th Conference.

Giampietro, M., Aspinall, R. J., Ramos-Martin, J. & Bukkens, S.G.F. (2014). *Resource accounting for sustainability assessment: The nexus between energy, food, water and land use.* London and New York: Routledge.

Jevons, W. S. (1879). *The theory of political economy.* London: Macmillan and Company.

Kahane, A. (2017). *Collaborating with the enemy: How to work with people you don't agree with or like or trust.* Oakland, CA: Berrett-Koehler Publishers.

Keynes, J. M. (1936). *The general theory of employment, interest and money.* A project Gutenberg of Australia eBook, eBook No.: 0300071h.html

Marglin, S. A. (2010). *The dismal science: How thinking like an economist undermines community.* Harvard: Harvard University Press.

McLeod, S. A. (2007). *Maslow's hierarchy of needs – Simply psychology.* Retrieved from www. simplypsychology.org/maslow.html

Monbiot, G. (2014). *Put a price on nature? We must stop this neoliberal road to ruin.* Retrieved from www.theguardian.com/environment/georgemonbiot/2014/jul/24/price-nature-neoliberal-capital-road-ruin (Accessed 14/04/2017)

Radermacher, W. (2006). 'The reduction of complexity by means of indicators: Case studies in the environmental domain', in OECD (Ed.) *Statistics, knowledge and policy: Key indicators to inform decision making.* pp. 163–173.

Ravetz, J. (1994). Economics as an elite folk science: The suppression of uncertainty. *Journal of Post Keynesian Economics* 17 (2): 165–184.

Ravetz, J. (2018). *Urban 3.0: Synergy for sustainable cities.* London: Routledge.

Rayner, S. (2012). Uncomfortable knowledge: The social construction of ignorance in science and environmental policy discourses. *Economy and Society* 41 (1): 107–125.

Schumacher, E. F. (1973). *Small is beautiful: A study of economics as if people actually mattered.* London: Abacus.

Stiglitz, J. E., Sen, A. & Fitoussi, J. P. (2010). *Report by the Commission on the measurement of economic performance and social progress.* Paris: Commission on the Measurement of Economic Performance and Social Progress.

18 Postscript

Heuristics for sustainability science

Jerome Ravetz

The challenge to transformative sustainability science

Transformative sustainability science can be seen to have emerged in response to fundamental challenges that relate to the world of established, conventional science. As is widely recognised now, conventional science is experiencing grave difficulties. The insights that define transformational sustainability science have emerged in response to recognition of these challenges, and as such they can help us to understand its problems and the way to their possible solution.

The crisis in mainstream science is seen as twofold. On the one hand there is a loss of credibility with the public, consequent on deep divisions within expert advice on crucial issues, as well as evidence of the manipulation of science by powerful vested interests (Goldacre, 2014). It is more difficult to defend science against such criticisms when it has been experiencing an internal 'replicability crisis' for some years (Ioannidis, 2005). That is, a significant proportion of published papers cannot be replicated, and therefore cannot be trusted. This results in part from a fragmentation of knowledge into sub-disciplinary silos, in the sciences as well as in the humanities, so that quality assurance is seriously impaired. On the other hand, there are also increased pressures on scientists to publish with 'impact' in order to advance their careers, which lead to practices that stress publicity over content (Saltelli et al., 2016). Of course, there have been many positive responses to these challenges at all levels within science (see also the framework for the conduct of ethical research in Chapter 2 of this book by Wals and Peters). But if the defects in scientific practice have risen from new structural features of the social activity of science, in the absence of deep structural change, their resolution remains problematic. These challenges can be traced to what has been called 'industrialized science' (Ravetz, 1971), characterized by huge increase in size of the total enterprise, scale of projects, and closeness to industry and commerce.

Transformative sustainability science is a new approach to science that takes into account these structural problems in the way that 'normal' or 'disciplinary science' is practiced and that institutionalised science cannot. It tries to guide the collaborative inquiry of scientific knowledge co-creation in such a manner as to overcome them. This is not trivial, as transformative science will in part have to emerge, interact, and mature while working partly within structures of traditional disciplinary science.

The label 'transformative sustainability science' has three parts, each meaningful in its own way. 'Science' seems the most familiar, but here we have been using it in a new sense. We do not imagine a lone discoverer exploring the secrets of nature. Rather, we have a process of social learning, achieving a co-creation of applicable knowledge intended for the solution of existential problems of living in the twenty-first century. Although detailed technical knowledge is an essential part of this sort of science, it is deployed by an engaged community in connection with their common problems. In this way our conception is closest to that of pragmatism in the American tradition. Then, 'sustainability' is what it is all about; we have learned that the concept itself grows in our understanding as we use it to guide our efforts. Finally, there is 'transformative'. Previously in our civilisation we have used science to transform our knowledge and control of the natural world around us. But we have imagined ourselves as standing outside the process, rather like in Descartes' phrase "masters and possessors of nature". Now we have come to appreciate the inadequacies and indeed dangers of that conception of ourselves and of science. Our engagement with sustainability issues has taught us that we ourselves need to be, and can be, transformed in the endeavour of sustainability science. This transformation arises from the work of co-creation, engaged on in a context of complexity, contingency, contradiction, and uncertainty and ignorance. Out of this experience we have learned the importance of self-awareness, and with that, of humility. This new understanding coheres well with the general awareness in society of the need of respect for nature along with other cultures and life forms. We are still at a very early stage of understanding and implementing this new conception of ourselves in the world. But we are sure that this is the way forward.

With this understanding of what we are about, we can ask: What are the challenges faced by transformative sustainability science and how will it meet them? There are the obvious challenges that confront any nascent social practice that aspires to wider adoption, not just within one but across very diverse and as yet bounded communities: survival, growth, and influence. Although all three elements of its title have familiar labels, together they define a concept that is new and demanding. One way to meet those challenges, along with gaining experience of teaching and practice, is to create our own stock of concepts and methods for guiding and assessing our practice. We will start on this task in this chapter, with a set of heuristics that together define this new sort of understanding and practice. Through living with them, our students and colleagues will enrich them and make them effective.

There are ways in which transformative sustainability science can offer insights and examples which can help mainstream science to understand and resolve its crises. First, there is the understanding of the objects of study and its goals. For us, the reality we study is complex, and our knowledge of it is contingent, situated, and varied depending on the perspective (shaped by expertise or interests) from which one looks at the problems, materials, and suggested solutions. This objective is fundamentally different from the traditional scientific objective of achieving timeless truths about nature.

The model for scientific discourse is not a simplifying mathematical analysis using quantitative data, out of which Truth somehow emerges. Rather it is a conversation among voices with a diversity of perspectives and interests. Our description of the debate on the Ecological Footprint (Chapter 15) is a good example of open, productive criticism, which is welcomed by leading statisticians and integrated in study programmes in higher education. Reflective practitioners know that no scientific result that depends on statistical inference can be properly described as 'true' or 'false'. Rather, it is only confirmed to a certain degree of 'probability', and that is usually accomplished by a sophisticated or even obscure process of reasoning and calculation. Indeed, in an age when 'p-hacking' is a very common technique, the traditional invocations of a Scientific Method can seem irrelevant or misleading. These insights need to be made common knowledge for competent citizens if we would like to work for a greater engagement in civic and citizen science.

Transformative sustainability science also offers a more realistic and fruitful image of the inquirer. We have not been conditioned by the traditional style of expression of school laboratory science, where the 'third person passive' is used to describe the actions of a disembodied agent. For us, research, practice, learning, and self-awareness are indivisible. Contradictions are not obstacles to knowledge, but are challenges for learning and growth. Again, in the practice of research in mainstream science, the contemporary researcher is constantly confronted with the need to make judgements about quality and ethics. There is no scientific rule that prevents cheating in science; in contemporary conditions, integrity is a hard-won virtue. The new generation of scientists is discovering that science does not provide a haven from the ills of society. Transformative sustainability science is showing how the messy world out there can be a locus of creative, indeed transformative, work by understanding scientific inquiry both as an expression of prevailing values and as a forum for their examination.

Heuristics

Transformative sustainability science is still taking shape, defining itself in every way. We need some aids to learning and understanding. For our purposes, the concept of 'heuristics' is quite fruitful. These are particular means or methods for organising information and action. The word comes directly from the classical Greek, where it means 'discovery'; thus, Archimedes shouted 'Eureka' to celebrate the discovery of 'his' principle in hydrostatics. So a 'heuristic' can be anything from a conceptual framework to a suggestive name. In mathematics, the line-and-circle of classical Greek geometry and Descartes' symbol for the unknown, x, were both powerful heuristics that shaped our civilization. Our understanding of contemporary science has been enhanced by the heuristic concept of 'paradigm' created by Thomas Kuhn to define what he called 'normal science'. Building on that there is the contrasting heuristic of 'post-normal science' which leads directly to sustainability science as we understand it. In economics, the graph of two intersecting curves representing 'marginal utilities' defines the mainstream theory, and a graph of 'stocks and flows' defines ecological economics. So the term is very general, and we use it to illustrate a very varied practice rather than to legislate for a particular approved methodology.

Our four general heuristic themes are complexity, contingency, contradiction, and uncertainty and ignorance. There is a fifth heuristic, a device we call the 'surprise checklists'. We will deal with them in turn.

Complexity

Somewhat paradoxically, we can say that 'complexity' is a good example of itself. Like any powerful organising idea, it has roots in a variety of forms of practice, each with its own history, techniques, and social organization and ideology. None of them is uniquely 'correct'; provided we have respect for other approaches to some extent we can play Alice's Humpty-Dumpty and choose our own definition. For clarity, we might even speak of 'deep complexity' to distinguish it from mere complication. That traditional concept is characterized by a large number of independent variables; by contrast this has irreducible structural elements. Complexity as we understand it is about certain sorts of systems. In these every element has systems above, below, and on the same level, related by any possible relationship: scale, inclusion, control, etc. Every system is then itself simultaneously a sub-system, a super-system, and a co-system. In consequence, there is no privileged perspective for an analysis of the whole system, which must be accepted as 'basic' or uniquely privileged. Nor is there a simple, linear causality among elements of the sort that is assumed in mainstream economics and in classical (but not contemporary) physics. Accordingly, there is no possibility of conclusive knowledge of the total system (Ravetz, 2006; König, 2017; Sonnleitner, 2017; Newell & Proust, 2017; Davila & Dyball, 2017).

The recognition of complexity can be a liberation; the scientist no longer needs to suppose that a reductionist (usually mathematical) scheme of analysis will yield the unique correct answer to their problem. Rather, the way is open to an appreciation of the variety of relationships among engaged parties and the adoption of a social learning, co-creating mode of inquiry. Most important, the legitimacy of diverse perspectives on the situation, even including those that are incompatible or in conflict, is provided with a scientific foundation. And solutions to problems are understood not as final, but rather as stages in a co-evolutionary process. An extreme case of complexity is expressed in the concept 'wicked problems', where everybody knows that there is something wrong but agreement stops there; the remedy is in 'clumsy solutions' aiming for a minimalist outcome (Verweij & Thompson, 2006).

Contingency

The next concept, the second 'C' in our list, is Contingency. In some ways this is of the most direct importance for our readers, for it requires them to do the most unlearning. There is an assumption that in order to be real, knowledge must be universal and permanent. This assumption is deeply embedded in our intellectual culture, so much so that we scarcely recognise it. For modern European civilization, the clarity and objectivity of mathematics sets it aside from all other sorts of knowledge. From Descartes onwards, mathematics has stood proud in contrast to the other disciplines; for him theology was irrelevant, philosophy futile, the humanities full of uncertainty, and the practical disciplines merely crafts. One of the most chilling passages in our intellectual tradition is the section of Descartes' *Discourse on Method* (Part 1) where he first praises and then assassinates the humanist curriculum. Then by a miracle he discovers geometry: clear, certain truth. Many practitioners who nowadays receive emotional security from the belief that their spreadsheet will tell them precisely what to do with a project or company are living with the consequences of Descartes' desperate grab for certainty.

We see the same phenomenon in the vast majority of university teaching programmes in the 'decision sciences', with economics setting the paradigm. There students live in a world where general theories are assumed to be both true and effective in practice. They are exemplified in exercises using the names of real-life situations, but where every problem has a unique correct solution, precise to three digits. Unknown to these learners, at least until they enter the messy world of practice, there is a tradition of passive resistance by those whose job is to keep the wheels from coming off the projects defined by those abstract disciplines. It is best documented in the 'Murphy's Law' literature. It is noteworthy that knowledge of how things can go wrong is generally relegated to collections of jokes. Even in what are called 'applied sciences', it is exceedingly hard to find university courses on how things can go wrong and what to do about them. There is a literature on these issues with headings like 'perverse incentives', 'unanticipated consequences', and (notoriously) 'tragedy of the commons', but those discussions relate more to social policy than to economic or organisational activity.

Another perspective on contingency is 'situated knowledge'. Again, this concept is rarely found in textbooks; but almost by definition, any knowledge that is actionable in any way will relate to a particular situation rather than to relationships among abstract concepts. Contingency reminds us that work on transformative sustainability projects will, above all, attempt to integrate and reconcile local, experienced knowledge with the general and abstract knowledge of the experts. There is a great challenge for mutual social learning; neither is sufficient by itself. 'Citizens' need to learn how to master the relevant technicalities of the scientific disciplines that will contribute to their framing of the problem, as well as the technicalities relevant to crucial areas of debate. On their part (and perhaps more difficult) 'scientists' (or experts) need to learn how to respect and make use of knowledge that is situated, experiential, and contingent.

Contradiction

Next we have 'contradiction'. Its history is not merely conceptual but also political, because Marx and Lenin made use of it in their writings. There is actually a theological dimension to this very complex concept, as it relates to creative change in the world. The classic study *Main Currents of Marxism* by Leszek Kolakowski (1978, 2005) starts, surprisingly, with a study of mystical writers of the classical, Medieval, and Renaissance worlds. Placing rational knowledge in the context of the super-rational and giving it a divine-developmental dimension lays the foundation for a dialectical, rather than a purely logical, conception of knowledge. Hegel, emphasising the creative aspect of contradiction, is the link between all that and the later 'dialectical materialism' that became notorious with Stalin. In that later tradition, a contradiction was the driving force that led to a transformation. For Hegel, it was conceptual (as between 'thesis' and 'antithesis' leading to 'synthesis'); for Marx's material dialectics, it was about social structures and power.

There is no sharp distinction between a complementarity, a tension, and a contradiction. We see in the biosphere how a commensual species can perform as a symbiote, a challenger, or a pest or pathogen (as do so many of our 'germs') depending on circumstances. And it can depend on circumstances whether the prey species can show sufficient resilience to meet the challenge, or if not then to die, individually or collectively. The 'panarchy' pattern of Holling (Gunderson & Holling, 2002), with its 'lazy-eight' cycle, can be seen as an example of a contradiction. Once the situation has gone into an excessively stable 'conservation' phase, it cannot smoothly evolve to meet its challenges. In the panarchy model it is fated to go into a sudden 'release' like a forest fire, which will be destructive or creative depending on your point of view. Our own conception of contradiction is that of a challenge to a system that cannot be resolved within its own terms.

The main use of contradiction in social theory has, of course, been in economics. Marx is the classic case, but the transformations of technology and its social matrix have also been studied, as by Kondratieff and Schumpeter. For ourselves, the 'contradictions' in the social system do not manifest so much in business cycles, as in the dysfunctional relations of the technological system with its context. Many of the challenges of the 'Anthropocene' can be understood as resulting from what can be understood as 'perverse incentives' of the economy and government that are very deeply embedded in the system. For Marx, the fundamental contradiction of the capitalist system was that between social production and individual appropriation. For ourselves, it can be that between the ecological burden of our 'affluent' society and the political imperative for ever more 'growth'. It is exacerbated by the post-colonial political requirement to share some of these unsustainable riches with the world's poor. And that is coupled with the contradiction of the need of the rich to export a uniform high-tech material culture in the interests of their own stability against the need for appropriate technologies in the 'majority' socio-technical-cultural milieux. In these ways the concept of contradiction can be very useful for clarifying the tasks of sustainability (Ravetz, 2006).

Uncertainty and ignorance

Now we come to the broadest and most ancient of our heuristics. Surprisingly, this has the deepest connection with sustainability. It goes back to Socrates. He is quoted as saying that his whole life was devoted to discovering his own ignorance. But this ignorance was not absolute; we might say that awareness, or rather self-awareness, is a special sort of knowledge, and it was crucial for Socrates. He said, "The unexamined life is not worth living". How does this principle relate to sustainability? It is a corrective, against all the pressures to desire scientific knowledge that is guaranteed to be true and effective, and further to believe that our particular sample is of just that sort. All those who depend on our work in some project will wish us to have success in our scientific studies and practical efforts. Who wants to be reminded of errors and failures? Yet projects for sustainability are, by definition, explorations into a hazardous unknown, where pitfalls abound. If we are not alert for things going in unexpected directions, we may come to realize them too late, when there is already a personal and institutions investment in the erroneous course. And most of our training in conventional science leads us away from this essential awareness.

There are well-developed sciences devoted to the management of uncertainty, which provide essential tools for the handling of empirical data. But none of these can eliminate uncertainty and ignorance, and it is all too easy to gain a false sense of security from these sophisticated techniques. Awareness of one's own ignorance also enhances the appreciation of the knowledge of others. This is particularly important when those others lack the technical knowledge that has hitherto seemed to define the problem and solution. This awareness has recently been described as a 'technology of humility' (Jasanoff, 2003). We could also call it 'transformation'.

The 'surprise checklists': images and devices

Finally, we introduce another heuristic, this one being more a tool than a concept. Checklists are a very common means of organising information and making it more vivid and comprehensible. Of course, they are always caricatures of a complex reality, but they can still be powerful caricatures.

The two checklists given here enable us to characterise the ways in which things can have 'unanticipated consequences'. They help us to anticipate events that would otherwise be surprises, usually unpleasant. That's why we call them 'surprise checklists'. The first relates to images of any sort. We can say that each image will simultaneously reveal a reality and suggest new possibilities, but also distort what lies behind and finally conceal some elements.

The easiest example is an x-ray, which shows solid structures and enables huge advances in diagnostics, but also displays only shadows of lesser or greater intensity and fails to show details of softer tissues. Other imaging techniques remedy some of these defects, but all introduce others of their own. That's why radiographers use a variety of imaging techniques, depending on the diagnostic problem. Another useful example is a map. If a map didn't reveal something usefully and reliably, it wouldn't be worth much. And by the interpretation of its images, the map enables all sorts of things that wouldn't be possible, or even conceivable, otherwise. Yet every map, particularly those where the sphericity of the earth is relevant, will necessarily distort. A comparison of the various projections is very instructive; somehow most of them unavoidably make Greenland as big as South America. And the concealment of many features on the surface is the price that any particular image pays. So is a map 'true'? Reflection on that question can help us appreciate that question in connection with science.

The other checklist relates to tools, or more generally devices. For these, we have intended use, creative new use, misuse from incompetence, and abuse from malevolence.

This checklist is particularly useful nowadays, when it becomes increasingly necessary to keep in mind that things can go wrong. Over previous generations there had been a comfortable assumption that when an invention has been made and applied, all will be well. We had learned that weapons of war could sometimes do more damage than intended, such as poison gas in the First World War and atomic weapons in the second. But it was only with Rachel Carson's book *Silent Spring* in 1962 about pesticides that the public became aware that even the most well-intentioned technology could be harmful. Since then, awareness has grown, but the general tendency is still to 'accentuate the positive' and hence to 'eliminate the negative'. Even with patently dangerous or vulnerable technologies like the 'Internet of things' or genetic engineering in the wild, the burden of proof is still on those calling for caution. The relation to sustainability needs no elaboration.

Conclusion

The usefulness of this set of four attributes for organizing teaching and research in transformative sustainability science is attested by the study programme in 'Sustainability and Social Innovation' at the University of Luxembourg. This was established in 2013, based on prior work in developing the individual courses that started in 2009.

Of course, there are potential weaknesses. As the 'surprise checklist for images' indicates, the possibility of incompetent misuse and malevolent abuse is always present. We can hope that under the conditions of transformative sustainability science, with social learning through open dialogue, such defects will be controlled better than in many other contexts of use.

Questions for comprehension and reflection

1 How can transformative sustainability science meet the challenges of trust and quality that now confront the mainstream sciences?
2 What are some examples of situations that are characterised by one of the four heuristics?
3 What are some examples of a surprise checklist for images and for devices?

References

Carson, R. (1962, 2002). *Silent spring*. Boston and New York: Houghton Mifflin Harcourt.

Davila, F. & Dyball, R. (2017). 'Food systems and human ecology: An overview', in König, A. (Ed.) *Sustainability science: Key issues in connecting learning, research, and practice*. Abingdon: Routledge. Chapter 10, pp. 183–210.

Goldacre, B. (2014). *Bad pharma: How drug companies mislead doctors and harm patients*. London: Macmillan.

Gunderson, L. H. & Holling, C. S. (Eds.) (2002). *Panarchy: Understanding transformations in human and natural systems*. Washington, DC: Island Press.

Ioannidis, J.P.A. (2005). Why most published research findings are false. *PLOS medicine* 2 (8): e124. doi:10.1371/journal.pmed.0020124

Jasanoff, S. (2003). Technologies of humility: Citizen participation in governing science. *Minerva* 41: 223–244. doi:10.1023/A:1025557512320

Kolakowski, L. (1978, 2005). *Main currents of Marxism: The founders, the golden age, the breakdown*. New York: W.W. Norton & Company.

König, A. (2017). 'Systems approaches for transforming social practice: Design requirements', in König, A. (Ed.) *Sustainability science: Key issues in connecting learning, research, and practice*. Abingdon: Routledge. Chapter 3, pp. 55–81.

Newell, B. & Proust, K. (2017). 'Escaping the complexity dilemma', in König, A. (Ed.) *Sustainability science: Key issues in connecting learning, research, and practice*. Abingdon: Routledge. Chapter 5, pp. 96–112.

Ravetz, J. (1971). *Scientific knowledge and its social problems*. Oxford: Oxford University Press.

Ravetz, J. (2006). Post-normal science and the complexity of transitions towards sustainability. *Ecological Complexity* 3 (4): 275–284.

Saltelli, A., Ravetz, J. & Funtowicz, S. (2016). 'Who will solve the crisis in science?', in Benessia, A., Funtowicz, S., Giampietro, M., Guimarães Pereira, Â., Ravetz, J., Saltelli, A., Strand, R. & van der Sluijs, J. P. (Eds.) *The rightful place of science: Science on the verge*. Tempe, AZ: Consortium for Science, Policy & Outcomes. Chapter 1, pp. 1–30.

Sonnleitner, P. (2017). 'Cognitive pitfalls in dealing with sustainability', in König, A. (Ed.) *Sustainability science: Key issues in connecting learning, research, and practice*. Abingdon: Routledge. Chapter 4, pp. 82–95.

Verweij, M. & Thompson, M. (Eds.) (2006). *Clumsy solutions for a complex world: Governance, politics and plural perceptions*. Basingstoke: Palgrave Macmillan.

19 Outlook

Citizens and science in the Anthropocene

Ariane König

The challenge: competent citizenship and reflective science for purposeful collaborative inquiry in the Anthropocene

The Anthropocene, the age in which humankind meets planetary boundaries, calls into question prevailing social practices and ways of knowing, including practices of how we produce science and technology (Maggs & Robinson, 2016). Transforming social practices such that they match planetary limits will require new ways of co-creating knowledge, social practices and technologies (Wiek & Lang, 2015; Schneidewind et al., 2016; Grunwald, 2016), as well as re-inventing how we conceive of citizenship at the local, national and planetary scales (Giddens, 2009). The transformation of society and the associated learning process can be conceived as complex process that brings along changes across different levels of social organisation, including in the personal, cultural, organisational, institutional and systemic spheres (O'Brien & Sygna, 2013). Transformation can thus be considered a social process with a Cognitive dimension that opens up new human potential for reconsideration of how we relate to ourselves, others and our environment. In such a process technologies can be conceived as an expression of prevailing values and worldviews. The recognition of how prevailing values can serve as ordering principles for attributing attention and resources at different levels of social organization (individual lifestyle decisions to whole nations or regions) should be an integral part of research and technological design and development. The networked society of the twenty-first century offers as yet untapped processes for co-design and co-creation that aim to better understand and express shared values in pluralist societies. Such processes for co-creation of new knowledge and technologies in processes of collaborative inquiry and design assume not only profound changes in how science may be practiced (see also Chapter 18 by Ravetz), but also a new notion of competent citizens.

In the remainder of this chapter we first explore in more detail what rights and responsibilities may be associated with citizenship in the Anthropocene, before we revisit implications for science in the Anthropocene as detailed in the various chapters of this book. The conceptual and methodological tools presented in this book have been selected as we consider them as particularly useful for the 'Anthropocene' – eventual drastic changes in our environment whose consequences are unpredictable. And with ecological awareness, the elements of the theoretical framework include layers of context, as society, culture and environment. Consideration of values, ethics and long-term consequences are integrated into the analysis. With transformative sustainability science, our theoretical framework includes ourselves as individuals engaged on social learning and personal growth. For this, it embeds us in a world of deep complexity and of uncertainty and ignorance. Subsequently, we present a research project designed to draw on a large range of conceptual tools and methods presented in this book, to engage citizens in collaborative inquiry for changing social practice for sustainable water governance. We conclude with an outlook for the practice of transformative sustainability science.

What is citizenship in the Anthropocene?

We conceive of citizenship in the Anthropocene in five dimensions. Building on Giddens (2009, p. 198) we consider civic, political, social and ecological citizenship, but we also add the fifth dimension of digital citizenship. Each dimension of citizenship holds rights and responsibilities with respect to contributing to a sustainable stewardship of our planet. *Civil citizenship* includes the respect for mutual property rights. Citizens can foster sustainability, for example, by holding producers or firms to account by engaging actively and critically in participatory democracy. *Political citizenship* includes, for example, active democratic participation by the exertion of voting rights and free speech. This in turn requires making and acting upon judgments on policies and possibly also seeking opportunities for participation in policy making. In order to foster sustainability, citizens can not only vote based on their judgments on political action plans for sustainability, but also engage for environmental and social justice, for example, by joining a social movement concerned with fair distribution. *Social citizenship* includes rights and obligations for collective provision of welfare and social benefits. Along with Giddens we add the fourth, less conventional dimension of *ecological citizenship* (Smith, 1998), which presents us with "new obligations to non-human animals, future generations of human beings, and maintaining the integrity of the natural environment. This requires a transformed human experience of nature and the self as tightly bound together" (Giddens, 2009, p. 198).

Last but not least, in this book we consider a fifth dimension – *digital citizenship* in a networked society also brings with it new sets of rights and responsibilities that will have profound implications on how we can cope in the Anthropocene. Wals and Peters in Chapter 2 highlight the need to reconsider prevailing ways of social coordination in democracies in the twenty-first century, including the role of social media in networked societies and its relation to the production and use of science and 'facts' in electoral politics. Formalizing and embracing new sets of rights and responsibilities as digital citizens may help to distribute responsibilities to contribute to and verify the digital commons and how it is ethically used for the common good and for coping in the Anthropocene. Digital citizenship could include, for example, rights of accessing quality information on the Web (for example, Wikipedia) and tapping into collective intelligence in virtual spaces (such as for the co-creative development of the Linux operating system described in the book *Reinventing Discovery* (Nielsen, 2012). Responsibilities include managing your digital footprint (information you place on the Web about yourself) and your digital shadow (what others place on the Web about you) and contributing to the respect of security and privacy (Negroponte, 1995). Resources such as Wikipedia will not work when only tapped; contributing to building the science commons will only work if it is a truly co-creative effort, to ensure the credibility and quality of information by contribution of one's micro-expertise where it matters, for verification and improvement. Some thought leaders even predict that judging and serving as a jury of peers will be something expected and drawn upon as part of a web-based commons (Schmidt & Cohen, 2013). Reporting in the digital age is made easy, be it on peers and perspectives, distributional issues or states and flows in the environment.

Citizenship in the Anthropocene may even be defined to include rights and responsibilities that combine those of ecological citizenship and digital citizenship described earlier, if we would like to become truly democratic and rely on distributed knowledge co-creation about the state of the environment and changes in social practice and how they relate to each other. In Chapter 3 we have learnt how vital new information flows and learning can be in designing balancing feedback regulatory loops – for example, in growth-driven runaway industrial resource- and energy-using material flows. Should citizens assume the responsibility to engage in collaboratively driven, participatory-sensing activities in order to contribute to regulatory processes towards a healthy earth system? How might we all contribute to co-creating situated knowledge on changes in the environments we are part of in order to understand local complexity and lived experiences in different groups? Actively and critically engaging with associated rights and responsibilities requires the capacity for self-directed learning also about civic affairs and competencies for effective political action, co-sensing and acting locally. What social processes and technologies are suitable? What are the social and technological challenges? These are, in our view, core research questions for the future of civic and citizen science in networked societies.

Science for the Anthropocene: methods to engage in systems- and future-oriented dialogues across different expertise and interests to transform social practice

The basic premise of this book is that in the Anthropocene, nature becomes a task for culture, not as something to dominate, but as a source of ways to organize life in view of boundaries, for transforming our societies' aspirations and lifestyles for their sustainability, in diversity and longevity. The emergence of new sociotechnical imaginaries,[1] based on a sophisticated understanding of our conceptions of the social, natural world and how they interrelate, will play an important role in transformative science and our ability to assume new responsibilities in view of contributing to shaping more desirable futures (Jasanoff, 2015). Most, if not all, chapters in Part I of this book on conceptual tools and methods provide examples on how such transformative dialogues can be structured and guided that promise to result in new powerful ideas, metaphors or prototypes from co-design that can serve to seed effective changes in social practice. Of the three chapters on understanding complex social-ecological-technological systems, a particularly effective approach towards this goal is described in Chapter 5 by Newell and Proust. Collaborative conceptual mapping of complex dynamic systems has as a main goal the production of simple, low-order models with a limited number of variables to characterize their relationships – one main outcome this research approach aims for is powerful metaphors that inspire concerted action in the face of shared challenges (Newell, 2012; Newell & Doll, 2015). Similarly, methods outlined in Chapter 6 by Drenth and colleagues to explore alternative futures are drawn upon in projects like Robinson's scenario projects (Vervort et al., 2015). The key is drawing together sophisticated understandings from natural and engineering sciences with a greater role for social science and humanities to explore human agency in diverse actors groups as creatively as possible (Maggs & Robinson, 2016; Castree, 2016; O'Brien, 2015). Based on such an understanding, human-centred co-design methods described in Chapter 8 by Gericke et al. then can serve to innovate, with the goal of developing prototypes for social practices and technologies that serve to express as well as live particular sets of values.

In this book we also point to citizen science as one more answer (Wals et al., 2014; Wals and Peters, Chapter 2). For example, Muki Haklay's definition also describes a new approach to science which aims to produce situated knowledge, embracing diverse values and worldviews, with the goal to transform: "Extreme Citizen Science is a *situated, bottom-up practice* that takes into account local needs, practices and culture and works with broad networks of people *to design and build new devices and knowledge creation processes that can transform the world*".[2] Again, the emphasis is also on the learning collective that produces a whole that is greater than the sum of its parts. For socially robust solutions, governance processes benefit from the input of diverse and conflicting viewpoints and interests. Citizen and civic science projects such as those by Haklay's group or described in Chapter 14 by Tran and König, and in the fifth section of this chapter, have been credited with a broad range of benefits (UBA, 2016). These include the more meaningful engagement of citizens when they are empowered and equipped to monitor data about their own environment. Also, there is an enhanced understanding of the nature of scientific knowledge and of the meaning of data (validity and reliability) when citizens are actively engaged in scientific inquiry. Capacity building of local expertise brings the discovery of how easy and quickly one can become an expert in a specific issue in their own local environment. Access to cheap information communications technology (ICT), with enormous monitoring and storing capacity, makes 'doing science' easier and more affordable. Enhanced understanding of complex systems by monitoring social, technological and environmental change in parallel for reflection about complex systems and how to better act upon them is in reach now. Last but not least, by self-monitoring the impact of one's own actions, the citizen can become more reflective and effective in bringing about change. These two chapters open up new spaces for knowledge co-creation in citizen science approaches with more distributed roles. This, however, can also be seen as problematic if science is seen as a 'certified' knowledge. An essential part of the transformative social learning that takes place in sustainability science, including economics, is that participants come to recognise, and come to terms with, the limitations of their own knowledge and the constraints of their own ignorance. This has been referred to by Sheila Jasanoff (2003) as 'technologies of humility'.

Chapter 17 by Manhart and Chapter 2 by Wals and Peters are key on how to conceive of learning in such social processes, aiming at dialogue across differences and learning we posit in this book as the ultimate goal. The design of powerful and engaging virtual learning environments will play a key role on whether this societal project will bear fruit and become scalable and extend its reach and influence, or whether it risks being nipped before it can blossom (Medema et al., 2014). The Internet and the networked society provide a learning environment that changes constantly through changing and dynamic participation. Meaning-making in such an environment relies fully on engagement in a social process that hopefully is not just taking place in the virtual realm but has complementary spaces in localities and social institutions. Emerging technologies shape the collective nature of participation with these media and reinforce peer-to-peer learning. Transformative learning by blogging is becoming part of everyday life, as shown in the following quote by a pupil: "I blog to learn about my views and how others receive them and take a different stance".

Science for the Anthropocene: heuristic tools

In this book, we present diverse approaches for co-creating meanings of sustainability in ways that draw on science and serve to transform social practice – in particular, sectoral challenges. Each chapter presents a set of conceptual tools in some cases, with a case study in which a particular approach to scientific analysis has contributed to this goal. Heuristics are particular means or methods for organizing information and then taking action; they illustrate a varied practice rather than aiming to legislate for an approved methodology. The chapters in Parts II and Part III of this book have provided a variety of heuristics. Some are expressed as graphs, others as lists or tables and still others are conveyed discursively in texts. In every case, they help us to organize our experiences and understandings so as to make them more effective. It is clear that behind every heuristic there is an assumption about reality.

In Chapter 9 by Dendoncker and Crouzat the concept of ecosystem services is put to use in research in human geography by focusing attention on development trajectories and path dependencies by looking at past agricultural revolutions. An associated research project embedded in a municipality has proven an effective space and process for directing attention to systemic interactions and re-framing values attributed to ecosystem services in an effective manner. The ecosystems services concept directs attention at human values and at environmental change and their interdependence by design. The question framed largely from economics on the internalization of externalities such as environmental impacts from diverse agricultural practices on ecosystems looks at 'values' and 'measures', but also what knowledge is relevant to think about emergent futures by starting with a historical analysis of past drivers of change. Chapter 16 takes up the question of economics and fundamental assumptions and conceptions as embedded barriers to systemic thinking and revisits the merits and limitations of the concept of ecosystem services under these aspects.

Chapter 10 on food systems by Davila and Dyball uses a graphical framework from human ecology (see also Dyball & Newell, 2015) to identify relations between environmental changes, human health and quality of life, prevailing cultures and powerful institutions. This framework is remarkable in having very few elements and no quantification beyond ±, and yet it is capable of illustrating quite complex situations. The framework was designed to be adapted for use in association with the visioning process and proved effective for fostering dialogue across differences, even in workshops with illiterate farmers and agronomist experts and policy makers in the Philippines. The use of this heuristic in this chapter shows by example that precise quantification of complex situations is not necessary for an effective analysis, and it opens up the possibility that it is not feasible anyway. This insight could be quite transformative of practice in studies of sustainability, the environment and more. In this respect, our approach is implicitly challenging the dominant assumption about reality and our knowledge of it. Each of these chapters provides salient tools and advice on how to implement methods that allow for the co-creation of at least one or several of the four types of knowledge in diverse groups.

Chapters 11 to 13 on energy transition focus on governance and technological change, also with reference to the Dutch conception of the multi-level perspective as heuristic for sociotechnical transitions. This conceptual tool helps to relate transition events across different scales and levels of social organisation. Limitations include necessary choices of boundaries (the focus is often on a 'system' at the national scale and on changes in the 'systemic structure'; research rarely manages to include a diversity mapping of felt experiences of diverse actors). Empirical studies based on this heuristic often used post-hoc analyses, in which, arguably, data can be selected in terms of their good fit to make a particular point of interest using this heuristic (how you define boundaries will direct attention and allow to develop a message and research insights by design).

Chapter 15 by Ravetz et al. on tracking change and evaluation in turn problematizes how we can learn and the need for tracking change with numbers, as well as complementary representations, changes of which are analysed over time.

The Luxembourg Nexus project as example of a project designed according to the guidance in this book

With the growing realisation of planetary limits and systemic interactions, several so-called 'Nexus projects' that seek to co-create knowledge on interacting water, food and energy systems, and considering their social, ecological and technological dimensions, have sprung up over the last five years (see, for example, Abson et al., 2017; Stirling, 2015).[3] In Luxembourg a similar project has been developed in parallel, which was designed based on the exact conception of transformative sustainability science presented in this book.

The project's overarching goal is to contribute to reconfiguring the science–policy–practice interface relating to the governance of water and food systems in Luxembourg; interactions with the energy system are also considered, but fewer resources can be attributed to more detailed research on the energy sector in the initial phases of the project. The policy and regulatory context for water systems is one of the most forward-looking sets of policies and laws in the EU. Related EU law creates an institutional openness to more transformative approaches to the practice of science, thus reducing potential barriers and offering pre-existing institutional arrangements for collaborations across different interests and expertise, such as river partnerships. The European Water Framework Directive (2000/60/EC) recognises that in view of the growing complexity, new approaches to water governance and knowledge processes informing water use are required. The definitions of 'water quality' and associated standards now include a wide range of human considerations beyond science. The law requires involvement of stakeholders in water governance, including citizens, at the EU, national and local levels. Related Sustainable Development Goals of the United Nations and associated targets and measures adopted as part of the Agenda 2030 also invite innovative governance approaches based on new forms of collaborations between diverse stakeholders, including public authorities, enterprises, research scientists and citizens. In Luxembourg, the EU Directive was transposed to national law in 2008 (Loi du 19 décembre 2008 relative à l'eau); it presents a legal basis for five river partnerships in which stakeholders make contractual commitments to improve water governance (www.flusspartnerschaft.lu). Government plans for adapting and implementing Agenda 2030 are being drawn up. Vision 2020 of the European Statistical System calls for the generation of data and statistics from more diversified sources.

Accordingly, the project aims to build communities engaged in more sustainable water governance that are networked across three spatial scales, including the level of the individual river partnerships, the collaborative development of futures at the national level, and the development of a citizen science tool in collaboration with the Non-Governmental-Organisation Earthwatch at the international level. Each community will have its set of indicators with associated quality criteria. The minimum set of indicators used across all communities are the technical indicators of phosphate and nitrate levels as well as turbidity and stream flow of the international fresh water watch project (more akin to contributory citizen science; see Chapter 14). The lower levels will be more co-designed sets of data for knowledge co-creation processes based on joint framing of what matters most and what might be actionable knowledge for each of the communities in Luxembourg. The national indicators will be a mix of mandated EU data and Luxembourg-specific indicators emerging from the scenario process.

More specifically, a series of workshops will engage stakeholders in two river partnerships to explore systems dynamics, as well as policy makers and experts on water and agriculture, to explore futures with a systems point of view at the national level. The series of workshops on systems and futures based on methods described in Chapters 5 and 6 of this book will then also serve to inform the co-design process of a citizen science tool and associated sets of indicators for monitoring changes in social practice, technologies and the environment. The tool will be structured with an agreed indicator set for participatory monitoring to create actionable knowledge for improved water governance.

The project serves to develop methods, including conceptual and computer-based tools, to structure social learning processes for transformative change for sustainability, with a focus on water governance. The methods to be further developed in the projects include collaborative conceptual systems mapping with the aim of developing simple low order conceptual systems models as basis for a shared understanding amongst diverse stakeholder groups. One main goal will be to generate shared representations of systems dynamics, including interdependencies between social, technological and environmental subsystems. Such shared representations on what matters most to present stakeholders will help to identify jointly feedback loops, reasons for 'lock-ins' in unsustainable social practices and leverage points for policy-making and changes in social practice as a basis for future concerted action and possible monitoring initiatives with the citizen science tool. Systems methods will build on approaches described in this book in Chapter 3 by König, Chapter 5 by Newell and Proust and Chapter 10 by Davila and Dyball.

A variety of methods will be used to collaboratively explore futures. Water governance and food production both today reveal clear interdependencies of how actions in the present can co-shape or restrict future spheres of activity in most places in the world, including in Luxembourg. The role of diverse ways to expect and conceive of different futures has a key role in arguments and motivations to change practice in the present (Grunwald, 2016). In particular, scenarios and visioning serve to structure collaborative processes for future-oriented explorations of systems and values, to better understand and discuss from diverse perspectives opportunities to collectively shape the future or constraints to future fields of actions. The scenarios will serve to explore alternative futures, risks, uncertainties and possible surprises. A vision, in contrast, is a normative collaboratively developed desired future for joint orientation and giving a direction to changes in social practice. The two methods are complementary, as visions without scenarios easily miss out on potential risks, uncertainties or threatening cliffs that might prove avoidable. Diverse methods for collaboratively deploring alternative futures that we will draw upon this project are described in Chapter 6 by Drenth et al.

The previous methods of structuring stakeholder dialogues to explore systemic relationships and alternative futures will inform the co-design of a citizen science tool for participatory monitoring and representing system dynamics and feedbacks as the basis for concerted action by stakeholders (see also Chapter 14 by Tran and König). The citizen science tool kit and associated indicators for monitoring and representations of data via web tools will be designed to service all communities based on insights from the collaborative mapping of systems dynamics. The scenario process for future orientation will potentially point at indicators relating to strategically important leverage points to change undesirable trajectories, taking account of interdependencies and feedbacks between social, technological and ecological changes. Co-design approaches for the citizen science tool and database structure and functionalities will be based on approaches described in Chapter 8 by Gericke et al. Co-design approaches for the citizen science tool and database structure and functionalities will be based on approaches described in Chapter 8 by Gericke et al.

The project will also develop new approaches to document and evaluate transformative learning, including based on assessing changes in communication and behaviour at the individual, organisational and systemic level that can be associated with the engagement of diverse stakeholders in these processes. The documentation and analysis of different discourses in diverse groups, areas of agreement and contradictions based on discourse analysis will play a central role in better informing judgment on acceptable and feasible actions. The conception and evaluation of learning will be that presented in Chapters 1 and 2 by Wals and Peters, and Chapter 17 by Manhart.

We will foster and evaluate learning from collaborative conceptual systems mapping and scenario development at the individual, organisational and systemic level in all stakeholders who engage in this process. Evaluation of impacts and outcomes of transformative research projects will include documentation of:

- **Changes in communication** indicating changes in 'expectations', 'conceptions and perceptions of realities', statements of purpose and goals over time, for example, with respect to sustainability and their evaluation and areas of accountability assumed by individuals and organisations, as well as transient or stabilised changes in social practice.
- **The emergence of new technologies**, including new uses and users of technologies, and social technologies such as measurement regimes will also be assessed and evaluated.
- **Evidence of transformative learning** at the individual, organisational and systemic level that is to be collected includes enriched systems dynamics understanding; improved value judgements; improved action judgments; and evidence for improved capacity to engage cognitive switching between viewpoints in dialogue and between past, present and alternative future worlds.
- **Environmental change and impacts from human action** will be assessed as far as this is possible in the given time frame. This will depend on existing baseline data and on the timeline for the development of the citizen science tool and its level of adoption and use in river partnerships. The need for research on learning from citizen science has been highlighted in research on environmental education, the learning sciences and community-based monitoring.

- **Scientific impact** will be evaluated based on the number and quality of publications and presentations at scientific conferences and eventually their citations.

There are three main innovative aspects of the project. First, the framework for collaborative conceptual systems mapping will direct attention to exploring inter-dependencies and feedbacks between social, ecological and technological change. Frameworks to date focus either on complex social-ecological systems or on human–environment interactions, often leading to a neglect of the influence of accelerating technological change on how humans relate to each other and the environment. Second, the project aims to build communities engaged in more sustainable water governance that are networked across three spatial scales (river partnerships, the national level and the global Earthwatch Initiative). Each commu-nity will have its set of indicators with associated quality criteria. Last but not least, at the start of the project we will conduct a detailed study relying on interviews and workshops on what is 'actionable knowledge' for different groups with stakes in water quality and security, including farmers, municipalities, households and firms, that will inform how we organise our workshops to explore systemic relationships and futures as well as the co-design process for the citizen science tool.

Outlook

Existential challenges in the Anthropocene will likely centre on distributional issues and raise questions on both environmental and social justice relating to the distri-bution of resources, as well as of knowledge, technologies and learning. At the same time, questions are raised on the adequacy of ways we produce science and technol-ogy. Established practices to produce science in particular fields of knowledge are experiencing difficulties from within science and its social institutions, in particular due to fragmentation and perverse career incentives for researchers (see also Chap-ter 18 by Ravetz). From the outside of science loom new demands for which the current organisation of knowledge production proves largely inadequate.

On top of that come new but related threats of democratic disengagement and social media–fuelled post-truth politics. Some scientists see this development as more reason for expert-driven technocratic and somewhat coercive approaches to regulating for sustainability. However, how well such approaches can really seize complexity, and in particular implications of the distributional issues, remains open to question. Such approaches to producing science for sustainability policy will not allow scanning the future for the possible unexpected, surprising or desirable implications of such changes in the short and long term from multiple perspectives in a way warranted in pluralist and societies which are becoming ever more diverse.

In the face of these possible future outlooks we join John Dewey in his con-viction that the only cure for a lack of democracy is more democracy (see also Chapter 2 by Wals and Peters). This book seeks to contribute to and to inspire the design social processes, conceptual and methodological tools, as well as supportive virtual spaces for the co-creation, tracking and evaluation of social transformation. In the face of existential challenges of sustainability we need to better understand complex circumstances in the face of accelerating and interdependent change in the social, technological and environmental spheres. We need more awareness and reflective attention to how we engage with the material world, and how scientific knowledge, technological objects and social orders are co-produced. We need to reflect and act upon a better understanding of how culture, and prevailing values are entangled with our conceptions of the material and the social. This understanding will help us to view technological design as an expression of prevailing values, and open our eyes in how design attributes of technological artifacts actually contribute to shaping how we interact with the world and attribute values.

The development and practice of transformative science for sustainability seems to open at least as many questions, including research questions, as it holds promise. Core questions include in particular those of quality criteria, quality control and validation of outcomes and impacts. Promises of the potential wider adoption of such practices include that they may contribute to reconfiguring and improving the science-policy-practice interface in a way that is appropriate for effectively networked societies. Emergent changes in expectations and social practice promise to be meaningful to many rather than to just a few. Furthermore, these practices may enable us to learn and reflect on progress for coping with the Anthropocene in a manner far more empowering than a mere flat belief in technological determinism. In sum, transformative sustainability science offers to all who choose to engage the opportunity to see the world with new eyes and to develop intentions and practices on how to transform it collaboratively.

Notes

1 Sociotechnical imaginaries are 'collectively held, institutionally stabilised, and publicly performed visions of desirable futures, animated by shared understanding of forms of social life and social order attainable through, and supportive of, advances in science and technology' (p.322). STI's can thus help to reveal topographies of power and morality as they interact with science and technology within hybrid networks.
2 Muki Haklay. www.ucl.ac.uk/excites/home-columns/full-what-is-extreme-citizen-science/ (emphasis added)
3 Examples include the UK Nexus Network organised from the University of Sussex www.thenexusnetwork.org, a joint project by Lang and Wiek and their colleagues largely based at Leuphana University and Arizona State University.

References

Abson, D. J., Fischer, J., Leventon, J., Newig, J., Schomerus, T., Vilsmaier, U., ... Lang, D. (2017). Leverage points for sustainability transformation. *Ambio* 46 (1): 30–39. doi:10.1007/s13280-016-0800-y

Castree, N. (2016). Broaden research on the human dimensions of climate change. *Nature Climate Change* 6 (8): 731. doi:10.1038/nclimate3078

Dyball, R. & Newell, B. (2015). *Understanding human ecology: A systems approach to sustainability*. London: Routledge.

Giddens, A. (2009). *Sociology, 6th edition*. Cambridge: Polity Press.

Grunwald, A. (2016). *Nachhaltigkeit verstehen: Arbeiten an der bedeutung nachhaltiger entwicklung*. München: Oekom Verlag.

Jasanoff, S. (2003). Technologies of humility: Citizen participation in governing science. *Minerva* 41 (3): 223–244. doi:10.1023/A:1025557512320

Jasanoff, S. (2015). 'Future imperfect: science, technology and the imaginations of modernity', in Jasanoff, S & Kim S.-H. (Eds.) *Dreamscapes of modernity: Sociotechnical imaginaries and the fabrication of power*. Chicago: University of Chicago Press. pp. 1–33.

Maggs, D. & Robinson, J. (2016). Recalibrating the Anthropocene: Sustainability in an imaginary world. *Environmental Philosophy* 13 (2): 175–194. doi:10.5840/envirophil201611740

Medema, W., Wals, A. & Adamowski, J. (2014). Multi-loop social learning for sustainable land and water governance: Towards a research agenda on the potential of virtual learning platforms. *NJAS – Wageningen Journal of Life Sciences* 69: 23–38.

Negroponte, N. (1995). *Being digital*. New York: Vintage Books.

Newell, B. (2012). Simple models, powerful ideas: Towards effective integrative practice. *Global Environmental Change* 22 (3): 776–783. doi:10.1016/j.gloenvcha.2012.03.006

Newell, B. & Doll, C. (2015). *Systems thinking and the cobra effect*. Tokyo: United Nations University.

Nielsen, M. (2012). *Reinventing discovery: The new era of networked science*. Princeton: Princeton University Press.

O'Brien, K. (2015). Political agency: The key to tackling climate change. *Science* 350 (6265): 1170–1171. doi:10.1126/science.aad0267

O'Brien, K. & Sygna, L. (2013). *Responding to climate change: Three spheres of transformation*. Proceedings of Transformation in a Changing Climate, 19–21 June 2013. Oslo: University of Oslo. pp. 16–23.

Schmidt, E. & Cohen, J. (2013). 'The future of identity, citizenship and reporting', in Schmidt, E. & Cohen, J. (Eds.) *The new digital age: Reshaping the future of people, nations and businesses*. London: John Murray Publishers. pp. 32–82.

Schneidewind, U., Singer-Brodowski, M., Augenstein, K. & Stelzer, F. (2016). *Pledge for a transformative science: A conceptual framework*. Wuppertal Papers, No. 191. doi:10.13140/RG.2.1.4084.1208

Smith, M. J. (1998). *Ecologism: Towards ecological citizenship*. Buckingham: Open University Press.

Stirling, A. (2015). *Developing 'Nexus Capabilities': Towards transdisciplinary methodologies*. Nexus methods discussion paper. Nexus Network Team, SPRU and STEPS Centre.

UBA. (2016). *Umweltpolitik für die Transformation fit machen: Neue Grundkonfigurationen für eine angewandte Umweltpolitik*. Dokumentation einer UBA-Kolloquiumsreihe. Heausgeber: Umweltbundesamt.

Vervoort, J. M., Bendor, R., Kelliher, A., Strik, O. & Helfgott, A.E.R. (2015). Scenarios and the art of worldmaking. *Futures* 74: 62–70.

Wals, A.E.J., Brody, M., Dillon, J. & Stevenson, R. B. (2014). Convergence between science and environmental education. *Science* 344 (6184): 583–584. doi:10.1126/science.1250515

Wiek, A. & Lang, D. J. (2015). 'Transformational sustainability research methodology', in Heinrichs, H., Martens, P. & Michelsen, G. (Eds.) *Sustainability science – An introduction*. New York: Springer. pp. 1–12.

Index